Environmental Change and African Societies

Climate and Culture

Series Editors

Carmen Meinert
Claus Leggewie

VOLUME 5

The titles published in this series are listed at *brill.com/clac*

Environmental Change and African Societies

Edited by

Ingo Haltermann and Julia Tischler

BRILL

LEIDEN | BOSTON

Cover Illustration: National Theatre of Ghana, Accra Ghana. Julien Lanoo © 2007.

Library of Congress Cataloging-in-Publication Data

Names: Haltermann, Ingo, editor. | Tischler, Julia, 1982- editor.
Title: Environmental change and African societies / edited by Ingo Haltermann and Julia Tischler.
Description: Leiden ; Boston : Brill, 2020. | Series: Climate and culture, 2213-0519 ; volume 5 | Papers originally presented at a workshop on "Environmental Change and African Societies" in Essen, Germany, June 2014. | Includes bibliographical references and index.
Identifiers: LCCN 2019024724 (print) | LCCN 2019024725 (ebook) | ISBN 9789004410831 (hardback) | ISBN 9789004410848 (ebook)
Subjects: LCSH: Africa, Sub-Saharan–Environmental conditions–Congresses. | Climatic changes–Africa, Sub-Saharan–Congresses. | Human ecology–Africa, Sub-Saharan–Congresses.
Classification: LCC GE160.A357 E59 2020 (print) | LCC GE160.A357 (ebook) | DDC 304.2/50967–dc23
LC record available at https://lccn.loc.gov/2019024724
LC ebook record available at https://lccn.loc.gov/2019024725

KULTURWISSENSCHAFTLICHES INSTITUT ESSEN
FORSCHUNGSKOLLEG DER UNIVERSITÄTSALLIANZ RUHR

Typeface for the Latin, Greek, and Cyrillic scripts: "Brill". See and download: brill.com/brill-typeface.

ISSN 2213-0519
ISBN 978-90-04-41083-1 (hardback)
ISBN 978-90-04-41084-8 (e-book)

This book is printed on acid-free paper and produced in a sustainable manner.

Contents

Foreword VII
Acknowledgements IX
List of Illustrations X
List of Abbreviations XII
Notes on Contributors XV

Introduction: Environmental Change and African Societies 1
Julia Tischler and Ingo Haltermann

PART 1
Ideas

1 To See or Not to See: on the 'Absence' of Climate Change (Discourse) in Maasailand, Northern Tanzania 23
Sara de Wit

2 Perspectives on Climate Change in Makonde District, Zimbabwe since 2000 48
Vimbai Kwashirai

PART 2
Past

3 Environmental and Climate Change in Africa: Global Drought and Local Environmental Infrastructure 73
Emmanuel Kreike

4 Shamba Forestry in Colonial Kenya: Colonial Dominance or African Opportunity? 98
Ben Fanstone

PART 3
Present

5 I'm Staying! Climate Variability and Circular Migration in Burkina Faso 123
Jonas Østergaard Nielsen

6 More of the Same: a Gender Lens on Life in a Changing Climate in Sub-Saharan Africa 149
Joy Clancy

7 Sustainable Mauritius? Environmental Change, Energy Efficiency, and Sustainable Development in a Small Island State in the Indian Ocean 177
Laura Jeffery

8 Transformative Learning for Global Change? Reflections on the WASCAL Master Programme in Climate Change and Education in the Gambia 199
Irit Eguavoen and Erick Tambo

PART 4
Prospects

9 Africa in Transition: What Role for the Environment? 227
Ton Dietz

10 Africa's High Modernism: Historical Ecologies of Climate Change and Hydrologies of Watersheds (Blue Nile and Zambezi) 250
James C. McCann

11 Increasing Urbanisation and the Role of Green Spaces in Urban Climate Resilience in Africa 265
Bertrand F. Nero, Daniel Callo-Concha, and Manfred Denich

Bibliography 297
Index 341

Foreword

Humans influence the environment and climate, with the consequences now felt across the globe. National or regional efforts to restrict or at least contain the damage are invariably insufficient: in principle, environmental and climate protection requires a global approach.

Paradoxically, the way we *perceive* environmental and climate change and respond to the harm that they cause is closely linked to local or regional patterns of perception. It is these particularistic perceptions that often lead to different, and in many instances conflicting, reactions to preventive and curative environmental and climate protection measures.

These local views are grounded not only in the different paths that socio-economic development has taken in specific regions of the world, but also in varying *cultural patterns*. Think, for example, of the vastly different ways in which current problems are perceived, or of how policy styles and politico-social environments differ. Also, the disturbance of the environment and climate causes relatively rapid social changes, in which the interpretation of symbols for the relationship between man and nature plays an important part. The history of climate and culture, patterns of perception of environmental and climate change, and an informed assessment of the future direction of environmental and climate policy in different parts of the world have to be taken into account in order to get to grips with the problem.

From a variety of angles, such as the history of ideas, historiography, the study of civilisation, and the political sciences, the monographs and edited volumes in *Climate and Culture* will all deal with the following questions:

- How do local and regional cultures perceive historical and contemporary changes in the environment and climate?
- How did and do they adjust to these changes?
- How do their various representatives and spokespeople introduce their respective views into the global debate and into emerging systems of international negotiation?

The following titles are included in the series:

VOLUME 1: *Nature, the Environment and Climate Change in East Asia*, edited by Carmen Meinert, 2013.

VOLUME 2: *Climate Change in South and Southeast Asia*, edited by Barbara Schuler, 2014.

VOLUME 3: *Cultural Dynamics of Climate Change and the Environment in Northern America*, edited by Bernd Sommer, 2015.

VOLUME 4: *Climate Change and Cultural Transition in Europe*, edited by Claus Leggewie and Franz Mauelshagen, 2017.

VOLUME 5: *Environmental Change and African Societies*, edited by Ingo Haltermann and Julia Tischler, 2019.

Carmen Meinert
Claus Leggewie

Acknowledgements

The editors would like to express their great appreciation to Carmen Meinert and Claus Leggewie for their patient guidance and support in enabling this volume to become a fruitful contribution to the *Climate and Culture* series. Further sincere gratitude goes to the series editors' team, namely Ben Müller, Sebastian Sponheuer, Thorben Pelzer and Nele Fabian, for their valuable help in ensuring that all contributions are in due form. At the University of Basel, Raphael Jenny and Zanele Leu helped to ensure that the publisher's guidelines were followed. The authors would also like to thank Patrick Grogan for his language editing and proof-reading.

We naturally owe the contributors to this volume a great debt of gratitude. Editing and re-editing this volume to ensure both its high quality and its compatibility with the overall series has proven far more time-consuming than we had anticipated. Thank you so much for your patience.

We also wish to thank all those who supported us in planning, preparing, implementing, and following up on the initial authors' workshop, especially the Fritz Thyssen Foundation, whose generous financial support rendered the workshop possible in the first place, and the Institute for Advanced Study in the Humanities (KWI) for providing the workshop venue as well as supportive staff and infrastructure. We are also especially grateful to Carmen Meinert, Steven Engler and Jan Schuster, who were excellent hosts of the panels, and Maria Klauwer and Britta Weber, who kept our administrative burden to a minimum. Andreas Eckert also helped initiate the workshop and provided guidance and assistance in its organisation.

Julia Tischler
Ingo Haltermann

Illustrations

Figures

2.1 Deforestation in the Obva area, Makonde; cutting down trees for crop fields and firewood. Photograph © Vimbai Kwashirai. 69

5.1 Rainfall anomalies in the Sahel. The figure captures the wet 1950s and 1960s and the dry 1970s and 1980s and the increased variability in precipitation since the early 1990s. Source: Lebel and Ali (2009). 127

5.2 Typical houses found in *non lotis* areas of Ouagadougou. Source: Google Earth. 132

5.3 Houses in *lotis* areas of Ouagadougou. Single-family homes consist of one or two rooms and are always mud-brick constructions. © Jonas Ø. Nielsen 133

8.1 Historical outline of the climate change adaptation policy process in the Gambia 210

8.2 Capacity building on climate change at West African universities 213

8.3 Areas of capacity building in the WASCAL Graduate Study Programme 213

8.4 Curriculum of climate change and education, February 2015 217

8.5 Educational background of students 219

9.1 Citations of the ICCD book between 2004 and 2014 246

11.1 Urban population growth in the African subregions 270

11.2 Distribution of green spaces in Kumasi, Ghana 281

11.3 People in Kumasi under a *Terminalia catappa* tree (coping with heat). Source: Bertrand Nero (author). 284

11.4 Building too close to the watercourse at Atonsu, Kumasi, Ghana. Source: Bertrand Nero (author). 285

Tables

5.1 Semi-structured interview distribution (N = 38) 135

5.2 Reasons mentioned by interviewees (N = 38) for moving to Ouagadougou since the year 2000 137

5.3 Reasons mentioned by interviewees (N = 38) for not returning to origin village 138

8.1 Transdisciplinary learning: Curation of an educational exhibition 217

8.2 Technology-enhanced learning in the context of climate change 218

9.1 Long-term population developments and prospects, Europe and Africa, 1650–2100 230

11.1 Changes in urban land use characteristics in mega, medium-sized, and small cities in Africa (average values) 274
11.2 Managerial actions/decisions, actors, and underlying reasons that favour and weaken UGS existence and maintenance in Kumasi, Ghana 289

Maps

1.1 Terrat in Northeastern Tanzania. © Monika Feinen, University of Cologne 39
2.1 Makonde District. Source: Wikipedia, "Districts of Zimbabwe." 58
5.1 The Sudano-Sahelian region of West Africa, showing the location of Burkina Faso and Ouagadougou 128
5.2 Satellite images taken in 2012 showing the difference between *lotis* and *non lotis* as seen from above. Source: Thorsen, D., "weaving in and out of employment and self-employment". 131
5.3 Interview sites and the spatial expansion of Ouagadougou in the period 1956–2012. Source: IGB, Institut Géographie du Burkina and Google Rarth. 136

Abbreviations

AD	*Anno Domini* (in the year of the Lord)
AER	Agro-Ecological Region
AIDS	Acquired Immune Deficiency Syndrome
BC	Before Christ
BMBF	German Federal Ministry of Education and Research
CDK	Climate and Development Knowledge Network
CEB	Central Electricity Board
CFA	Community Forest Associations
CFU	Commercial Farmers' Union
CO_2	Carbon Dioxide
COP	Conference of Parties
COPD	Chronic Obstructive Pulmonary Disease
CSC	Climate Service Centre
DALYs	Disability-Adjusted Life Years
DRC	Democratic Republic of the Congo
ECOWAS	Economic Community of West African States
EIA	Environmental Impact Assessment
EMA	Environment Management Act
ENSO	El Niño Southern Oscillation
ESAP	Economic Structural Adjustment Programme
EU	European Union
FPE	Feminist Political Ecology
FTLRP	Fast Track Land Resettlement Programme
GAIA	Global Anti-Incinerator Alliance
GDP	Gross Domestic Product
GHGs	Greenhouse Gases
GIS	Geographic Information Systems
GPS	Global Positioning System
IAP	Indoor Air Pollution
ICDPs	Integrated Conservation and Development Projects
ICT	Information and Communication Technology
IEEA	Indigenous Economic Empowerment Act
IFAD	International Fund for Agricultural Development
IMF	International Monetary Fund
INDC	Intended Nationally Determined Contributions
IPCC	Intergovernmental Panel on Climate Change
IPPs	Independent Power Producers

ITCZ	Inter-Tropical Convergence Zone
KFD	Kenya Forest Department
KLFA	Kenya Land and Freedom Army
KNA	Kenya National Archives
LAA	Land Acquisition Act
LED	Light-Emitting Diode
LPG	Liquefied Petroleum Gas
LRA	Labour Relations Act
MCTP	Mauritius CT Power
MID	Maurice Ile Durable/Sustainable Mauritius
MIMES	Multi-Scale Integrated Model of Ecosystem Services
MoESD	Ministry of Environment and Sustainable Development
MW	Megawatts
NGO	Non-Governmental Organisation
NPK	Nitrogen, Phosphorus, and Potassium
PELIS	Plantation Establishment and Livelihood Improvement Scheme
PMO	Prime Minister's Office
REDD	Reducing Emissions from Deforestation and Forest Degradation in Developing Countries
ROR	Run-Of-The-River
SADC	Southern African Development Community
SIDS	Small Island Developing States
SIS	Small Island State
SMS	Short Message Service
SSA	Sub-Saharan Africa
SST	Sea Surface Temperatures
TNA	The National Archives of the United Kingdom
UGS	Urban Green Spaces
UHI	Urban Heat Island
UK	United Kingdom
UN	United Nations
UNDP	United Nations Development Programme
UNESCO	United Nations Educational, Scientific and Cultural Organization
UNFCCC	United Nations Framework Convention on Climate Change
UNU-EHS	United Nations University Institute of Environment and Human Security
USA	United States of America
UTG	University of The Gambia
VOC	Dutch East India Company
WASCAL	West African Science Service Center on Climate Change and Adapted Land Use

WB	World Bank
WCED	World Commission on Environment and Development
WGIAR5	Working Group I to the Fifth Assessment Report of the Intergovernmental Panel on Climate Change
WHO	World Health Organisation
WMO	World Meteorological Organization
WtE	Waste-to-Energy
ZANU–PF	Zimbabwe African National Union–Patriotic Front

Notes on Contributors

Daniel Callo-Concha

is a Senior Researcher at the Center for Development Research (ZEF), University of Bonn. He has fifteen years of experience in the interface between agriculture and natural resource management in Latin America and Africa. His academic approach is based on the systems thinking method for handling complexity in social-ecological systems. Callo-Concha's recent research explores farmers' adaptation to climate change, food and nutrition security in developing countries, urbanisation in the Global South, and sustainable mining and agriculture. His most significant publications include: Callo-Concha et al., "Lessons for Research, Capacity Development and Policy in Agroforestry for Development," *Agroforestry Systems* 91.5 (2017): 795–798. doi:10.1007/s10457-017-0085-6; and Callo-Concha, "West African Farmers' Climate Change Adaptation: From Technological Change Towards Transforming Institutions," in *Implementing Climate Change Adaptation in Cities and Communities*, ed. Leal Filho et al. (Berlin: Springer, 2016), 253–265.

Joy Clancy

a member of the Department of Governance and Technology for Sustainability (CSTM), University of Twente, is a professor in Development Studies who specialises in gender and energy. For more than thirty years, Joy's research has focused on small-scale energy systems for developing countries, including the technology transfer process and the role that energy plays as an input for small businesses and the potential it offers entrepreneurs, particularly women, through the provision of new infrastructure services. Joy is currently the Principal Investigator in a five-year £4.5 million research project by ENERGIA, funded by the UK's Department for International Development (DfID): "Building the Evidence Base for Improving Energy Investments' Effectiveness by Taking a Gendered Approach" (http://energia.org/research/). Two of her recent publications are: Clancy, Joy, *Biofuels and Rural Poverty* (Abingdon: Routledge, 2013); and "Unpacking the Discourse of Social Inclusion in Value Chains," in *Value Chains, Inclusion and Endogenous Development: Contrasting Theories and Realities*, ed. A.H.J. Helmsing, and Sietze Vellema (Abingdon: Routledge, 2011).

Manfred Denich

is Senior Researcher at the Center for Development Research (ZEF), Department of Ecology and Natural Resources Management, University of Bonn, There he conducts research on land use and food security, renewable energy, environmental and climate change, ecosystem services, sustainable use of natural resources, biodiversity, and urbanisation. He is part of the West African Science Service Center on Climate Change and Adapted Land Use (WASCAL) and Project Director of the GlobE—BiomassWeb Project at the University of Bonn. His publications include: Callo-Concha D. et al., "Bridging Science and Development: Lessons Learnt from Two Decades of Development Research," *Agroforestry Systems* 91 (2016): 799–810. doi 10.1007/s10457-016-0008-y; and Atela, Joanes O. et al., "Agricultural Land Allocation in Small Farms around Maasai Mau Forest, Kenya, and the Implications on Carbon Stocks," *Journal of Ecology and the Natural Environment* (*JENE*) 4.4 (2012): 98–108.

Sara de Wit

is a Fellow at the Institute of Science Innovation and Society at the University of Oxford and part of the Forecasts for Anticipatory Humanitarian Action (FATHUM) project. Trained in anthropology and African Studies, de Wit has a strong empirical orientation, with long-term fieldwork experience in southeast Madagascar, the Bamenda Grassfields in Cameroon, and Maasailand in northern Tanzania. She has carried out "ethnographies of aid"—at the intersection of Science and Technology Studies, development theories, environmental anthropology, and postcolonial studies—in which she broadly focused on how globally circulating ideas, such as climate change and notions of development, travel, and what happens when they are translated by various actors along the translation chain.

Ton Dietz

is Emeritus Professor of African Development at Leiden University and former director of the African Studies Centre, Leiden, where he currently works as a Senior Researcher. His most recent publications include: Dietz, Ton, *Africa: Still a Silver Lining* (Leiden: Leiden University, 2017), http://www.ascleiden.nl/sites/default/files/afscheids-college_dietz_web.pdf; and Akinyoade, A, Ton Dietz, and C.U. Uche, ed. *Entrepreneurship in Africa*, African Dynamics Series, vol. 15 (Leiden: Brill, 2017). Dietz is also a guest professor at the University of Amsterdam. He has a PhD degree from the University of Amsterdam and an honorary degree from Moi University, Eldoret, Kenya, and was awarded a *Lauréat d'Honneur* by the International Geographical Union as well as an *Order of Merit* by the Dutch Ministry of Foreign Affairs.

Irit Eguavoen

is a social anthropologist who works at the Center for Development Research (ZEF), University of Bonn. She was part of the WASCAL staff from 2011 to 2017 and has coordinated the research group "Politics of Adaptation." The focus of Irit's work is on environmental themes in West Africa and Ethiopia. Her empirical research has turned towards questions of sustainable urbanisation and urban citizenship in the "Waterfront Metropolis Abidjan," an individual research project running from 2017 to 2019. Her most recent publications include: Lauer, Hannes, and Irit Eguavoen, "Mainstreaming Climate Change Adaptation into Development in The Gambia: A Window of Opportunity for Transformative Processes?" in *Innovation in Climate Change Adaptation*, ed. Walter Leal Filho (Berlin: Springer, 2016), 87–98; and Tan, Jiaxin, and Irit Eguavoen, "Digital Environmental Governance in China: Information Disclosure, Pollution Control, and Environmental Activism in the Yellow River Delta," *Water Alternatives*, 10.3 (2017): 910–929.

Ben Fanstone

graduated from the University of Stirling's Centre for Environment, Heritage and Policy with a PhD in 2017. His thesis entitled "The Pursuit of the 'Good Forest' in Kenya, c.1890–1963: The History of the Contested Development of State Forestry within a Colonial Settler State" examines the often discordant relationships that state forestry officials had with other colonial actors, especially the European settler population, as well as exploring the diverse forestry personnel and their evolving scientific ideals. A particular emphasis is the development of an agroforestry labour system within the colony and its relationship to protest, food security, and the Mau Mau uprising. His other research interests include colonial forestry in Nigeria, environmental conflict, indigenous adaptation to colonial land use regimes as well as colonial and postcolonial policies and their impact on land use today.

Ingo Haltermann

is a graduate geographer who also works as a freelance editor. From 2009 to 2016 he was Junior Fellow at the Institute for Advanced Study in the Humanities (KWI) in Essen, Germany, where he was part of the Research Unit "Climate and Culture," and Research Assistant at the Centre for Global Cooperation Research in Duisburg. After gaining his Diploma in Geography from the University of Münster in 2007, he studied Peace and Conflict Studies at the FernUniversität in Hagen. His particular research interest lies in the perception of and response to climate change at an individual and household level, but he also addresses corresponding cultural and political processes. Among other publications, he has written on "The Perception of Natural Hazards in the Context of Human (In-)Security," in *Negotiating Disasters: Politics, Representation, Meanings*, ed. Ute Luig (Frankfurt am Main: Peter Lang Verlag, 2012), 59–78;

and on "The Attachment to Place in the Aftermath of Disasters," (original title: "Woven Together: Perspectives from Four Continents.") in *Listening on the Edge: Oral History in the Aftermath of Crisis*, ed. Mark Cave, and Stephen M. Sloan (New York: Oxford University Press, 2014).

Laura Jeffery
is Senior Lecturer in Social Anthropology at the University of Edinburgh, where she held an ESRC Research Fellowship (RES-063-27-0214) with a particular focus on debates about environmental knowledge in the context of the Chagos Archipelago in the Indian Ocean. Her relevant recent publications include articles on ecological restoration and human-environment relations: "Ecological Restoration in a Cultural Landscape: Conservationist and Chagossian Approaches to Controlling the 'Coconut Chaos' on the Chagos Archipelago," *Human Ecology* 42.6 (2014): 999–1006; and " 'We are the True Guardians of the Environment': Human-Environment Relations and Debates about the Future of the Chagos Archipelago," *The Journal of the Royal Anthropological Institute* 19.2 (2013): 300–318.

Emmanuel Kreike
is Professor of African and Environmental History at Princeton University. He studied African History at Yale University (PhD, 1996) and Tropical Forestry at Wageningen University (Dr.Sc., 2006). His current research focuses on the impact of war and population flight on rural societies and environments in Southern Africa as well as in global and comparative perspective. Kreike's most recent books include: *Deforestation and Reforestation in Namibia: The Global Consequences of Local Contradictions* (Leiden, Boston: Brill, 2010); and *Environmental Infrastructure in African History: Examining the Myth of Natural Resource Management in Namibia* (Cambridge: Cambridge University Press, 2013).

Vimbai Kwashirai
is a Zimbabwean, Oxonian and independent scholar. He is a visiting lecturer in the Department of Economic History at the University of Zimbabwe and is also associated with the Centre for World Environmental History at the University of Sussex. His research interests are in economic and environmental issues, specifically in modern Zimbabwe and Africa more generally. The author of *Green Colonialism in Zimbabwe, 1890–1980* (Amherst: Cambria Press, 2009), Vimbai has published extensively on these themes as well as on social and political developments in Zimbabwe, as reflected in his forthcoming monograph *Election Violence in Zimbabwe: Human Rights, Politics and Power 1980–2018* (Cambridge: Cambridge University Press). He has taught at several

universities in Zimbabwe and the United Kingdom, including Durham and Liverpool. Vimbai has received many academic awards and fellowships in the UK and Germany. He is a leading expert on Zimbabwean human rights, land rights, poverty, and ecological transformations. His current research is concerned with the sustainability of drought coping strategies in Southern Africa, in which he seeks to find lasting solutions to climate change drought-induced socio-economic problems.

James C. McCann
is Chair of the Department of History and Professor at Boston University. He won the 2014 Distinguished Scholar Prize from the American Society of Environmental History. In 2012/2013 he was awarded a Guggenheim Fellowship for a study of the ecology of malaria in a region of unstable transmission: *The Historical Ecology of Malaria in Ethiopia: Deposing the Spirits* (Athens, OH, 2014). He has been a scholar in residence at Harvard, Yale, the National Humanities Center, Addis Ababa University, Hiob Ludolf Centre, University of Hamburg, the National University of Lesotho, and the University of Khartoum and has served as a consultant to many NGOs and international organisations. He is the author of six books, including *Green Land, Brown Land, Black Land: An Environmental History of Africa* (Munich: Heinemann, 1999) and *Maize and Grace: Africa's Encounter with a New World Crop* (Harvard: Harvard University Press, 2005).

Bertrand F. Nero
is a Ghanaian postdoctoral fellow in the BMBF-funded research project GlobE—BiomassWeb at the Institute for Food and Resource Economics at the University of Bonn. He is a former junior researcher at the university's Center for Development Research (ZEF), Department of Ecology and Natural Resources Management, and works for the Ghanaian Ministry of Petroleum in Accra. One of his recent publications is: "Urban Green Space Dynamics and Socio-Environmental Inequity: Multi-Resolution and Spatiotemporal Data Analysis of Kumasi, Ghana," *International Journal of Remote Sensing* 38.23 (2017): 6993–7020.

Jonas Østergaard Nielsen
is Professor in Integrative Geography at Humboldt-University's Department of Geography and at IRI THESys, where he leads a research group. Together with his team he is investigating the emerging framework of telecoupling in land use science, considering how distal environmental, social, political, and economic drivers of change influence local land use decisions. Jonas' research

is also concerned with human dimensions of global climate change and how climate change opens up for explorations around global–local interactions in an increasing connected world. Since 2007, Jonas has conducted extensive ethnographic fieldwork in Burkina Faso. His publications include: Friis, Cecilie, and Jonas Ø. Nielsen, "Land Use Change in a Telecoupled World: The Relevance and Applicability of the Telecoupling Framework in the Case of Banana Plantation Expansion in Laos," *Ecology and Society* 22.4 (2017): 30. DOI: https ://doi.org/10.5751/ES-09480-220430; and Niewöhner, Jörg et al., *Land Use Competition: Ecological, Economic and Social Perspectives* (Dordrecht: Springer, 2016). http://www.springer.com/in/book/9783319336268.

Erick Tambo
is a lecturer at the West African Science Service Center on Climate Change and Adapted Land Use (WASCAL) at the University of The Gambia. He graduated in Computer Science from the Technical University of Dortmund, Germany, and holds a PhD in Computer Science from the FernUniversität in Hagen. He is an Associate Academic Officer at UNU-EHS, where he is in charge of the conceptualisation, organisation, and management of e-learning activities. Tambo leads and coordinates UNU-EHS contributions towards the implementation of the Higher Education Cooperation with the African Union's Pan African University Institute for Water and Energy including Climate Change. His publications include: Tambo, Erick R.G. et al., *Management of Environmental Risks and Natural Disasters: How Can eLearning Help?* (Bonn, 2012).

Julia Tischler
is Tenure Track Assistant Professor of African History at the University of Basel, where she focuses on Southern Africa, development, and labour, environmental, and global history. After completing her doctorate at Cologne University, she worked as a postdoctoral researcher at Bielefeld University and at the International Research Centre "Work and Human Lifecycle" (re:work) in Berlin. Her first monograph, *Light and Power for a Multiracial Nation: The Kariba Dam Scheme in the Central African Federation* (Basingstoke: Palgrave Macmillan, 2013), examined the Kariba Dam on the border between today's Zambia and Zimbabwe as a case study of late colonial nation-building. Her interest in environmental history is further reflected in the volume *Grounding Global Climate Change: Contributions from the Social and Cultural Sciences* (Berlin: Springer; 2015, co-edited with Heike Greschke) as well as her ongoing second book project on agricultural development in South Africa (c.1900–1950).

Introduction: Environmental Change and African Societies

Julia Tischler and Ingo Haltermann

1 The Vulnerable Continent?

On Friday, 13 June 2014, we anxiously awaited the arrival of the attendees of our author's workshop on "Environmental Change and African Societies" in Essen, Germany. We had not heard from several participants who were coming from overseas; others called to report that they were stuck in various remote corners of western Germany, waiting for a bus to take them a bit closer to their destination. Even participants from relatively close to Essen struggled to reach the workshop venue, being forced to find a lift or other resourceful means of travel. A few days earlier, a major thunderstorm, a result of the cyclone Ela and the anticyclone Wolfgang, had begun to ravage through parts of western and central Germany, causing massive damage. The storm destroyed sections of railway line, uprooted trees, blocked major highways, and shattered cars. Tragically, six people lost their lives, and over sixty were injured.

Essen's main train station was entirely cut off from the railway network for several days. We were left with no idea as to how our participants would reach the venue, if at all. Eventually, however, everybody made it, many after quite an odyssey. This was a nerve-racking yet fitting and insightful start for a conference on climate change in Africa. In current climate change discourse, vulnerability is mainly cast as a characteristic of the Global South, affecting societies that lack the infrastructure and financial means to cope adequately with extreme weather events, long-term changes in the environment, and the consequences thereof. Yet, having just experienced a thunderstorm that had paralysed large parts of Germany, we became very aware of how even one of the richest countries in the world was vulnerable to nature's capriciousness.

Emmanuel Kreike, who also participated in the workshop, has previously taken issue with the Western trope of the 'conquest of nature'. While high-income societies are commonly perceived to be characterised by technology, science, management, and elaborate infrastructure, all of which help to transform nature into culture, Africans are said to live with nature and (still) depend on it. However, none of our modern accomplishments has allowed industrial and postindustrial societies to leave nature in their wake. As Kreike points out, their entire infrastructures depend on one finite natural resource,

 | DOI:10.1163/9789004410848_002

hydrocarbons.[1] While we certainly do not wish to question the fact that societies across the African continent are suffering the negative consequences of climate change to an extent that may in many cases exceed its experienced effects on Western societies, our violent thunderstorm did remind us of the many tropes in climate change discourse on 'developing' nations that are often left unquestioned.[2]

The present volume is the fifth in the series *Climate and Culture*, which seeks to make a contribution to current debates on global climate change at the local or regional level and from the perspectives of the social sciences and the humanities. These disciplines entered the stage of climate change research as latecomers, carving out a space in a research field contoured by the natural sciences. In the past decade, however, an increasing number of studies on the social or cultural sides of climate change, including aspects of adaptation, vulnerability, and mitigation, as well as broader questions of human-environment relations, have emerged. A first catalyst of such research was the 1992 Rio Earth Summit, while a further steep increase in the number of social science publications on global environmental change has been recorded since 2005.[3] The present volume is based on the premise that the ways in which communities will respond to the effects of climate change cannot be apprehended without addressing the various 'climate cultures' that have evolved historically around the world.[4] The study of climate change as a social phenomenon must be carefully contextualised in relation to existing conceptions of nature and environmental history and to people's experiences in response to environmental change.

1 Kreike, Emmanuel, Environmental Infrastructure in African History: Examining the Myth of Natural Resource Management in Namibia (Cambridge: Cambridge University Press, 2003), 1–6.

2 The cover image, taken by the French photographer Julien Lanoo in Accra, Ghana, has been chosen in an attempt to counter this alarmist-interventionist discourse, which often finds its figurative expression, for example, in pictures of cracked soils and solar panels on mud huts. By choosing Lanoo's photograph, we reproduce iconic elements of the Western climate change discourse (i.e. bus and bike), but in an African context. This choice does not relate to climate change itself, but to how we translate it for a wider audience.

3 Compare the figures and graph in Caillods, Françoise, "Regional Divides in Global Environmental Change Research Capacity," in *World Social Science Report 2013: Changing Global Environments*, ed. ISSC and UNESCO, 126, accessed August 3, 2017. doi:10.1787/9789264203419-en.

4 Leggewie, Claus and Harald Welzer, *Das Ende der Welt, wie wir sie kannten: Klima, Zukunft und die Chancen der Demokratie* (Frankfurt: Fischer, 2009); Greschke, Heike, "The Social Facts of Climate Change: An Ethnographic Approach," in *Grounding Global Climate Change: Contributions from the Social and Cultural Sciences*, ed. Heike Greschke and Julia Tischler (Dordrecht: Springer, 2015), 121–138, here 122.

The present volume is organised into four parts, analogous to the existing and forthcoming volumes in the series. The first section, "Ideas", inquires into local perceptions of the environment. Rather than taking the social consequences of climate change as a given, we take a step back to ask how specific African communities conceptualise nature, whether and how they perceive changes therein, and, if they do, to what they attribute these changes. Climate change is commonly seen as a threat to the long-established ways in which groups of people interact with the environment and how they make sense of these interactions. Similarly, the two contributions in this section depict highly dynamic and tension-ridden situations; however, instead of defining these as a simple confrontation between 'traditional' knowledge and global climate change (see also below), de Wit and Kwashirai describe the perceptions that they trace on the ground as fluid and creative bricolages, assembled from various sources that include long-standing local cosmologies and current climate change discourses.

The section "Past" discusses historical cases of environmental change and state regulation. The chapters by Kreike and Fanstone help us to contextualise ongoing climate-related changes within longer historical trajectories of how different communities have dealt with disturbances in their use of environmental resources, including droughts and state-induced restrictions. In addition, many contributions in the other sections are also historically informed and examine contemporary problems in the light of diachronic developments. While popular climate change discourse often casts African communities as victims of global climate change, past experiences also reflect the existing knowledge and practical skill repertoires that have allowed various groups to counter environmental stress. Without implying that past responses to environmental change can be transposed to the present or future in a linear manner, this volume does make a strong plea for climate change research that is historically grounded, exploiting the rich body of research on African environmental history.

The section "Present" addresses decision-making and agenda-setting processes that relate to current representations and/or predicted effects of climate change. Østergaard Nielsen discusses individual adaptations to climate change by illustrating how the common coping strategies applied by Sahelian farmers in the face of interseasonal and interannual precipitation variability alter under the influence of a changing climate. Climate change is predicted to increase inequalities, including those based on gender, as Clancy's contribution shows. Men and women are affected by climate change to different degrees and display different levels of vulnerability and resilience. This is, for one, conditioned by the unequal provision of economic, cultural, and social capital between genders. Moreover, perceptions of environmental change often vary between men and women. As adaptation is in many regards a collective process, unequal power relations and cultural conditions may further lower women's ability to adapt. Central to adaptation is communication, which thus calls for an intersubjective understanding of climate change and its meanings for the respective community, as Jeffery highlights. Describing the conflicts between different actors in the Mauritian sustainable development programme "Maurice Ile Durable", Jeffery shows how disparate interests as well as divergent interpretations of sustainability, development, and environment can result in such initiatives achieving reduced levels of efficacy and social inclusion. This demonstrates the need for effective communication on climate change, which is the main focus of the section's third article. Therein, Eguavoen examines an African university as an arena of knowledge transfer and depicts multiple levels of climate change communication. While such communication occurs naturally in the lecturer-student relationship and through the desired transfer of knowledge from graduates to wider society, the difficult communication process between the university and its sole foreign donor exposes how imbalances within agenda-setting processes remain an important issue.

The section "Prospects" is concerned with contemporary African megatrends. The challenges of a changing climate impinge on a continent characterised by rapid economic and population growth and unprecedented levels of urbanisation, all corresponding with massive shifts in land use patterns. The interrelations between the causes and effects of climate change, its local and regional implications, the required adaptation and mitigation measures, and the above-cited megatrends are ambiguous. In his overarching chapter on "Africa in Transition", Dietz identifies the opportunities as well as the risks that climate change brings with it, thereby contesting the "images of doom and gloom" that still dominate climate change writing on Africa. As he observes, multiple factors may "reinforce or counter the risks brought about by climate change" (see Chapter 9 of this volume). One example thereof is provided by McCann in his contribution on Blue Nile and Zambezi hydrologies. As he notes, Africa is currently experiencing a renewed wave of dam-building projects. On the one hand, this will contribute to a decoupling of the continent's growing energy demands from its rising CO_2 emissions and provide African farmers with irrigation facilities to render them less dependent on natural precipitation. On the other hand, the construction of dams requires massive interventions into river hydrologies, which may already be adversely affected by changing rainfall patterns in their catchment areas. Besides adaptation, mitigation and resilience have played a major role in recent debates about Africa's climate future. As the continent's population and economy become increasingly urban, the question of climate resilience is also turning into more and more of an urban challenge. With reference to the example of Kumasi, Ghana, Nero et al. define the opportunities and constraints for sustaining and enhancing green cover in African cities, showing that this is to a large extent a bottom-up process driven by customary and private interests.

In line with the previous volumes in the *Climate and Culture* series, the structure of this volume implies that environmental change in Africa should be approached from a broad perspective. Although we cannot aim for a complete discussion in any sense, we do surmise that insights from a multitude of approaches and disciplines provide us with a more multifaceted understanding of the key issues at hand. By juxtaposing past cases of environmental change, adaptation, and contestation, we counter the presentist bias that has marked much of the ongoing climate change debate. In addition, the authors in this volume engage with climate change on different scales. Meso- and macro-scale analyses and prognoses (see, for instance, Dietz and McCann) are combined with interpretations that build on highly localised ethnographic data (see, for example, de Wit and Østergaard Nielsen).

2 Climate Change in Africa

In public discourse, Africa is often referred to as the continent that is most adversely affected by climate change while contributing the least to its causes. A range of detrimental environmental factors in conjunction with socio-economic stressors such as poverty, unstable political circumstances, or infrastructural constraints are seen to render many regions in Africa particularly vulnerable. Droughts and hunger catastrophes, as experienced in the 1980s in the Sahel and Ethiopia or in 2011/2012 in the Horn of Africa, have a long-lasting effect on global perceptions of the continent.

In general terms, African climates are primarily characterised by variable levels of moisture, predominantly rain, and less by varying temperatures as in Europe or North America. Most significant is the annual cycle of wet and dry seasons. While variations occur from year to year, as well as on a decadal time scale,[5] annual weather patterns are usually predictable. With the onset of the rainy season, dried-out landscapes turn green within a very short period of time, while fields ripen for harvest in the dry season. With their Mediterranean

5 The latter are mainly influenced by global climate anomalies such as the El Niño Southern Oscillation (ENSO) or fluctuations in the Atlantic Meridional Overturning Circulation (AMOC).

climates, the Cape and the North African littoral form an exception to these patterns. From a long-term perspective, Africa's climate has become drier and warmer, with desert zones in the Sahara expanding and the Central African forest zone shrinking. Some changes, like the increase in frequency and intensity of drought in the West African Sahel from 1951 to 2010, have already been attributed to anthropogenic climate change.[6] Today, Africa is about 1.1°C warmer than it was in the world's preindustrial era. The continent's six warmest years of the 20th century all occurred in the final decade, with 1998 being the warmest.[7] This trend has continued throughout the first two decades of the 21st century, with the years 2015 and 2016 being the warmest ever recorded in Africa (and worldwide).[8]

Human beings have always adapted to changing environments throughout the course of human history, making the capacity to adapt a key attribute of humanity.[9] "Adaptation", as Carmen Meinert has argued in a previous volume in this series,

> [...] is not a new phenomenon at all; climatic vicissitudes throughout human history have acted as significant stimuli for social and technological innovations and for the establishing of new institutions on a local, regional or national level, which in turn have proven the possibilities of adapting social practices towards changing climates.[10]

This is especially true for Africa. As James McCann has pointed out, most African communities have adapted to long-term changes, for instance through agricultural systems and livestock economies that were mobile enough to follow the path of moisture rather than by seeking to transform their environments.

---

6 McCann, James, *Green Land, Brown Land, Black Land: An Environmental History of Africa, 1800–1990* (Portsmouth, Oxford: Heinemann, and James Currey, 1999), 15–19; IPCC, Climate Change 2013: The Physical Science Basis. Contribution of Working Group I to the Fifth Assessment Report of the Intergovernmental Panel on Climate Change [Stocker, Thomas et al] (Cambridge: Cambridge University Press, 2013), 7, 50, 215.

7 Hulme, Mike et al., "African Climate Change: 1900–2100," *Climate Research* 17 (2001): 145–168, here: 150.

8 World Meteorological Organisation (WMO), "WMO Statement on the State of the Global Climate in 2016," 2017, accessed August 2, 2017. https://public.wmo.int/en/resources/library/wmo-statement-state-of-global-climate-2016.

9 Adger, W. Neil et al., "Adaptation to Climate Change in the Developing World," *Progress in Development Studies* 3.3 (2003): 179–195, here 186.

10 Meinert, Carmen, "Introduction: Climate and Culture in East Asia," in *Nature, Environment and Culture in East Asia: The Challenge of Climate Change*, ed. Carmen Meinert (Leiden: Brill, 2013), 1–20, here 2.

Throughout history, climate patterns, particularly the seasonality of rain, have affected food supply as well as the spread of disease, migration, the timing of military campaigns, political cycles, and religious practices.[11] In the future, climatic conditions in Africa, like elsewhere in the world, will undergo profound changes, some unprecedented in their extent and/or in the speed by which they take effect, which will place the range of local experience-based coping mechanisms under severe pressure. Furthermore, not all existing coping strategies can be successfully maintained in times of globalised markets and conflicts, mass urbanisation, population growth, and land use change.

In the proclaimed age of the Anthropocene, it is important to remember that African landscapes—despite often being cast as a last bastion of 'original nature'—have long been anthropogenic. Historians and archaeologists have shown how the use of tools and technology, labour, draught animals, and fire have shaped landscapes on the continent for centuries. In the past two hundred years, however, the extent to which human activities have impacted the natural environment has increased significantly, primarily due to changing technology, demographics, and patterns of trade.[12] The question remains to what extent and at what pace human beings will continue to transform the natural environment in Africa and beyond. While Africa still accounts for the smallest portion of the world's cumulative net CO_2 emissions,[13] this may soon change considering the continent's growing population, rising levels of economic development, rapid urbanisation, and corresponding shifts in its patterns of land use and resource consumption. For instance, energy consumption is predicted to increase dramatically, but Africa has the potential to leapfrog the traditional centralised utility model for energy provision and may base its future energy supply on renewables.[14] Ambitious programmes, like the Africa Renewable Energy Initiative launched by the African Ministerial Conference on the Environment, have been formulated at a national and intergovernmental level

---

11 McCann, Green Land, Brown Land, Black Land see above, 15–19.

12 McCann, Ibid., 1–51.

13 In the period from 1900 to 2012, Africa was responsible for only 1.8 per cent of global energy-related CO_2 emissions (0.6 per cent if South Africa is excluded). See International Energy Agency (IEA), *Africa Energy Outlook: A Focus on Energy Prospects in Sub-Saharan Africa, World Energy Outlook Special Report* (Paris: IEA, 2014), 117. Emissions from land-use change are also dominated by regions outside of Africa, including deforestation in the mid-northern latitudes prior to the 1980s and in the tropical Americas and Asia thereafter, with only small contributions from tropical Africa. See IPCC, *Climate Change 2013*, 491.

14 International Renewable Energy Agency (IRENA), *Africa 2030: Roadmap for a Renewable Energy Future* (Abu Dhabi: IRENA, 2015), 6.

to promote renewable energy sources.[15] However, their regulatory frameworks often remain patchy and inconsistent. Additionally, as Dietz notes in this volume, recent oil and gas field discoveries, such as those along the coast of southeastern Africa or in the Gulf of Guinea, will tempt political leaders to meet the growing energy demand by utilising such resources instead of harnessing Africa's vast potential for renewable energy. Ultimately, much will depend on political will.

As of yet, we still do not fully understand the physical drivers of many African climate systems.[16] Most significantly, there are still massive deficiencies in current climate model simulations for the Central African convective region and the West African monsoon, which each influence the global climate as one of the world's three major convective or monsoon systems respectively.[17] Both regions demonstrate the complexity that arises from the reciprocal effects of a dominant climate system interacting with regional climate drivers and more distant processes. For example, West African precipitation is influenced significantly by teleconnection with global sea surface temperatures (SSTs). ENSO and the changing SSTs in the Gulf of Guinea, the Mediterranean, and even the Indian Ocean were also found to have a strong influence on the West African rain belt,[18] but confidence in any projected change of specific SSTs and related regional phenomena for the 21st century remains low.[19]

Many regions in Africa continue to be severely understudied, while we also face a lack of resources and expertise within Africa for addressing climate-related issues.[20] Moreover, the density and coverage of weather stations, which is much lower in Africa than in other parts of the world and falls beneath the standards recommended by the WMO, renders rather difficult the compilation

15 IASS, The Future of Africa's Energy Supply, 10.

16 Senior, Catherine et al., "Improving Climate Modelling for Africa," in *Africa's Climate: Helping Decision-Makers Make Sense of Climate Information*, ed. Future Climate for Africa (Cape Town: Future Climate for Africa, 2016), 38–43, here 39.

17 Washington, Richard, Mike Harrison, and Declan Conway, *African Climate Report: A Report Commissioned by the UK Government* (London, Oxford: 2004), 13. For the Congo Basin, see Washington, Richard et al., "Congo Basin Rainfall Climatology: Can We Believe the Climate Models?" *Philosophical Transactions B* 368.1625 (2013), accessed July 24, 2017. doi: 10.1098/rstb.2012.0296; Creese, Amy, and Wilfried Pokam, "Central Africa's Climate System," in *Africa's Climate*, ed. Future Climate for Africa, 4–10. For the West African Monsoon, see: IPCC, *Climate Change 2013*, 1219, 1234.

18 IPCC, Climate Change *2013*, 803; Hartley, Andrew et al., "A Century of Climate Change: 1950–2050," in *Africa's Climate*, ed. Future Climate for Africa, 33–38, here 33.

19 IPCC, *Climate Change 2013*, 23, 106, 1229, 1240–1243.

20 Washington, Richard et al., "African Climate Change: Taking the Shorter Route," *Bulletin of the American Meteorological Society* 87.10 (2006), 1355–1366, here 1358, 1361, 1363.

of well-founded reports on historical, current, or likely future climates.[21] Future climate modelling under these constraints produces contradictory scenarios and thus leads to a greater degree of uncertainty. Nonetheless, some trends and predictions can be identified that are at least of medium confidence. The "Fifth Assessment Report by the IPCC's Working Group I" (WGIAR5) concludes that it is very likely that the whole African continent will continue to warm during the 21st century. While temperature increases will occur within a 2°C range in most geographical areas and under most scenarios, the biases are much larger for some locations.[22]

In comparison to the predicted increases in temperature, projected changes in precipitation are much less consistent across the whole continent, although some changes can be forecast as likely. In the Southern African region, reduced precipitation in the austral winter and an increase in surface temperatures throughout the year are expected, which will lead to the incremental drying of soil moisture towards the end of the century.[23] The AR5 authors also predict that it is very likely that the Sahara will remain very dry. For East Africa, however, rainfall patterns are expected to change, with rainfall likely to increase during the short rainy season.[24] Meanwhile, the long rainy season, which lasts from March to May and has experienced decreasing rainfall and a series of devastating droughts over the past thirty years, is predicted by the majority of climate models to see a reversal of this trend, with an increase in rainfall by the end of the 21st century likely.[25] Furthermore, extreme rainfall events will, with

---

21 In Africa, 1,152 weather stations report to the WMO, with the average station covering an area of more than 26,000 square kilometres. In comparison, Germany hosts 287 operational WMO weather stations covering an average of 1,244 square kilometres each. Even more unfortunate is that not all of the stations in Africa actually record observations and transmit these to the international network. See African Climate Policy Centre (ACPC), *Assessment of Africa's Climatic Records and Recording Networks Including Strategies for Rescuing of Climatic Data* (United Nations Economic Commission for Africa, 2011), 3.; Washington et al., *African Climate Change*, 12; Washington et al., *Congo Basin Rainfall Climatology*, 2.

22 IPCC, *Climate Change 2013*, 761. One of these hotspots is West Africa, particularly the West African drylands, where warming is not only likely to exceed average global warming (1.5–4°C by mid-century), but projections also indicate "that unprecedented changes in climate will occur earliest in these regions, by the late 2030s to early 2040s"; see Hartley et al., "A Century of Climate Change," 34; Climate & Development Knowledge Network (CDKN), *The IPCC's Fifth Assessment Report: What's in it for Africa?* (CDKN, 2014) 18.

23 IPCC, *Climate Change 2013*, 7, 45, 1079.

24 Ibid., 1267–1268.

25 Araujo, Julio et al., "East Africa's Climate: Planning for an Uncertain Future," in *Africa's Climate*, ed. Future Climate for Africa, 11–16, here 12; IPCC, *Climate Change 2013*, 1266; IPCC, *Climate Change 2014: Impacts, Adaptation, and Vulnerability. Part B: Regional Aspects.*

a high likelihood, become more intense and more frequent in the region by the end of the century.[26] However, predictions about the probability and intensity of dry spells are contradictory.[27] For West Africa, general statements about future drying or wetting are of low confidence.[28] Monsoonal rainfall is expected to become more intense, while probable localised effects remain "complex and more uncertain", the AR5 authors point out.[29] It seems likely that there will be an increase in monsoon-related interannual rainfall variability. As in other parts of Africa, the intensity of rainfall events is very likely to increase, although their general frequency is projected to decrease.[30] If correct, the result will be an increase in West Africa's already high interseasonal rainfall variability. Accordingly, Washington et al. note that for Africa in general

> climate change will manifest itself in part, and possibly largely, as a change in the frequency of events that are currently experienced within current climate variability. Consequently, [...] climate change and climate variability are [...] closely coupled in the complicated evolution of the climate system. From a practical perspective, [...] some of the largest impacts of climate change could arise through the superposition of more intense forms of existing modes of variability on the underlying change.[31]

These broad-brush patterns, however, tell us very little about how the weather and changes therein are experienced by individuals in concrete situations. In many ways, it seems presumptuous to speak of 'climate change in Africa' in the first place given the continent's vast dimensions. As Jane Carruthers has pointed out, Africa's extreme diversity of climates, topographies, soil qualities, fauna, and flora defy generalisation.[32] In line with previous research on

---

Contribution of Working Group II to the Fifth Assessment Report of the Intergovernmental Panel on Climate Change, ed. Vicente R. Barros et al. (Cambridge, New York: Cambridge University Press, 2014), 1199–1265, here Chapter 22, 1209–1210.

26 Araujo et al., "East Africa's Climate," 13; IPCC, *Managing the Risks of Extreme Events and Disasters to Advance Climate Change Adaptation: A Special Report of Working Groups I and II of the Intergovernmental Panel on Climate Change*, ed. C.B Field et al. (Cambridge: Cambridge University Press, 2012), 147; IPCC, Climate Change 2013, 106–107, 1268.

27 Araujo et al., "East Africa's Climate," 13. IPCC, *Climate Change 2013*, 968, 1079, 1081.

28 Hartley et al., "A Century of Climate Change," 34.

29 IPCC, *Climate Change 2013*, 1228.

30 Ibid., 105, 1218–1219, 1234, 1268.

31 Washington et al., *African Climate Change*, 1360.

32 Carruthers, Jane, "Environmental History in Africa," in *A Companion to Global Environmental History*, ed. J.R. McNeill (Chichester: Wiley-Blackwell, 2014), 96–115.

human-environment relations, this volume thus does not seek to provide large syntheses, but rather focuses on particular cases in specific parts of Africa.

Together, the chapters in this volume offer a broad, and at times diverse, perspective on climate change. In its direct effects on the ground, climate change qua changing long-term weather patterns is not necessarily any different from more fleeting alterations in the weather, or even other forms of environmental change. For this reason, and in order to harness the "rich cultural knowledge" of people in different parts of Africa,[33] we regard it as neither necessary nor fruitful to limit ourselves to cases that are direct manifestations of climate change; instead, we focus on shifting human-environment relations more generally. Moreover, as far as we do engage with climate change as a subject of research, we need to be precise about what the concept means in a social science context. As has been argued in previous publications, climate change is not directly perceptible, and knowledge about its causes and consequences has to be mediated. It is an abstraction based on aggregated scientific data and only becomes socially relevant in distinct manifestations, such as changing rainfall patterns or extreme weather events. Local perceptions of the weather and unexpected changes therein must be related to scientific discourse in order for the latter to be seen as 'climate change'.[34] Climate change can therefore not be taken as a given, but must be disaggregated. Some of the contributions in this volume thus take distinct manifestations of climate change—such as increasing interannual and interseasonal precipitation variability (Østergaard Nielsen), perceived extreme weather events (Clancy), or a growing likelihood of their occurrence (Nero et al.)—as a starting point to examine coping and mitigation strategies or to reveal to what degree, if any, conventional strategies are modified under climate change conditions. Others, by contrast, adopt a self-reflexive stance, interrogating some of the basic tropes in climate change discourse, like the nature-culture dichotomy (see Kreike), or reflecting critically on the epistemological and ontological implications of climate change in a social science context, as de Wit does in her deliberations on the *absence* of climate change discourse in her research field. Further chapters explore a broad array of human-environment relations and changes therein, including those that have been state-induced (see Fanstone).

In studying the adaptations and environmental perceptions of local residents in various settings on the African continent, the chapters in this volume

33 Hulme, Mike, "Meet the Humanities," *Nature Climate Change* 1 (2011): 177.

34 Cf. Stehr, Nico, and Hans von Storch, *Klima, Wetter, Mensch* (München: Beck, 1999); Strauss, Sarah, and Ben Orlove, "Up in the Air: The Anthropology of Weather and Climate," in *Weather, Climate, Culture*, ed. Sarah Strauss, and Ben Orlove (Oxford: Berg, 2003), 3–14.

adopt a bottom-up approach. Even if we dissect climate change in terms of actual occurrences on the ground, it is still difficult to assess what these represent for affected groups of people. Climate change is generally interpreted as a threat to African communities; nevertheless, as Nancy Jacobs has argued, changing environments are not inevitably perceived in a negative light. Discussing bush encroachment in Kuruman, South Africa, she illustrates how the local population did not interpret the increasing density of vegetation as a problem, but used the bush as fodder. Instead of assuming a tacit acceptance of an optimum ecological status quo that is thrown out of balance when it is changed, social scientists should pay close attention to the—often divergent—interpretations of and within affected communities (compare Clancy in this volume).[35] Similarly, de Wit (in this volume) highlights the importance of the specific interpretative context in which research takes place. Since climate variability is the norm, rather than an exception, in her study region in northern Tanzania, many residents find it difficult to connect with a notion of climate change that presupposes a stable climate as standard. For the Burkinabe informants in Østergaard Nielsen's article, climate change is framed as the dissolution of 'normal' climate variability, thereby illustrating the inextricable manner in which climate change and variability are often interwoven.

3 Climate Change and Causality

The contributions in this volume make a case for not isolating climate or environmental change, but paying close attention to their embedded nature. Environmental factors always interact with several other variables that impact on people's lives—including economic, political, and cultural ones—among which they are not necessarily the most dominant. Østergaard Nielsen (in this volume) has found that circular labour migration—a common and well-documented technique applied by Sahelian farmers to cope with interseasonal and interannual precipitation variability for decades—has become a more and more permanent feature of their lives in recent times. While persisting climate variability and the corresponding difficulties that this creates for farmers looking to resume their agricultural practices have been acknowledged as drivers of this trend, a variety of other factors have also been listed. These include push factors, such as social constraints in the respective villages of

35 Jacobs, Nancy, "Grasslands and Thickets: Bush Encroachment and Herding in the Kalahari Thornveld," *Environment and History* 6.3 (2000): 289–316.

origin compared to the liberties offered by a more anonymous urban lifestyle, as well as pull factors, including increasing job opportunities in the capital, Ouagadougou. Rather than trying to identify sole factors for decision-making, Østergaard Nielsen asks how coping mechanisms are used individually or in combination with one another to manage the risks associated with changing environmental conditions. In a similar vein, Kreike argues that "the environment and climate are embedded in and fractured through social, economic, and political factors" (see Chapter 10 of this volume). The near "killer famine" that Kreike discusses in this volume was the result of a global climate event, a worldwide drought, that hit Southern Africa in the 1920s and 1930s as severely as it affected the USA by causing the Dust Bowl crisis. However, the famine was also the outcome of a worldwide economic crisis as well as regional political and demographic developments. Furthermore, the effects of the famine were not homogenous, but highly context-dependent, as reflected by the fact that some of Kreike's informants remember the crisis as "the enriching famine". Such cases warn us not to resurrect environmental determinism, but to raise the question of causality in an open-ended manner. Contributions in this volume thus treat the environment as a context that "contours" historical events, but "not as a discrete historical actor" in its own right.[36]

Moreover, there can be wide discrepancies between the ways in which different groups of actors experience environmental changes. Climate change is a global phenomenon, but it manifests in very particular ways when analysed from a bottom-up perspective. As Demeritt argues, climate change may hold "little meaning for developing nations and the poor people in them struggling daily [...] with more basic and immediate needs of sanitation, health, and hunger".[37] Further discrepancies can be identified in relation to gender and age. As Clancy (in this volume) has found, there are differences in the ways in which women and men receive information about changing weather patterns and encounter the scientific concept of 'climate change'. While in some cases men obtained their information primarily from radio broadcasts, women tended to gather information from official meetings, enabling them also to ask questions. Similar disparities in how information is received exist between the old and the young.

---

36 McCann, James C., *Green Land*, 47–48. See also McCann, James C., "Climate and Causation in African History," *The International Journal of African Historical Studies* 32.2/3 (1999): 261–279.

37 Demeritt, David, "The Construction of Global Warming and the Politics of Science," *Annals of the Association of American Geographers* 91.2 (2001): 307–337.

4 Knowledge and the Environment

Social aspects of environmental change primarily concern the generation and communication of knowledge, which is a key topic explored by several contributions in this volume. In interrogating different types of knowledge, this volume connects with a growing body of research that addresses historical and contemporary interactions between what have been termed 'local' or 'indigenous' and 'Western' or 'scientific' systems of knowledge. Anthropological and historical research on African human-environment relations has drawn out local cosmologies as well as cultural practices that, intentionally or unintentionally, have resulted in the protection of soils, forests, water, or wildlife. These show how African farmers, pastoralists, and fishermen developed a range of practices—including fallowing and the banning of fishing or hunting for specific time periods—that ensured that natural resources were not overused. In some cases, people consciously reflected on conservation and lobbied for the implementation of corresponding measures at government level.[38] Much of this research has underlined the preservationist ethics of such practices and cosmologies, as well as the ways in which African societies have seen humans as an integral part of nature rather than as occupying a separate sphere. One such example is provided by Jeffery in this volume. Curious about relevant environmental aspects that were rarely mentioned in her interviews on the Maurice Ile Durable sustainability programme, she identified an understanding common to all of her interviewees of the environment as "the totality of one's environs, milieu, surroundings, or physical context, including the social, political, or cultural circumstances therein". Her informants saw "nature and society as interconnected constituent parts of an anthropocentric lived environment" (see Chapter 7 of this volume).

Colonialism and capitalism, it has been argued, have upset these intertwined physical, cultural, and spiritual worlds by introducing a rigid Western

38 See, for instance: Akeampong on fishermen in the colonial Gold Coast and modern Ghana: Akeampong, Emmanuel, *Between the Sea and the Lagoon: An Eco-Social History of the Anlo in Southeastern Ghana, c. 1850 to Recent Times* (Athens: Ohio University Press, 2001); Mulwafu on conservation practices in Malawi: Mulwafu, Wapulumuka, *Conservation Song: A History of Peasant-State Relations and the Environment in Malawi, 1860–2000* (Cambridge: The White Horse Press, 2011); and Khan on conservationist thought and lobbying among African farmers in South Africa: Khan, Fareida, "Soil Wars: The Role of the African National Soil Conservation Association in South Africa, 1953–59," *Environmental History* 2.4 (1997): 439–459; Khan, Fareida, "Rewriting South Africa's Conservation History: The Role of the Native Farmers Association," *Journal of Southern African Studies* 20.4 (1994): 499–516.

nature-culture dichotomy.[39] As a result, parts of such research project a meta-narrative about how African traditional beliefs ensured a sound balance in human-environment relations that was later destroyed under external pressure. This degradation narrative, which highlights deforestation, soil erosion, and the loss of biodiversity, has itself been met with criticism. In particular, some scholars have taken issue with (and partly invalidated) the romanticised notion of a pre-colonial or pre-capitalist past when humans and nature co-existed in harmony.[40] Declensionist narratives were also prevalent in colonial discourses. As several historical studies have shown, colonial officials and experts often identified Africans as being culpable for the degradation of nature that they witnessed. By claiming that Africans' supposedly irrational and wasteful land use—a prime example being overstocking—had destroyed the continent's pristine nature, officials legitimised colonial intervention.[41] After independence, similar discourses were often sustained by governments and international aid agencies alike. The notion of a human-induced decline of nature also resonates with current climate change discourse, although now Africans are cast as the victims of a global process rather than as its perpetrators. Whichever way the blame is shifted, both perspectives are too simplistic. In contrast, the present volume seeks to show how African societies interact with the environments that they inhabit in reciprocal ways and with outcomes that are contingent and multi-causal.

Furthermore, the category of indigenous knowledge itself has been subjected to serious critique. While the notion has become widely adopted among policymakers and activists in the field of development, among whom it enjoys considerable political appeal, scholars have cautioned against slipping into simple dichotomies. Gordon and Krech, among others, have warned that the concept of a sealed package of knowledge belonging to "an unchanging group of indigenes" is a "romantic projection of our modern imaginations to

---

39 Kolkman, Harold, "Inequity and Strife in Community Based Natural Resource Management," in *Culture, Organization and Management in South Africa: In Search of Equity*, ed. Marja Spierenburg, and Harry Wels (New York: Nova Science Publishers, 2004), 111.

40 See, for instance, Luig, Ute, and Achim von Oppen, "Einleitung: Zur Vergesellschaftung von Natur in Afrika," in *Naturaneignung in Afrika als sozialer und symbolischer Prozess*, ed. Ute Luig, and Achim von Oppen (Berlin: Das Arabische Buch, 1995), 5–27, here: 12–16.

41 Compare, for instance, Leach, Melissa, and Robin Mearns, ed. *The Lie of the Land: Challenging Received Wisdom on the African Environment* (Portsmouth: Heinemann, 1996); McCann, James C., "The Plow and the Forest: Narratives of Deforestation in Ethiopia, 1840–1992," *Environmental History* 2.2 (1997): 138–159.

the past".[42] This applies equally to the term 'local knowledge', as even relatively localised bodies of knowledge are diffused and appropriated elsewhere.[43] Also, studies of 'Western' (colonial) science in Africa have provided ample evidence of how it relied heavily on and incorporated the knowledge of Africans.[44] Similarly, Weiser et al. describe the concept of "Adaptation to Climate Change" as a travelling idea that was "conceptually developed at an international level", fed into national programmes and projects, and ideologically and socially negotiated at the local level, thereby undergoing "reinterpretation, modification and appropriation" to meet the different stakeholders' experiences and interests.[45]

Rather than pitting different categories of knowledge against each other, contributions in this volume therefore focus on their interactions. In several chapters, climate change or related notions like "sustainable development" (see Jeffery) appear as a "contact zone",[46] in which different bodies of knowledge meet and are renegotiated. Arguing that research and policy need to "rely on a mutual understanding of numerous and diversely interpreted frameworks on climate change", Eguavoen underlines the need for "translation work to be done between these different systems of knowledge as well as between different interest groups and across various sectors and scales" (see Chapter 8 of this volume).

---

42 Gordon, David, and Shepard Krech III, "Indigenous Knowledge and the Environment," in *Indigenous Knowledge and the Environment in Africa and North America*, ed. David Gordon, and Shepard Krech III (Athens: Ohio University Press, 2012), 1–24, here 1.

43 Gordon, and Krech, "Indigenous Knowledge," 8; see also: Agrawal, Arun, "Dismantling the Divide between Indigenous and Scientific Knowledge," *Development and Change* 26.3 (1995): 413–439.

44 As twoexamples from a vast body of research, see Tilley, Helen, *Africa as a Living Laboratory: Empire, Development, and the Problem of Scientific Knowledge, 1870–1950* (Chicago: University of Chicago Press, 2011); and Tilley, Helen, "Global Histories, Vernacular Science, and African Genealogies; or, Is the History of Science Ready for the World?" *Isis* 101.1 (2010): 114–115.

45 Doevenspeck, Martin et al., "Translations of the 'Adaptation to Climate Change' Paradigm in Eastern Africa," accessed July 26, 2017. http://www.ias.uni-bayreuth.de/de/forschung/forschungsprojekte/projektdetails/index.php?id=108. Weiser, Florian et al., "Translating the 'Adaptation to Climate Change' Paradigm: The Politics of a Travelling Idea in Africa," *The Geographic Journal Volume* 180.2 (2013): 111–119.

46 Compare Pratt, Mary L., "Arts of the Contact Zone," *Profession* (1991): 33–40.

5 Environmental Regulation and Power

Another major theme in this volume are the ways in which climate and environmental change play out politically at the local level. Studies on conservation policy—both historical and contemporary—have shown how environmental concerns, or the pretext thereof, motivated (colonial) state actors to intervene in the lives of African populations. In order to protect soils from erosion, numerous colonial states imposed restrictions on the ways in which local residents farmed their land and kept their livestock—including, for instance, the mandatory building of contour ridges or the compulsory culling of cattle—often provoking fierce resistance in the process.[47] Moreover, white settlers, government officials, and foreign experts cast themselves as stewards of nature, thus claiming authority over contested landscapes, such as game parks or nature reserves, and banning African communities therefrom.[48] Building on such literature, Fanstone (in this volume) discusses *shamba* agroforestry in Kenya, illustrating how the distribution of environmental resources was a key arena in which state-peasant relations were negotiated. *Shamba* allowed landless Kikuyu to farm in forests, while at the same time providing the British colonial authorities with a cost-effective form of plantation forestry. In this way, *shamba* represented both a means of colonial domination and a livelihood for impoverished farmers. The case of state food for work programmes in colonial Namibia, discussed by Kreike in this volume, reveals the similarly ambivalent dynamics of environmental politics. In response to a series of droughts and famines in the late 1920s and early 1930s, the colonial state devised several labour schemes—mainly dam construction projects and other programmes to improve water infrastructure—through which women,

---

47 See, for instance: Carswell, Grace, "Multiple Historical Geographies: Responses and Resistance to Colonial Conservation Schemes in East Africa," *Journal of Historical Geography* 32.2 (2006): 398–421; Anderson, David, and Richard Grove, ed. *Conservation in Africa: People, Policies and Practice* (Cambridge: Cambridge University Press, 1999); McCracken, John C., "Conservation and Resistance in Colonial Malawi: The 'Dead North' Revisited," in *Social History and African Environments*, ed. William Beinart and JoAnn MacGregor (Athens: Ohio University Press, 2003), 155–174; Beinart, William, "Soil Erosion Conservationism and Ideas about Development: A Southern African Exploration, 1900–1960," *Journal of Southern African Studies* 11.1 (1984): 52–83. See also Kreike in this volume.

48 See, for instance: Carruthers, Jane, *The Kruger National Park: A Social and Political History* (Pietermaritzburg: University of Natal Press, 1995); Gißibl, Bernhard, *The Nature of German Imperialism. Conservation and the Politics of Wildlife in Colonial East Africa* (New York, Oxford: Berghahn Books, 2016); Mackenzie, Fiona, Land, Ecology and Resistance in Kenya, 1880–1952 (Edinburgh: Edinburgh University Press, 1998).

children, and older men could earn food. As Kreike shows, environmental distress enhanced the power of the state, which would step in to offer the assistance that local patronage networks had ceased to provide. Simultaneously, the colonial administration, lacking the staff to supervise food distribution, had to rely on local elites, whose authority among the population it thus bolstered. In both Kreike's and Fanstone's case studies, the power derived from the control of environmental resources was never monolithic, but was instead negotiated between the state and the inhabitants. Such dynamics have not been confined to the colonial era. As Kreike argues, post-colonial environmental discourses have displayed a similar tendency to attribute resource problems—such as famines or 'overpopulation'—to specific aspects of African culture. Similarly, David Hughes has drawn analogies between the ways in which European colonisation reshaped rural politics in the Zimbabwe-Mozambique border zone and the ongoing efforts of states and investors to commodify African land under the banner of environmental protection and development.[49] De Wit (in this volume) offers a further telling example of how climate change can be instrumentalised by the state as a one-size-fits-all explanation for local problems, thereby obfuscating state responsibility. The Tanzanian government, she argues, employs climate change to naturalise and depoliticise problems of poverty that are primarily the result of state neglect and discrimination.

At the same time, such tensions should not be conceptualised in terms of a simple state-population dichotomy. By highlighting the agency of individuals, research on environmental policy has demonstrated how people do not merely succumb to state-expert discourses, but appropriate and challenge them. De Wit argues that climate change can also be utilised as a counter-discourse in this manner, for example in the context of her case study by constituting a means for local NGOs to criticise the Tanzanian government and its top-down style of planning. More generally, scholars have shown that the designation of knowledge as 'indigenous' may also represent "an intervention in power relations". During colonial indirect rule as much as in the context of modern development initiatives, which claim to be sensitive to local perspectives, claims to indigeneity have been able to add weight to local groups' demands. At the same time, this should not obscure the vastly asymmetrical power relations in which claims to environmental knowledge and resources are upheld. Even development initiatives that are inspired by the perspectives of the marginalised

49 Hughes, David McDermott, *From Enslavement to Environmentalism: Politics on a Southern African Frontier* (Seattle: University of Washington Press, 2006).

have been shown to have been "more neoliberal than previous state-oriented strategies".[50] Whether a critical or more optimistic reading of the environmental politics negotiated between states and their populations is adopted, we deem it important to contextualise current climate change discourse. It is imperative for scholars as much as for developers or international and state bodies to be aware of the ways in which narratives of degradation and decline have—for a century and more—been used, or even exploited, by state actors, settlers, businesses, and development experts alike to claim authority over vast spaces and resources, cast themselves as saviours, and intervene in people's lives.

At the same time, the African continent has been associated with weak states and limited governmental power, which is as true for Africa's past as it is for the contemporary age. Contemporary adaptation and mitigation strategies are often hampered by an inability to implement them due to systemic institutional malfunction and a lack of technical, managerial, and financial means on the part of state actors. Consequently, other bodies, such as NGOs or community-based organisations (CBOs), or private citizens are left to take the initiative. For example, by highlighting the roles of traditional leaders and individual landowners, Nero et al. (in this volume) show how de-facto bottom-up processes drive the creation and management of urban green spaces in Kumasi, Ghana.

6 African Exceptionality?

This volume is the fifth in a series covering different regions of the world. The question of whether there is anything specific to the geographic region known as Africa that would justify such a structure is a difficult one. In our understanding, the answer must be pragmatic. Our volume offers snapshots, all of which may share commonalities, but which together do not amount to a broad synthesis of this highly diverse continent. In climate change discourse, what sets 'Africa' apart from other world regions is its particularly high vulnerability. This resonates with debates in environmental historiography, which have tended to highlight the hostility of African environments. As one of the few historians who has adopted a Braudelian long-term perspective on environmental and demographic change, John Iliffe has famously referred to Africans as "the frontiersmen of mankind", "who have colonized an especially hostile

50 Gordon, and Krech, "Indigenous Knowledge," 1–24.

region of the world on behalf of the entire human race".[51] Africans, he claims, were confronted with an environment—"poor soils, fickle rainfall, abundant insects, and unique prevalence of disease"—that rendered human existence more difficult than in any other part of the globe.[52] While subsequent research has not fundamentally disputed Iliffe's analysis, there is a danger of essentialising African environments and societies and of reintroducing an alarmist-interventionist discourse. As Kreike argues in this volume, the perception of Africa as the continent most vulnerable to climate change is based on the cliché that African societies are highly dependent on their natural environments and not protected by a modern technological infrastructure in ways in which Western societies are (at least whenever the latter are not reminded of the contrary, as we have seen at the beginning of this introduction). There may be some justification in regarding Africa as 'special' for its difficult environment and relative lack of mitigating capacities, including technology, infrastructure, and financial power; however, what is most certainly unique about the continent in comparison to Europe is the extent to which outsiders—colonisers, foreign enterprises, international organisations—have defined Africa's environmental problems and sought to regulate them.

51 Iliffe, John, *Africans: The History of a Continent* (Cambridge: Cambridge University Press, 2007), 1.

52 Ibid.

PART 1

Ideas

CHAPTER 1

To See or Not to See: on the 'Absence' of Climate Change (Discourse) in Maasailand, Northern Tanzania

Sara de Wit

Abstract

This chapter explores how climate change as a global idea travels to and is (only marginally) translated in Simanjiro, Maasailand, in northern Tanzania. It examines the possible reasons for why the idea of climate change is largely rejected in the village of Terrat. While the Maasai are said to be among the most vulnerable communities to the future effects of climate change, their own understandings of climate and environment bear little resemblance to this travelling idea. This chapter interrogates a largely neglected question within the climate change research agenda: how to approch the absence of climate change discourses and an apparent lack of the typically all too conspicuous concerns about experiential climate change realities? By doing so, it critically engages with the current climate change research agenda in the social sciences and the humanities and questions the general tendency to 'see' and thus construct climate change as an overall dominating reality and single determinant of lifeworlds 'on the ground'. It thus explores the taken-for-grantedness of the supposedly universal, all-embracing, and threatening biophysical effects of climate change, while questioning the bases for and effects of these knowledge claims.

1 Introduction

This chapter explores the possible reasons for both why the idea of climate change is sometimes rejected as well as why—in some contexts—it is not a commonly articulated concern. More specifically, it interrogates the reasons as to why the idea of climate change is wholeheartedly embraced by most actors along the translation chain from international to regional actors until it reaches the village of Terrat in Maasailand, where it is to some extent questioned and rejected, or only hesitantly adopted. During my fieldwork in Tanzania, climate change discourses were omnipresent throughout policymaking circles and regularly featured on the radio, in newspapers, at conferences and workshops,

 | DOI:10.1163/9789004410848_003

and in daily conversation. Yet, I encountered few traces of climate change narratives in Terrat. While some informants gave accounts of the lack of rain that they experience nowadays, others perceived climate variability as being part and parcel of their normal lives. This 'climate change void' has fascinated me in my fieldwork ever since I was confronted with this absence of what I was looking for, at least as far as Maasailand was concerned. The data that is presented here is based on 14 months of multi-sited ethnographic fieldwork, which I carried out between 2012 and 2014.[1] In this research, I moved between (inter)national and local platforms where the idea of climate change was translated by different stakeholders and policymakers before reaching Maasai communities in northern Tanzania. The central argument that I put forward in this chapter is a simple one, namely that we have to take the absence of data—in this case, the lack of climate change narratives and observations—seriously. How does one write about and give meaning to something that is not overtly present?

This chapter interrogates some theoretical concerns related to the question of why ideas sometimes do *not* travel or are not adopted.[2] Against this background, the absence of climate change 'realities' will be juxtaposed with the presence of other concerns, including those influenced by the ways in which earlier environmental paradigms have entered Maasailand. By contextualising older discourses that have profoundly shaped human-environment relations in Maasailand, such as conservation, this chapter sheds light on how the idea of climate change is entangled with current socio-political structures in both old and novel ways. I begin by discussing some epistemological considerations on the notion of absence, before continuing with more in-depth analysis of my research findings.

1 This research formed part of the DFG-funded priority programme SPP 1448, 'Adaptation and Creativity in Africa—Technologies and Significations in the Production of Order and Disorder'. I wish to thank the DFG for providing financial support. This is a reworked chapter from my dissertation: *Love in Times of Climate Change: How an Idea of Adaptation to Climate Change Travels to Northern Tanzania* (Cologne: University of Cologne, 2017).

2 While the focus of this paper is mainly on Terrat village, the rejection of climate change was prevalent among educated Maasai too. During several workshops that I attended the adaptation to climate change discourse was at times (emotionally) contested by Maasai participants from the grassroots and NGO representatives, on the basis of the claim that they have other things to worry about than climate change, such as land alienation in relation to conservation issues. I have written about the politics of adaptation elsewhere: "A Clash of Adaptations—How Adaptation to Climate Change is Translated in Northern Tanzania," in *A Critical Approach to Climate Change Adaptation. Discourses, Policies, and Practices*, ed. Silja Klepp and Libertad Chavez-Rodriguez (New York: Routledge, 2018), 37–54.

2 On Seeing Climate Change

> We tend to find what we are looking for. This does not mean that there is no real basis to what we find. Rather, it is just that reality has a tendency to reveal itself in accordance with the perspectives through which it is engaged.
>
> GARETH MORGAN[3]

Since climate change's rise as a key topic for anthropological analysis and ethnographic inquiry over the last two decades, it has, perhaps to nobody's surprise, been accompanied an anthropological lens that captures and encounters the phenomenon almost everywhere across the globe. Of course, anthropologists do not see or feel climate change with the naked eye or their bare senses, for it is a phenomenon that is—unlike its manifestations—not perceptible in unmediated form.[4] However, we ask questions about it, and very frequently we receive satisfying answers that by and large testify to the fact that climate change is real and that it is already being experienced by 'vulnerable', or 'frontline', communities on the ground. As we all do, the communities that are generally under anthropological climate change scrutiny (i.e. those in the Global South) have a tangible relationship with the weather, for they feel and experience short-term weather patterns as much as they give meaning to longer-term climatic conditions based on their experiential realities. It is often assumed and stated that communities that are more 'nature-bound' and whose livelihoods are directly dependent on nature have a much more detailed, attached, and thorough understanding of their environment, thus making them climate change witnesses par excellence.[5] Nonetheless, we should not forget that the ways in which we give meaning to the weather and the climate, and the knowledge that is generated in the process, is always the result of a mediated process that is given life through the categories we have at hand and the language that we use.[6]

3 Morgan, Gareth, *Images of Organization* (London, Delhi: Sage, 2006), 338.

4 I am aware that this is not an uncontested statement, but it should principally be understood as a philosophical position that holds that as soon as we claim that a certain climatic event is the result of climate change, this assertion is always mediated by language.

5 Some scholars have also questioned this dichotomy pitting a more nature-dependent 'South' against a detached and industrialised 'North'. They argue that this distinction is exaggerated as both are as dependent on the environment as the other; see, for example, Emmanuel Kreike (Chapter 3 of this volume).

6 By proposing a theory of direct perception, Tim Ingold has criticised such a 'cognitivist account of perception'—which holds that people can only know and act upon their

Furthermore, it is not only impossible to isolate natural climate variability from its anthropogenic counterpart through observation, but it is all the more difficult to translate a global statistical abstraction into a personal and tangible reality.[7] Yet, climate change realities and testimonies abound within anthropological studies and social science research. Only recently have anthropologists advanced a more critical position that scrutinizes the bases for such knowledge claims.[8] In a similar vein, this chapter seeks to build upon this increasing epistemological 'climate change sensibility' by arguing that reflexivity towards the concepts that we employ[9] could prove a welcome corrective to the overwhelming climate change realities that we bring into being with narrations of climate change and crisis as each other's cognate.

By way of reflecting on concepts, a crucial distinction must be made between direct (sensory) *observation* of changing environmental and climatic phenomena, on the one hand, and having access to climate change information and knowledge, or the so-called *reception* of secondary sources, on the other.[10] This distinction is important, since having access to the idea of climate change—in the form of scientific knowledge or discourse—inevitably informs and shapes the ways in which people give meaning to and speak about their

---

environment indirectly through the medium of cultural representations—whose roots lie, according to Ingold, in a Western, dualistic worldview. He instead proposes the notion of mutualism between people and the environment, an idea that I embrace. See: Ingold, Tim, *The Perception of the Environment: Essays on Livelihood, Dwelling and Skill* (London: Routledge, 2000), 39–40. Nevertheless, the point here is related to knowing *climate change* directly, which is inevitably a second-order observation.

7 Cf. Jasanoff, Sheila, "A New Climate for Society," *Theory, Culture & Society* 27.2–3 (2010): 233–253.

8 Hastrup, Kirsten, "Comparing Climate Worlds: Theorising Across Ethnographic Fields," in *Grounding Global Climate Change: Contributions from the Social and Cultural Sciences*, ed. Heike Greschke and Julia Tischler (Dordrecht: Springer, 2015), 139–154; Greschke, Heike, "The Social Facts of Climate Change: An Ethnographic Approach," in *Grounding Global Climate Change: Contributions from the Social and Cultural Sciences*, ed. Heike Greschke and Julia Tischler (Dordrecht: Springer, 2015), 121–138; Rudiak-Gould, Peter, " 'We Have Seen It with Our Own Eyes': Why We Disagree about Climate Change Visibility," *Weather, Climate and Society* 5 (2013): 120–132; Rudiak-Gould, Peter, "The Influence of Science Communication on Indigenous Climate Change Perception: Theoretical and Practical Implications," *Human Ecology* 42.1 (2014): 75–86.

9 Cf. Uekötter, Frank, "You Ain't Seen Nothing Yet: A Death-Defying Look at the Future of the Climate Debate," in *Grounding Global Climate Change: Contributions from the Social and Cultural Sciences* (Dordrecht: Springer, 2015), 175–181.

10 Rudiak-Gould, Peter, "Climate Change and Anthropology: The Importance of Reception Studies," *Anthropology Today* 27.2 (2011): 9–12; Rudiak-Gould, Peter, "Promiscuous Corroboration and Climate Change Translation: A Case Study from the Marshall Islands," *Global Environmental Change* 22.1 (2012): 46–54.

(changing) environments and lifeworlds.[11] Yet, as anthropologist Peter Rudiak-Gould has demonstrated, the treatment of these two distinct sources of information in anthropological climate change research has not come without epistemological confusion. Firstly, the majority of research has largely focused on observation rather than reception, while studies on the interplay between the two are largely missing. Secondly, and of more relevance to the argument that will be put forward in this chapter, climate change accounts by local or 'indigenous' communities are predominantly treated as being solely the fruit of direct observation.[12] However, in this ever more interconnected world in which people increasingly have access to global information flows, local climate change accounts are often the result of the interaction between direct observation and translated information.[13]

In a similar vein, Julie Cruikshank, among other anthropologists, has queried the category of 'local knowledge', emphasising that local knowledge is not something static or timeless that can be 'discovered', but that it is instead continuously being brought into being via human encounters and could possibly also be used to inform science.[14] This attests to the idea that knowledge does not emerge in isolation, but is always in motion and a co-production involving many different people and manifesting itself in many different forms. Hastrup shares such a perspective on climate knowledge, which allows for an understanding of the ways in which perceptions and knowledge continuously come into being by focusing on the entanglement of places and epistemic practices. According to Hastrup the task of anthropology is "to show how such potentially incongruent sources and knowledge practices fuse into a shared knowledge space, upon which people may act. In the process, traditional cosmologies may bend and twist, incorporating new national ideologies as well as scientific

11 The asymmetries regarding media coverage and people's concern for global environmental problems also came to the fore in the ethnographic research carried out by the research team of the *Climate Worlds* project. See: Greschke, "The Social Facts of Climate Change"; De Wit, Sara, *Global Warning: An Ethnography of the Encounter Between Global and Local Climate-Change Discourses in the Bamenda Grassfields, Cameroon* (Bamenda, Leiden: Langaa & African Studies Centre, 2015).

12 Rudiak-Gould, "The Influence of Science Communication on Indigenous Climate Change Perception," 75.

13 De Wit, Sara, Arno Pascht, and Michaela Haug, "Translating Climate Change: Anthropology and the Travelling Idea of Climate Change—Introduction," *Sociologus* 68.1 (2018): 1–20.

14 Cruikshank convincingly demonstrates in her collection of oral histories that glacier stories both speak of geophysical changes as well as depict human encounters with European colonials. See: Cruikshank, Julie, "Melting Glaciers and Emerging Histories in the Saint Elias Mountains," in *The Anthropology of Climate Change: An Historical Reader*, ed. Michael Dove (Boston: Wiley Blackwell, 2014), 263.

knowledge from many sources."[15] In this chapter, the term 'climate change realities' refers to this complex set of interactions between knowledge that constantly takes form through both sensory observation and the reception of information as practices that take (and make) place in a particular 'knowledge space'. Two theoretical propositions inform my overall argument.

First, it is posited that if we wish to make sense of local or 'located' climate change accounts, while acknowledging their biophysical and sensory underpinnings we also have to be attentive to the ways in which climate change travels as an idea that entangles with, shapes, and is shaped by varying perceptions and lifeworlds in an array of intricate ways.[16] Second, environmental or climatic changes often coincide with societal changes;[17] hence climate change discourses can provide "the 'imaginative grist' for making sense of broader societal changes."[18] This awareness is important for the following analysis, in which a conundrum will be addressed: what do we do when—considering both reception and observation as sources of information—our findings do not immediately match global climate change narratives, in which a looming crisis is assumed for vulnerable populations in the Global South? Furthermore, how can we make sense of the fact that the communities under study are, to

15 Hastrup, "Comparing Climate Worlds," 142.

16 Weisser, Florian, Michael Bollig, Martin Doevenspeck, and Detlef Müller-Mahn. "Translating the 'Adaptation to Climate Change' Paradigm: the Politics of a Travelling Idea in Africa," *The Geographical Journal* 180.2 (2014): 111–119, accessed January 23, 2015. doi: 10.1111/geoj.12037; Eguavoen, Irit, Karsten Schulz, Sara de Wit, Florian Weisser, and Detlef Müller-Mahn, "Political Dimensions of Climate Change Adaptation: Conceptual Reflections and African Examples," in *Handbook of Climate Change Adaptation,* ed. Walter Leal Filho (Berlin, Heidelberg: Springer, 2015), 1183–1198; Hulme, Mike, "Geographical Work at the Boundaries of Climate Change," *Transactions of the Institute of British Geographers* 33 (2008): 5–11; Hulme, Mike, *Why We Disagree About Climate Change: Understanding Controversy: Inaction and Opportunity* (Cambridge: Cambridge University Press, 2009); Hulme, Mike, "The Idea of Climate Change: Exploring Complexity, Plurality and Opportunity," *GAIA* 19.3 (2010): 171–174; Hulme, Mike, "Climate and its Changes: a Cultural Appraisal," *Geo: Geography and Environment* 2.5 (2015): 1–11; Bollig, Michael, and Sara de Wit, "Commentary on: Orlove, Ben, Heather Lazrus, Grete K. Hovelsrud, and Alessandra Giannini, Recognitions and Responsibilities: On the Origins and Consequences of the Uneven Attention to Climate Change around the World," *Current Anthropology* 55.3 (2014): 262–264; De Wit, Sara, "Victims or Masters of Adaptation? How the Idea of Adaptation to Climate Change Travels Up and Down to a Village in Simanjiro, Maasailand Northern Tanzania," *Sociologus* 68.1 (2018): 21–42.

17 Dove, Michael R., "Introduction," in *The Anthropology of Climate Change: An Historical Reader,* ed. Michael Dove (West Sussex: Wiley-Blackwell, 2014), 1–36; Boersema, Jan J., *Beelden van Paaseiland: Over de Duurzaamheid van een Cultuur* (Amsterdam, Antwerp: Atlas), 2011.

18 Cruikshank, "Melting Glaciers and Emerging Histories," 266.

varying degrees, cognisant of the scientific rationale underpinning climate change, yet either reject the idea or do not embrace it? These questions will be explored on the basis of the fieldwork that I carried out among the Maasai (agro)pastoralists on the Simanjiro plains in northern Tanzania.

More specifically, the subsequent sections address the following questions: How is climate change translated in Terrat, northern Tanzania, and how do different truth regimes fuse or confuse as they encounter one another in a particular place? How can we make sense of the overall absence and rejection of climate change discourses as well as the lack of otherwise all too conspicuous experiential climate change realities? Based on findings (or rather the lack thereof) that are similar to my own, Greschke and her research partners have raised questions that I deem relevant in this context: "Are they [the informants] all 'sceptics' or ignorant?" And: "Do we primarily have to enlighten our research fields about the 'real' causes and dynamics of the global socio-ecological system [...]?"[19] These questions bring us back to old anthropological concerns about the relationship between the researcher and the subject of the research—and the models of abstraction that we use and generate—that deserve more critical attention within current climate change debates.

3 Blinded by Sight[20]

Anthropology and the social sciences have embarked rather late upon climate change research, which was first considered a topic of primary interest for the natural sciences and was thus largely confined to technical and positivist inquiry.[21] Nowadays we can speak of a new climate change paradigm that has progressively set the stage for the anthropological research agenda. This disciplinary engagement is to be embraced as a re-examination of climate change by the more interpretative sciences, contributing to a critical understanding of the global and universaling tendencies that climate change discourses can

19 Greschke, "The Social Facts of Climate Change," 124.

20 I took the liberty of borrowing the title of this subheading from Francis Nyamnjoh's excellent article on postcolonial power structures in academia. See: Nyamnjoh, Francis, B., "Blinded by Sight: Divining the Future of Anthropology in Africa," *Africa Spectrum* 47.2–3 (2012).

21 Buttel, Frederick H., Ann P. Hawkins, and Alison G. Power, "From Limits to Growth to Global Change: Constraints and Contradictions in the Evolution of Environmental Science and Ideology," *Global Environmental Change* 1.1 (1990): 57–66; Hulme, Mike, "Meet the Humanities," *Nature Climate Change* 1 (2011): 177–179.

bring about[22] and yielding insights into the discordances between locally grounded approaches and global models and discourses.[23] Yet, this climate change focus similarly bears the risk of engendering blind spots, for it increasingly serves as an all-embracing explanatory framework. I contend that we should not lose sight of the reasons why climate change realities are not to be found everywhere we look, as is imagined and encapsulated in global climate change narratives that by and large evoke a sense of crisis.

Given the burgeoning global awareness of human-induced climate change, not only anthropologists but also climate-cognisant communities across the globe are increasingly prone to explaining environmental changes through a climate change lens. This should in turn lead us to reflect on the local accounts that we as anthropologists assemble and that we tend to accept as direct proof of climate change all too easily. Mike Hulme has criticised the way scholars have elevated climate change to become the dominant predictor variable in a complex matrix of interdependencies as a new form of climate determinism or 'reductionism'.[24] Drawing on this epistemological critique, this chapter is an ethnographically based discussion of *absence*, a phenomenon that has been given scant attention within the wider climate change literature.

As such, we must also allow for the possibility that the increasing 'visibility' of climate change on the ground could be (at least partially) the consequence of a new research focus and corresponding paradigm, or, as Gareth Morgan has framed it, our tendency "to find what we are looking for."[25] By exploring or positing this alternative view, one clearly enters a slippery ontological road, for fear of being accused of neglecting the severity of climate change impacts across the globe. Yet the point here is not (merely) a constructivist one—in the sense that language is our access to reality—but primarily addresses issues of representation, namely the relationship between the anthropologist and the

---

22 Hulme, Mike, "Problems with Making and Governing Global Kinds of Knowledge," *Global Environmental Change* 20 (2010): 558–564; Wisner, Ben et al., "Let Them Eat (Maize) Cake: Climate Change Discourse, Misinformation and Land Grabbing in Tanzania" (paper presented at "The International Conference on Global Land Grabbing II," New York, October 17–19, 2012).

23 Greschke, and Tischler, "Grounding Global Climate Change"; Hastrup, Kirsten, ed., *The Social Life of Climate Change Models: Anticipating Nature* (New York, London: Routledge, 2014).

24 Hulme, Mike, "Reducing the Future to Climate: A Story of Climate Determinism and Reductionism," *Osiris* 26.1 (2011): 247. For an example of climate reductionism and the construction of Maasai vulnerability in Kenya see: Filho, Walter, Nzengya, Daniel, Muasya, Gladys, Chemuliti, Judith, and Wanzuu Kalungu, Jokastah, "Climate change responses among the Maasai Community in Kenya," *Climatic Change* 145.1–2 (2017), 71–83.

25 Morgan, Images of Organization, 338.

empirical reality that he or she seeks to describe. For example, Edmund Leach, among others, criticised the idea that the social structure of a society is directly observable, insisting instead that the analyst created an abstract model that was the product of a particular way of looking at the world. As such, his criticism was directed at how anthropological analyses create 'entities' by lifting them out of the space and time of social interaction.[26] It is in this context that Greschke,[27] drawing on Albert Schütz, has pointed out the underlying difference between physical and social facts, thereby reminding us that natural and social climate scientists begin from fundamentally different realities. Whereas natural scientists focus on the physical facts, social scientists deal predominantly with the social facts of global warming, and hence with "second-order observations" as opposed to "first-order observations" or the biophysical effects of climate change.[28] By drawing attention to climate change as an explanation of, and primary determinant for, social realities and vulnerabilities on the ground, anthropological climate change models risk concealing as much of human-environment interactions as they seek to reveal.

Anthropologist Kirsten Hastrup has rightfully criticised the manner in which a climate change lens has been used to push interpretations that focus on the devastating impacts of global warming on local communities, as "[...] climate is no longer seen to *make* places but rather mostly to *destroy* them, with anthropologists called upon to mediate local understandings through their incomparable method of fieldwork."[29] This apocalyptic eye has by and large found ways of seeing and talking about communities in the Global South—with particular urgency in the case of sub-Saharan Africa and small island developing states—in terms of climate vulnerability, or an "adaptation deficit."[30] One consequence of global narratives on climate change that regard frontline communities in the Global South as victims is that it deprives them of their agency, thereby "redirecting their fate out of their hands."[31] The agropastoral

---

26 Leach, Edmund, 1970, in *Anthropology in Theory: Issues in Epistemology,* ed. Henrietta L. Moore, and Todd Sanders (West Sussex: Wiley-Blackwell, 2014), 1–18: here 6.

27 Greschke, "The Social Facts of Climate Change."

28 Albert Schütz, 1953, in Greschke, "The Social Facts of Climate Change," 129.

29 Hastrup, "Comparing Climate Worlds," 146.

30 IPCC, "Summary for Policymakers," in *Climate Change 2013: The Physical Science Basis. Contribution of Working Group I to the Fifth Assessment Report of the Intergovernmental Panel on Climate Change*, ed. T.F. Stocker et al. (Cambridge, New York: Cambridge University Press, 2013).

31 Cf. Farbotko, Carol, and Heather Lazrus, "The First Climate Refugees? Contesting Global Narratives of Climate Change in Tuvalu," *Global Environmental Change* 22.2 (2012): 382–390.

Maasai communities in Tanzania are considered to be among the most vulnerable in the country to the consequences of climate change.[32] However, NGOs representing the Maasai in Tanzania have a different view of the pastoralists' vulnerability. Instead of seeing them as climate change victims, they consider them to be "masters of adaptation,"[33] for their ways of life not only boast a long history of adapting to climatic fluctuations,[34] but historical evidence shows that their nomadic lifestyles have even sprung therefrom.[35]

Finally, scholars have pointed to the danger of 'naturalising' when using the climate change lens as an explanatory framework. This occurs in particular by erasing the global political economy from the focus of analysis, with poverty and vulnerability increasingly being explained by the natural and (seemingly) apolitical forces of the climate.[36] In the case of the agropastoral Maasai, who have had a longstanding problematic relationship with the government of Tanzania, climate change discourses are indeed embraced by the latter as the ultimate scapegoat to explain away the deplorable socio-economic conditions experienced by the Maasai.[37] As will be demonstrated below, by retrofitting older conservation discourses the government seeks to obscure ongoing practices of land alienation and the fact that thousands of Maasai are increasingly losing their rights in the name of conservation. In this context the discourse on adaptation to climate change is also embraced by regional NGOs as an opportunity

32 United Republic of Tanzania (URT), *National Adaptation Program of Action* (NAPA for the UNFCCC), 2007.

33 De Wit, "Victims or Masters of Adaptation?"

34 Tanzania Natural Resources Forum (TNRF), "Integrating Pastoralist Livelihoods and Wildlife Conservation? Options for Land Use and Conflict Resolution in Loliondo Division, Ngorongoro District," *Policy Paper*, 2011.

35 Saringe, Emanuel, "Impact of Climate Change to Pastoralist and Hunter-Gatherer Communities of Tanzania: A Case Study of Kiteto and Simanjiro Districts," *Pastoralist Indigenous Non Governmental Organizations (PINGO's) Forum*, 2011.

36 See also Ferguson's "Anti-Politics Machine" on development: Ferguson, James, *The Anti-Politics Machine. 'Development', Depoliticization, and Bureaucratic Power in Lesotho* (Cambridge: Cambridge University Press, 1990); Eguavoen et al., "Political Dimensions of Climate Change Adaptation"; Gesing, Friederike, Johannes Herbeck, and Silja Klepp, ed., *Denaturalizing Climate Change: Migration, Mobilities and Space.* Artec Paper 200, 2014, accessed November 3, 2015. http://www.uni-bremen.de/fileadmin/user_upload/single_sites/artec/artec_Dokumente/artec-paper/200_paper.pdf; Wisner et al., "Climate Change Discourse, Misinformation and Land Grabbing in Tanzania"; Smucker, Thomas A et al., "Differentiated Livelihoods, Local Institutions, and the Adaptation Imperative: Assessing Climate Change Adaptation Policy in Tanzania," *Geoforum* 59 (2015): 39–50; De Wit, "Denaturalizing Adaptation, Resocializing the Climate."

37 Bollig, and De Wit, "On the Origins and Consequences of the Uneven Attention to Climate Change around the World."

to bring about new pathways of development for the Maasai. However, the idea of "masters of adaptation" should also be seen as a counter-discourse that at the same time has also provided ground for criticising the top-down nature of government planning, competing land uses, and incompatible policies that are limiting the communities in their traditional planning mechanisms.[38]

4 Methods

From 2012 to 2014, I carried out fourteen months of ethnographic fieldwork. During this research I moved between (inter)national and local platforms, where the idea of climate change was translated among different stakeholders and policymakers before reaching Maasai communities in northern Tanzania. Trying to understand how climate change realities came about in Terrat, I employed participant observation as a basic ethnographic method, which mainly consisted of walking, herding cows, fetching water, going to the market, spending time with the local men and women, cultivating the land, and attending church services. Moreover, I employed a set of mixed (yet predominantly qualitative) methods, such as in-depth interviews, focus group discussions, and a survey. All interviews were conducted with a research assistant who translated between English, the Maa language, and occasionally Swahili (depending on the informant and the concepts that we were using). The majority of the local residents speak Swahili, but the main language in Terrat is Maa. However, since many educational activities also occur in Swahili, knowledge and information about climate change was often communicated in this lingua franca. Officially, there is no accepted translation for 'climate change' in Maa yet, but some attempts have been made by the local radio station (a point to which I will return below). My choice of interlocutors aimed at a balance between both female and male respondents as well as different age groups, since Maasai society is structured along age sets (with each person staying within his or her own age set throughout his or her life). Furthermore, I talked to informants from different social layers, such as Christians and non-Christians (although the majority of society is Christian), richer and poorer families, and families that both cultivate crops and keep cattle and families that only engage in pastoralism. In order not to bias my informants, I adhered to a methodology that Rudiak-Gould

38 Tanzania Natural Resources Forum (TNRF) and International Institute for Environment and Development (IIED), "Implications of Climate Change for Drylands Planning in Tanzania at District and National Levels: Opportunities and Challenges" (report of workshop held in Arusha, May 29–31, 2012).

has framed as "following best practices in measuring climate change perceptions."[39] I therefore did not announce myself as a climate change researcher, but explained that I was doing research about the environment.[40] Moreover, I always began interviews with general questions about people's livelihoods, and only towards the end, if they would not talk about it themselves, did I gradually address more weather-specific topics. This enabled me to understand whether my informants would spontaneously speak about changes in the weather, seasons, or climate.[41]

Moreover, in addition to 'being there', I also followed trajectories of 'getting there', tracing the idea of climate change along its varied and multiple trajectories from international platforms to national translation forums and ultimately to Terrat itself. I analysed several media sources such as radio programmes, newspaper articles, NGO talks, and educational video material. Furthermore, I attended workshops, conferences, and sensitisation meetings where Maasai pastoralists were exposed to (new) climate change information. While my stay in Terrat was primarily intended to be devoted to grasping processes of translating climate change, due to its apparent absence, I was ultimately, and naturally, bound to pursue the task of making sense of what my informants actually conveyed to me. As we shall see below, Terrat is inhabited by a society that finds itself enmeshed in a complex web of historically situated political discourses and environmental paradigms that have profoundly shaped human-environment relations.

5 Study Site: Terrat Village and the Interpretive Context

As the map below indicates, the village of Terrat lies in the midst of Tanzania's Maasailand, on the northeastern outskirts of Tarangire National Park. The village is home to some 6,000 agropastoral Maasai people, who predominantly live in shared enclosed compounds (*boma* in Swahili, *engang* in Maa) scattered

39 Rudiak-Gould, "The Influence of Science Communication on Indigenous Climate Change Perception."

40 In Swahili, the word *mazingira* means environment. It is a well-known word for people in Terrat who understand Swahili, because the radio station that is situated in the village (which is the only radio station that broadcasts programmes in both the Maa language and Swahili) has a weekly programme that discusses topics related to the environment. The word signifies much the same as what we mean by 'the environment', including wildlife.

41 Cf. Rudiak-Gould, "The Influence of Science Communication on Indigenous Climate Change Perception," 76–77.

across the vast plains of Simanjiro. In the case of a good year, during the long rainy season, the area is exceptionally beautiful and blessed with an abundance of lush pastures, blossoming flowers, and sufficient crops like maize or beans to feed whole families. When the rains arrive on time, if they arrive at all, people and cattle alike will have the chance to recover from and regain their strength after the harsh dry season. In this period of abundance, there is no need for long-distance movement, since grass and water are usually found in the local vicinity. The contrast with the dry season could not be sharper, during which the area turns into a semi-desert, with dust clouds covering the pastures and land, water sources drying up, and food for both people and cattle gradually diminishing. This is the time of the year when the *illmuran* (circumcised boys, or young men often glossed as warriors) begin their search for greener pastures and water sources.

It might go without saying that the Maasai are inextricably bound to their cattle, since this symbiotic relationship is at the heart of how they sustain their livelihoods and forms the fabric of socio-cultural life. Cattle provide the Maasai with milk, butter, meat (blood, while still perceived to have been a good adaptation strategy in the past, is only consumed rarely nowadays as it is prohibited by most churches), and hides. Furthermore, their other livestock, including goats and sheep, also form part of the lifeblood that sustains social relationships (bride wealth, gift exchanges, loans or exchanges to build patron–client relations or to repent for injustices) and feature in ritual practices and feasts (births, circumcisions, weddings, sacrifices, rain prayers). Moreover, human-animal relationships form part and parcel of the ways in which the Maasai adapt to a continuously changing environment and climate. Many informants referred to their herd as fulfilling the role of a bank that provided them with money to buy maize and other foodstuffs, pay their children's school fees, purchase medication, or cover unforeseen expenses in the case of a drought or other (environmental) calamity. While the Maasai have a history of semi-nomadism, they are increasingly diversifying their livelihoods through the adoption of agriculture and labour migration.[42] This is in part due to a long history of policies in Tanzania that have not been favourable for the Maasai, for they have increasingly been marginalised by their own government. All in all, the crafting of new sustainable livelihoods, such as their engagement with cultivation over the last fifty years, has been a response to rising population

42 McCabe, Terrence J., and Paul W. Leslie, and Laura DeLuca, "Adopting Cultivation to Remain Pastoralists: The Diversification of Maasai Livelihoods in Northern Tanzania," *Human Ecology* 38.3 (2010): 321–334.

pressure, a reduction in grazing areas, and the increasing emphasis on and pressure from a monetary economy.[43]

6 History, Politics, and the Construction of Nature

While in the early 19th century, Maa-speaking people coexisted with different economic specialties, by the late 1870s the ethnonym 'Maasai' had shifted to indicate economic specialisation, and its meaning gradually coalesced with pastoralism.[44] The fixation of Maasai identity and their consolidation into a 'tribe' of pastoralists began to take root when the British took control of Tanganyika from Germany in 1916. In order to control the scattered, 'dangerous warriors', and to find orderly alignments of culture, political system, and modes of production, the British administrators decided to isolate the Maasai in a newly created reserve that formed part of the driest and most desolate land in Tanganyika.[45] The Simanjiro District (Map 1.1), of which Terrat forms a part, has been home to the Kisongo Maasai, who have dwelled here since the mid-19th century and began to practise transhumant pastoralism around the the beginningof the 20th century on the Simanjiro Plains. Their livelihood was based upon seasonal mobility patterns, in which families were on the move with their herds in search of water and green pastures, following a migratory pattern similar to that of the wildlife that can be found in the area.[46] More specifically, during the dry season (July to October) both people and animals concentrated around permanent and reliable water sources, particularly along the Tarangire River. With the onset of the short rainy season (October or November) they

43 McCabe, Terrence J., "Sustainability and Livelihood Diversification among the Maasai of Northern Tanzania," *Human Organization* 62.2 (2003): 100–111.

44 Hodgson, Dorothy L., " 'Once Intrepid Warriors': Modernity and the Production of Maasai Masculinities," in *Gendered Modernities,* ed. Dorothy L. Hodgson (New York: Palgrave, 2001), 105–145. The term Maasai is by no means a homogeneous or static category as it refers to a large and heterogeneous group of Maa-speaking people who have followed distinct historical trajectories and adopted a diverse array of systems of production. Contrary to common interpretations, not all Maasai are (or were originally) pastoralists. For a historical overview of shifting Maasai identities, see: Spear, Thomas, and Richard Waller, ed., *Being Maasai: Ethnicity & Identity in East Africa* (London: Villiers Publications, 1993).

45 Hodgson, "Modernity and the Production of Maasai Masculinities," 115–116.

46 Igoe, Jim, "National Parks and Human Ecosystems: The Challenge to Community Conservation. A Case Study from Simanjiro, Tanzania," in *Conservation and Mobile Indigenous Peoples: Displacement, Forced Settlement and Sustainable Development,* ed. Dawn Chatty, and Marcus Colchester (Oxford: Berghahn Books, 2002), 77–96: here 80.

would disperse onto the Simanjiro Plains.[47] My informants pointed out that shifting far away with their herd has been until the present day their most vital strategy to adapt to a changing climate, a practice that is increasingly impeded by a reduction in grazing areas. In addition, a large herd size and good social relations based on trust and reciprocity are crucial to adapt to environmental changes and are particularly important during drought.[48] Maasai life in general and human-environment relationships in particular have undergone drastic changes from the early days of colonialism to the neo-liberal political regime of today, which allows private investors to lease land at the expense of local communities. One particular continuity in political efforts, which began around the 1920s, has largely been driven by discourses that have perceived the pastoral mode of living as environmentally destructive. These misconceptions have largely resulted in policies that have impacted upon precisely the Maasai's most important adaptation strategies. The Tanzanian state's attempts to promote the adoption of agriculture and to relocate the pastoralists in order to force them to live a sedentary life date back to the era of British colonial rule.[49] However, even after independence, when President Nyerere established the socialist-inspired villagisation programme known as *Ujamaa* ('familyhood' in Swahili, *Operation Imparnati* in Maa) that reached Maasailand around 1977–1978, the pastoralists were forced into permanent settlements and compelled to engage in farming and reduce their herd size. This villagisation programme was perceived by the Maasai as yet another attempt by the state to subjugate them and alienate them from their grazing lands.[50]

A dominant paradigm that has shaped historical trajectories of the Maasai in Tanzania and persisted throughout different political regimes has been the idea that pastoralists have an 'irrational' attachment to the environment and their livestock. At the beginning of the 20th century, Herskovits posited the assumption that pastoral people maximise herd size regardless of the carrying capacity of their rangelands.[51] Similarly, Hardin's 1968 article on "The Tragedy

47 Igoe, "National Parks and Human Ecosystems," 80.

48 Goldman, Mara, and Fernando Riosmena, "Adaptive Capacity in Tanzanian Maasailand: Changing Strategies to Cope with Drought in Fragmented Landscapes," *Global Environmental Change* 23 (2013): 588–597.

49 Hodgson, Dorothy, *Being Maasai, Becoming Indigenous: Postcolonial Politics in a Neoliberal World* (Indiana: Indiana University Press, 2011); De Wit, "Victims or Masters of Adaptation?"

50 Ndagala, D.K., "'Operation Imparnati': The Sedentarization of the Pastoral Maasai in Tanzania," *Nomadic People* 10 (1982): 29.

51 Herskovits, Melville J., "The Cattle Complex in East Africa" (PhD diss., Columbia University, 1926).

of the Commons" was buttressed by the supposition that pastoral systems are fundamentally unsustainable.[52] These scientific paradigms and Western conceptions about pastoral inefficiency continued to shape rangeland policies that advocated for the total abandonment of pastoralism or for a sedentary lifestyle, a reduction in herd sizes, or the privatisation of rangeland resources.[53] Yet, while being externally imposed, over time these scientific theories about grazing capacities were perpetuated by the Tanzanian government and continue to persist into the present day.[54] The pastoral inefficiency paradigm has been so perseverant in part because it has served the conservationist agenda so well. As a result, supposedly objective scientific arguments have been misused to justify political ends rather than to support sound management.[55] What is important to bear in mind is that this part of northern Tanzania is well-known worldwide for its scenic beauty and spectacular wildlife. Hence, this region has a particular history of being subjected to Western images of nature, which are based on the ideology of separating 'nature' and 'culture'.[56]

Conservationist ideology had already started to play out in Tanganyika during the onset of the German colonial period, which coincided with a devastating crisis for wildlife populations due to the rinderpest epizootic that entered East Africa in 1889 and caused ninety per cent of ruminant ungulates to die.[57] Hence, to protect these animals the first hunting regulations were implemented in the colony in 1891, after which protected areas for wildlife followed. When the British took control of Tanganyika from the Germans after the First World War, these game reserves were re-gazetted with varying levels of restricted access and wildlife use.[58] In the 1930s the first national park was created, for which pastoral people needed to be evicted from near Lake Manyara. This

52 Hardin, Garret, "Tragedy of the Commons," *Science, New Series* 126.3859 (1968): 1243–1248.

53 McCabe, "Sustainability and Livelihood Diversification among the Maasai of Northern Tanzania"; Igoe, "National Parks and Human Ecosystems"; Sachedina, Hassanali. T., "Wildlife is Our Oil: Conservation, Livelihoods and NGOs in the Tarangire Ecosystem, Tanzania" (PhD diss., University of Oxford, 2008).

54 Sachedina, "Conservation, Livelihoods and NGOs in the Tarangire Ecosystem, Tanzania"; Bollig, and De Wit, "On the Origins and Consequences of the Uneven Attention to Climate Change around the World."

55 Sandford, 1983, in Homewood, Katherine M., and W.A. Rodgers, *Maasailand Ecology: Pastoralist Development and Wildlife Conservation in Ngorongoro, Tanzania* (Cambridge: Cambridge University Press, 1991), 69.

56 Ibid.

57 Nelson, Fred, Rugemeleza Nshala, and W.A. Rodgers, "The Evolution and Reform of Tanzanian Wildlife Management," *Conservation and Society* 5.2 (2007): 232–261.

58 Ibid.

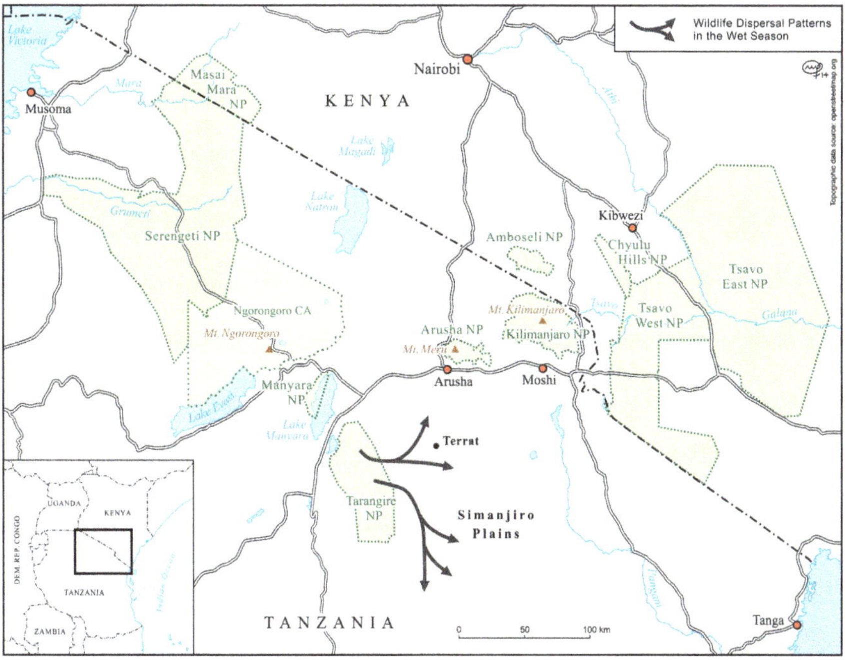

MAP 1.1 Terrat in Northeastern Tanzania

trend continued throughout the 1950s and 1960s and gained momentum after independence.[59] Today, land evictions in the name of conservation continue to take place, while the Maasai still receive the blame for environmentally destructive practices. Very recently, an old land conflict came to the fore after thousands of Maasai were violently evicted from their lands in Loliondo. This conflict dates back to the year 1992 and resurfaced after former President Kikwete leased the land to a private investor from Dubai for hunting purposes. The government initiated a police operation and violently evicted thousands of Maasai from a disputed area east of the world-famous Serengeti National Park, leaving more than 3,000 people homeless and more than 50,000 cattle without access to grazing lands. Allegations of human rights abuses followed,

59 McCabe, Terrence J., "Giving Conservation a Human Face? Lessons from Forty Years of Combining Conservation and Development in the Ngorongoro Conservation Area, Tanzania," in *Conservation and Mobile Indigenous People: Displacement, Forced Settlement, And Sustainable Development*, ed. Dawn Chatty, and Marcus Colchester (New York, Oxford: Berghahn Books, 2002), 66.

while economic losses to the communities, such as burnt houses, the death of livestock, and property loss, were reported.[60]

The Tarangire National Park in the vicinity of Terrat also became a game reserve in 1957 before being upgraded to a national park in 1970, which remains a painful memory for all those who were evicted from it.[61] Since the most important and reliable water sources are found inside the park, local herders have claimed that it has negatively impacted upon their traditional herding systems.[62] As two older men explained to me:

> When we lost this area to the national park we did not know that in the long run this would have such drastic consequences for us and our cattle. But now since life has become difficult because we lack pastures, we realise that this land was really important for our cattle and for our own survival.

One striking similarity between the older conservation discourses and the new climate change idea is the artificial separation between nature and culture on which both narratives are premised. The idea of climate change that is largely translated as a secular discourse finds little resonance within Maasai society for people who generally do not detach changes in the weather and climate from altered values, eroding morals, and general ways of being and believing (see the section on values and culture below).

In brief, if we wish to understand the broader interpretive context of which Terrat forms part, we need to historicise the village's exposure to the travelling ideas of the construction of nature, thereby revealing the fear and actual problem of large-scale land alienation that has haunted Maasailand over the past decades and continues to do so today. It is within a neo-liberal political landscape that Maasai grazing land is being sold to private investors, leading to extensive and violent mass evictions, thus depriving many pastoralists of their

60 TNRF, "Integrating Pastoralist Livelihoods and Wildlife Conservation?" Since the time of writing the Loliondo conflict has been been the subject of much controversy since violent land evictions and human rights abuses have continued. As such, the land conflict gained increasing international attention and remains unresolved. For a chronological overview of the events, see: https://www.arcgis.com/apps/Cascade/index.html?appid=8a472140d55e48aaaaab554625a5558b. Accessed 19 November 2018.

61 Igoe, Jim, and Dan Brockington, "Pastoral Land Tenure and Community Conservation: A Case Study from North-East Tanzania," *IIED Drylands Programme: Pastoral Land Tenure Series* 11 (1999).

62 Cf. Igoe, "National Parks and Human Ecosystems," 80–82.

most basic rights and reinforcing their marginalised position within Tanzania. It is therefore not surprising that this political context has incited a certain reluctance to accept foreign (environmental) discourses as well as an opposition to externally imposed policies. Not only in Terrat, but also in surrounding villages in the vicinity of Tarangire National Park, there exists an overwhelming perception that there is a strong risk of losing land due to government-imposed conservation plans.[63]

The restriction of grazing areas has not only resulted in changing livelihoods among local residents, but also in a deeply altered identity. An account by an informant, a middle-aged man, gives a good idea of how shifting human-environment relations and changing perceptions of the environment go hand in hand: "It could be that the rains are less nowadays, but it is hard to tell because we are also more dependent on rain because we live and eat differently. Our mentality has changed." Similarly, an elderly woman commented: "Perhaps the weather has changed, but we have changed too. We used to follow the clouds, nowadays we have settled." Thus while 'following the clouds' used to guide the Maasai's relationship to their environment and the climate, due to drastic changes in their ways of living it has become impossible to disentangle climatic and societal changes.[64]

My informants' most widely shared concerns were their lack of access to drinking water, good schools for their children in the area, basic healthcare facilities, medication, and good roads and general infrastructure. Other major issues were the lack of rain and pasturage for cattle (mainly during the dry season), population growth, diseases spread by wildebeest that give birth in the area and posed a risk to cattle, the lack of dipping facilities, and the high costs of vaccinations and medication for treating animals. Some informants mentioned the challenges of cultivating in the area due to the unpredictable rainfall and lack of sufficient rain as well as to wildlife (mainly zebras and impala) that destroy and eat their crops. Finally, most informants mentioned the fact that there were too many wild animals grazing in their area and finishing off the grasses before the onset of the dry season, thus not leaving enough pasture for their herds to survive until the next significant rains. One underlying problem is the fact that while wildlife can enter the lands of Terrat, the Maasai cannot enter the grazing area of Tarangire with their herds, which contains the most fertile dry season pastures.

63 Leslie, Paul W., and Terrence J. McCabe, "Response Diversity and Resilience in Social-Ecological Systems," *Current Anthropology* 54.2 (2013): 122.

64 De Wit, "Victims or Masters of Adaptation?"

7 On Climate

The second feature of the interpretive context and one of the reasons why the idea of climate change does not resonate in Terrat has, at least partly, a climatic grounding. As much as the semi-nomadic Maasai have faced rapid socio-cultural, political, and economic changes in the face of globalisation, they have also been confronted with an environment that is known to be characterised by strong, and often unpredictable, ecological fluctuations.[65] The Simanjiro ecosystem forms part of a semi-arid environment, or Savanna drylands, and is characterised by bimodal rain patterns, with an annual average rainfall of 500–800 mm. Rains are unreliable, and the area is drought-prone.[66] There is commonly a long period of rains (*Masika* in Swahili and *Alari* in Maa) that occurs approximately between February and April; in some years the dry season is first followed by a short period of rains (*Vuli* in Swahili and *Irkisirat* in Maa). However, in Simanjiro precipitation is limited and is spatially and temporally highly variable,[67] while droughts occur frequently. The annual occurrence of (partial or complete) crop failure is testimony to the ecosystem's inherent climatic variability and to the fact that—at odds with the government's push to stimulate agriculture—the area is not very well suited for cultivation. One informant expressed the insecurity of cultivation in a vivid, metaphoric way: "planting crops is like gambling, sometimes you win but more often you lose." Similarly, the chairman of the village mentioned that "it is more common not to harvest than to harvest. You cannot know in advance if you will lose or not because the environment is very unpredictable here. But we cannot lose hope so we keep on trying, even though we lose a lot of money and effort."

In other words, irregular rainfall patterns and environmental hazards like severe droughts are (to some extent) part and parcel of normality. The droughts of the past are ingrained in people's memories and still spoken about, have been given special names, and feature prominently in warrior songs. For example, the year 1997 has been called *alari lengolong* ('year of the sun') and is remembered as an extraordinarily dry year in which all pastures dried up. Another very bad year was 2005, when numerous cows died and many Maasai arrived from Kenya in search of pastures. It is remembered by the name *alari leunoto oorkoryanga,* referring to the circumcision ceremony that initiated a

65 Leslie and McCabe, "Response Diversity and Resilience in Social-Ecological Systems," 116.

66 Igoe, Jim, and Dan Brockington, "Pastoral Land Tenure and Community Conservation: A Case Study from North-East Tanzania," *IIED Drylands Programme: Pastoral Land Tenure Series* 11 (1999): 4.

67 Leslie and McCabe, "Response Diversity and Resilience in Social-Ecological Systems," 120.

new age group of *illmuran* that took place in that year. The unpredictability of the climate makes climate change a confusing idea, as emphasised by the traditional leader: "When I heard about climate change for the first time I thought it was just a repetition of the past. We had many drought spells in our history."

Nevertheless, many villagers in Terrat continue to cultivate crops, as this forms an important livelihood diversification strategy. Their accounts reveal that the benefits of a somewhat rare successful harvest may outweigh the risks and costs of crop failure.[68] Furthermore, this attests to the fact that irregular rainfall and climate variability—and thus unpredictable wealther pattners have been accepted as the norm. It is possible to imagine that in a relatively stable climate with regular seasonal variation, manifestations of climate change might be more distinctive and thus easier to observe. By contrast, the climate in Terrat is characterised by such pronounced variability that the idea of climate change arrives as a notion that appears initially all too common, but ultimately becomes alien for its construction of a stable climate as the norm. As one older female informant put it: "Of course the rains are changing these days, they have never been the same in this locality." In understanding the interpretive context in which climate change is translated it is thus crucial to consider both the climatic as well as the environmental context and the consequent production of nature. Climate itself is not so much spoken of in terms of fear, threat, or crisis; what is at stake rather are dwindling resources and restricted mobility patterns. In a similar vein, the broader network of NGOs representing pastoralist groups in Tanzania holds the opinion that climate change is just one factor among many that exacerbate the Maasai's already marginalised position. In fact, regional NGOs argue that the pastoral mode of living—as a complex livelihood system based on (seasonal) mobility—is inherently very well adapted to changing climatic conditions. Since mobility features among their most important adaptation strategies, their decreased access to traditional grazing land forms an increasing impediment to pastoralists' livelihood security.[69]

---

68 However, in 2013 many people in Terrat (as opposed to in other villages) did not plant maize because of successive failed harvests in the preceding years. According to my informants, these failures were due to wildlife that destroyed the crops.

69 Allegretti, Antonio, "Mainstreaming Climate Change Adaptation in Drylands Development Planning in Tanzania: Traditional and Government Planning Processes," Research Report Tanzania Natural Resources Forum (TNRF) & International Institute for Environment and Development (IIED), 2012; Saringe, "Impact of Climate Change to Pastoralist and Hunter-Gatherer Communities of Tanzania"; Hakikazi Catalyst, *A Future for Pastoralism: A Guide to the African Union's Policy Framework for Pastoralism*

8 On Culture, Values, and Believing

The conversations that I had about climate and weather in Terrat almost always ended up in a general discussion about a changing society, globalisation, and eroding cultural values, as a fragment of a focus group discussion held in Terrat in December 2011 reveals:

> Researcher: "Do you think that the rains nowadays are different from the past?"
>
> Informant A: "Yes, in the past my grandfathers, they went to large trees to pray to their god, then they made a sacrifice. And at the same time they went to see a witchdoctor to ask for rain. [...] But then we realised that they are not really gods because they die as normal people. So when the Christians came and told us about Christ we decided to follow them. [...] Now in church we are happy to have found the real God, and we believe that when we pray everything will be possible."
>
> Informant B: "Nowadays people don't respect each other anymore. Our society has changed; we don't have faith in each other any longer."

The final explanation why climate change as a scientific (and supposedly secular) discourse is rejected in Terrat is both a cultural and religious one. The great majority of my informants described an intrinsic relationship between morality, social ties, and the climate. For instance, an old man who is known for his extensive environmental knowledge and weather prediction skills stated: "The rains have changed because we have changed. We are not good people anymore, so the rains are not good either." Especially when asking elderly people about changes in the environment, seasonal changes, or altered rainfall patterns, they began to exclaim their discontent over the behaviour of the younger generation, whom they accused of no longer greeting properly nor showing any respect for others. When speaking about changes in the climate, the Maasai in Terrat—like many other societies worldwide—told tales of general changes in their culture, belief system, and way of life. In other words, the idea of 'climate' cannot be detached from themselves and the world they inhabit, and, as such, climate discourse must be understood as a meta-commentary on society's moral well-being.[70]

---

in Africa: Securing, Protecting and Improving the Lives, Livelihoods and Rights of Pastoral Communities (Arusha, 2011).

70 De Wit, *Global Warning*; Van Beek, Walter E.A., "Echoes of the End: Myth, Ritual and Degradation," *Focaal* 35 (2000): 29–51.

It should be emphasised here that the majority of Maasai have converted to Christianity. Yet, the 'traditional' Maasai religion is a monotheistic belief system in which people worship the (predominantly female) deity *Eng'ai*; while many different names exist for *Eng'ai*, there is only one supreme being, and therefore a certain continuity between the traditional way of believing and Christianity can be observed. Nevertheless, religious practices such as ritualistic rain prayer and other customary religious rituals underwent drastic changes due to the establishment of churches in the area. Interestingly, the Maasai were initially suspicious of Christianity (and other features that they considered to be part and parcel of 'modernity's project'). The mass conversion took place relatively late and occurred predominantly among women, despite concerted efforts by Catholic missionaries to evangelise men.[71]

For the villagers of Terrat, talking about the climate and the weather is a commentary upon a rapidly changing society in which God expresses his discontent with people, as manifested in the tangible realities of the sky. For example, similar to many other religious traditions in the world, the climate is seen as a mirror of God's judgment of a society in which drought is perceived to be a curse and rain as a blessing. Some climate- cognisant informants asked me the question of "whether it is true that these people from Europe and America who call themselves scientists do not believe in God?" Other cognisant informants wondered "why these scientists, who are talking about climate change on the radio, say that God has nothing to do with it?" However, even informants who were unaware of climate change asked me this question frequently, such as the chairman, who enquired: "Is it true that white people like you don't trust God?" The fundamental transcendental and moral connection that exists between God, society, and the weather is revealed by the most significant word in the Maa language: *Eng'ai,* which concurrently means God, rain, and the sky (or heaven). Remarkably enough, the first attempts to translate 'climate' into the Maa language by the radio station were contested by the traditional leader of Terrat, who stated that the climate could only be captured by the holy trinity of *Eng'ai*.[72] According to the traditional leader, who understood that climate

71 Although the question of gender is beyond the scope of this paper, it would certainly be worthwhile to explore whether the Maasai's initial rejection of Christianity bears similarities to the ways in which current climate change discourses are advanced or rejected, since I found that women were more prone to embracing the idea of climate change than men, Hodgson, Dorothy L., *The Church of Women: Gendered Encounters Between Maasai and Missionaries* (Indiana: Indiana University Press, 2005).

72 For an elaboration on linguistic translation difficulties from English to Swahili to Maa, see: De Wit, "Victims or Masters of Adaptation?"

change is the result of industrial pollution, the domains of rain and sky are—ultimately—in the hands of God. This nascent scientific story about climate change with its secular causality appeared for many informants to be a genuine attempt to disprove the existence of God, to deny his power—a disavowal of their understanding of the world.

9 Concluding Reflections

This chapter has dealt with a largely underexplored phenomenon within the climate change research agenda in anthropology: what to do with an apparent lack of both experiential as well as discursive climate change realities? In exploring the socio-cultural consequences of climate change across the globe, we as researchers also have to make sense of the *absence* of that which we were initially looking for, rather than imposing our own climate change lens and crisis narrations. The abstract models that we create are the product of a particular way of looking at the world; juxtaposing them with what is found on the ground might prove a necessary epistemological corrective to the climate change focus that is currently dominating the research agenda. In a similar vein, I have argued that one consequence of the current discourse on climate change—which almost exclusively manifests itself in terms of crisis—is that it can create epistemological blind spots. As many other researchers have also demonstrated, if we turn our research subjects into victims of climate change we risk depriving them of their agency. The case of the large-scale land evictions in Maasailand has illustrated that the idea of climate change also serves as a scapegoat for obscuring power relations (for those in power). This is not to say that climate change is not there, or that it is not (or will not become) a real concern for the Maasai. However, we must take seriously the reasons why both the *idea* of climate change as 'the ultimate crisis' as well as biophysical experiential realities do not always resonate with the socio-cultural horizons and knowledge practices that emerge in a certain locality.

Rather than reproducing climate determinism, social science climate research should allow for the fact that—at least at times—climate change might not be the supreme or sole determinant of contemporary lifeworlds of supposedly 'vulnerable' societies in the Global South. For the Maasai pastoralists in Terrat (and presumably elsewhere in Maasailand),[73] primary concerns are the lack of access to basic facilities and services such as schools, potable water,

73 See the work of Hodgson, *Being Massai*.

medication, and pastures for their cattle as well as the ways in which the construction of national parks has limited their access to natural resources. Bringing the complexity entailed by the politics of climate change knowledge back into our analysis means doing justice to local concerns and processes of marginalisation, whose root causes may reach beyond climate change while being politically and historically produced.[74]

Finally, the point has been made that climate change is also an idea that travels, entangles, and takes on different shapes along its manifold journeys. In order to understand the ways in which this novel idea takes root—or *does not*—a thorough insight into the interpretive context is needed. Against this background, arguing for or against the visibility of climate change is perhaps not so much the task of anthropologists. What remains for our field then is to stay true to the old anthropological merit of seeking not to enlighten our research subjects, but rather searching for a reflexive lens that stands receptive to the intricacies and complexities of what we encounter in the field—even if this means that we acknowledge that what we are looking for, or expect to find, is said to be not there.

74 Cf. Smucker et al., "Assessing Climate Change Adaptation Policy in Tanzania."

CHAPTER 2

Perspectives on Climate Change in Makonde District, Zimbabwe since 2000

Vimbai Kwashirai

Abstract

The following chapter discusses the acquisition and transmission of climate change ideas, knowledge, and perceptions in the Makonde District in Zimbabwe. It argues that knowledge of weather and climate in Makonde Shona society has primarily been acquired through informal pathways and is inextricably linked to agricultural practices. Climatic transformation is understood and explained through cultural beliefs, ideas, folklore, and riddles concerning agrarian knowledge. Climate change in Makonde is not just thought of in terms of the periodic acute and severe drought occurrences, but is generally conceptualised as a broader weather issue very much connected to rising temperatures, erratic or plentiful rains, wind patterns, the central role of ancestral spirits, and God (or *Mwari*). Indigenous and scientific knowledge of the climate tends to be a mixture of teachings from parents and external authorities, including government agencies, churches, NGOs, and school teachers, hence encompassing both cultural and scientific perspectives.

1 Introduction

This chapter focuses on the acquisition and transmission of climate change ideas, knowledge, and perceptions in the Makonde District in Zimbabwe (see Map 2.1) between 2000 and 2015. It argues that knowledge of the climate in Makonde Shona society has been acquired and experienced through both formal and informal pathways, but especially the latter. The weather and climate are conceptualised and experienced mainly through agricultural practices; notions of the climate and agrarian life are two sides of the same coin. Climatic transformation is further understood and explained through cultural beliefs, ideas, folklore, and riddles concerning agrarian knowledge. Climate change perspectives are shaped by agrarian livelihood activities relating not only to crop and livestock production, but also to the attendant environmental ramifications arising from plant cultivation and animal husbandry—the two major

 | DOI:10.1163/9789004410848_004

pillars of the Makonde region's economy. Climate change in Makonde is not just thought of in terms of the periodic acute and severe drought occurrences, but is generally conceptualised as a broader weather issue very much connected to rising temperatures, erratic or plentiful rains, wind patterns, the central role of ancestral spirits, and God (or *Mwari*). Since ideas about these concerns are mostly introduced informally to children from a very tender age, indigenous and scientific knowledge of the climate tends to be a mixture of teachings from parents and school teachers. Similar traditions of acquiring and transmitting climate knowledge also applied to white settler farmers in Makonde and Zimbabwe more generally. When Makonde people think and talk about the weather, they generally do so from both cultural and scientific perspectives. The different actors in the dissemination of climate change information are children, parents, teachers, the government, churches, religious leaders, and a few non-governmental organisations.

2 Inequalities in Zimbabwe's Agricultural Sector

Zimbabwe endured ninety years of British colonial capitalism from 1890 to 1980. Since independence, much of the literature on agriculture in modern Zimbabwe has remained preoccupied with analyses of colonial injustices in the area of land rights, ownership, and distribution. Populist narratives by Moyo and Yeros[1] and Mamdani[2] maintain that colonially induced land inequality in Zimbabwe has been the biggest obstacle to development. Important as they are, such debates have been at the expense of equally topical questions regarding agricultural development that take environmental factors, including climate, into closer consideration. Socialist rhetoric from the Zimbabwe African National Union–Patriotic Front (ZANU–PF) during its 38 years in power has neglected the significance of customary and scientific knowledge of climate change. For example, the late ceremonial President of Zimbabwe, Canaan Banana, portrayed colonial and post-colonial capitalism as a Western evil that inculcated selfishness among Zimbabweans that distorted their traditional egalitarianism—perhaps an apology for wealth accumulation by black political elites like himself after independence.[3] This explains why the

1 Moyo, Sam, and Paris Yeros, ed., *Reclaiming the Land: The Resurgence of Rural Movements in Africa, Asia and Latin America* (London: Zed Books, 2005).

2 Mamdani, Mahmood, "Lessons of Zimbabwe," *London Review of Books* 30.23 (2008): 17–21.

3 Banana, Canaan, *Towards a Socialist Ethos: Socialism without Socialists is Capitalism* (Harare: College Press, 1987).

structure of production and capitalist relations of production in agriculture, manufacturing, mining, money and banking, and foreign trade remained unaltered until 2000.[4] Agriculture remained dualised between blacks and whites, and over seventy percent of the capital stock in the above sectors was foreign-owned.

To be sure, political and economic factors, both in the colonial and post-colonial eras, account for many of the problems that Zimbabwean agriculture currently faces. As Sachikonye observes, the ideological and material basis of law in capitalism in Zimbabwe remained a major battleground of farmer- and working-class struggles for emancipation.[5] The new black bourgeoisie designed the law to enhance its efficacy as an instrument of control and regulation and for capital accumulation. Provisions of post-colonial agrarian and labour legislation, particularly those of the 1985 Labour Relations Act (LRA), clearly demonstrated the alliance between the government and capital as well as their shared material and ideological preoccupations. The LRA hamstrung farmworkers' freedom to organise into viable trade unions, to participate in nationalist politics, and to undertake industrial action to back up their legitimate demands for improvements in working conditions. Draconian state bureaucratic powers of intervention in industrial relations and repressive labour legislation strangled agrarian, public, and private sector workers alike. Mazingi and Kamidza have shown how pseudo-socialism during the 1980s disguised capitalism, irrespective of the tangible gains in social service provision (including in housing, education, and health).[6]

The Economic Structural Adjustment Programme (ESAP) of the 1990s sponsored by the World Bank (WB) and International Monetary Fund (IMF) wiped out the social gains of the 1980s.[7] The UN envoy Tibaijuka (2005) and the UNDP (2008) not only reported on these social losses, but also concluded that the violent state-sponsored operation *Murambatsvina* had widened inequalities. Indeed,

---

4 Roussos, Peter, *Zimbabwe: An Introduction to the Economics of Transformation* (Harare: Baobab Books, 1988).

5 Sachikonye, Lloyd M., *Labour Legislation in Zimbabwe: Historical and Contemporary Perspectives* (Harare: Zimbabwe Institute of Development Studies, 1985).

6 Mazingi, Lucy, and Richard Kamidza, "Inequality in Zimbabwe," in *Tearing Us Apart: Inequality in Southern Africa*, ed. Herbert Jauch, and Deprose Muchena (Open Society Initiative for Southern Africa (OSISA), 2010), 322–383, accessed July 2, 2017. http://www.osisa.org/sites/default/files/sup_files/chapter_5_-_zimbabwe.pdf.

7 Mlambo, Alois S., *A History of Zimbabwe* (Cambridge: Cambridge University Press, 2009).

Alexander,[8] Johnson,[9] Moore,[10] and Campbell[11] not only attribute inequality in Zimbabwe to skewed land distribution, but also to protracted authoritarianism and governmental inefficiency. Helliker[12] disagrees and explains this inequality with reference to an extremely chaotic land reform process in 2000—a view dismissed by Scoones et al.[13] However, Scoones and other scholars fail to acknowledge that fragile tropical environments prone to drought and severe climate change contribute immensely to economic difficulties and social inequality. New debates have emerged on the increasingly negative climatic and environmental impacts associated with capitalist agrarian and mining activities.[14] Shortcomings in land reform are believed to have been reflected in natural resources destruction that ignored environmental ethics.[15] Environmental concerns in agrarian societies provide salutary reminders of the interlocking relationships between capitalism, politics, production, property, poverty, conflict, and the environment.[16]

These studies provide critical context to understand the impact of extreme weather patterns that pose great challenges to food security in Zimbabwe's agriculture-based economy and in a country with seventy percent (9.8 million) of its total population living in widely dispersed agrarian communities. At independence in 1980, Zimbabwe's agricultural sector was dominated by 6,000 white and a few emergent black large-scale farmers affiliated to the

8 Alexander, Jocelyn, *The Unsettled Land: State- Making and the Politics of Land in Zimbabwe 1893–2003* (Harare, Oxford: Weaver Press, and James Currey, 2006).

9 Johnson, David, "Mamdani, Moyo and 'Deep Thinkers' on Zimbabwe," *Pambazuka News*, February 12, 2009, accessed June 8, 2011. http://www.pambazuka.org/en/category/comment/54039.

10 Moore, David, "Marxism and Marxist Intellectuals in Schizophrenic Zimbabwe: How Many Rights for Zimbabwe's Left? A Comment," *Historical Materialism* 12.4 (2004): 405–425.

11 Campbell, Horace, "Mamdani, Mugabe and the African Scholarly Community," *Pambazuka News*, December 18, 2008, accessed June 8, 2011. http:// www.pambazuka.org/en/category/features/52845.

12 Helliker, Kirk, "Dancing on the Same Spot: NGOs," in *Contested Terrain: Land Reform and Civil Society in Contemporary Zimbabwe*, ed. Sam Moyo, Kirk Helliker, and Tendai Murisa (Pietermaritzburg: S&S Publishers, 2008), 239–274.

13 Scoones, Ian, et al., "Livelihoods after Land Reform in Zimbabwe: Understanding Processes of Rural Differentiation," *Journal of Agrarian Change* 12.4 (2012): 503–527.

14 Chitiga, Gibson, and Percyslage Chigora, "An Analysis of the Implications of the Fast Track Land Reform Program on Climate Change and Disaster Management in Zimbabwe: A Case of Chegutu District," *Journal of Sustainable Development in Africa* 12.2 (2010): 124–143.

15 Scoones et al., "Livelihoods after Land Reform in Zimbabwe."

16 Kwashirai, Vimbai C., "Ecological and Poverty Impacts of Zimbabwe's Land Struggles: 1980 to Present," *Global Environment: A Journal of History and Natural and Social Sciences* 2.3 (2009): 222–253; Gaidzanwa, Rudo B. "Women and Land in Zimbabwe" (paper presented at the conference "Why Women Matter in Agriculture," Stockholm, April 4–8, 2011).

Commercial Farmers' Union (CFU). The Makonde branch of the CFU comprised a powerful rural bourgeoisie who controlled the region's economy, led by wealthy landowners, including the Nicole Syndicate, which owned ten plantations, Little England, and Anglo-American tobacco estates. These titled large-scale commercial concerns had strong financial (collateral) ties with foreign-owned banks, such as Barclays or Standard Chartered, operating in Chinhoyi, Makonde District's capital city. Makonde's land elites in the CFU accumulated large profits from sales of beef, maize, wheat, sugar cane, tobacco, tea, cotton, horticulture, timber, and game products. They enriched themselves from the creation of farm wealth by exporting cash and food crop commodities to world capitalist markets, including in South Africa, Europe, and North America.[17]

White minority privileges and ostentatious lifestyles were made possible mainly because Makonde CFU employers exploited ultra-cheap illicit farmworkers, who were alienated not only from the fruits of their labour but also from the means of production. Farm wage earners have stood out as the most exploited class in post-colonial Zimbabwe. Both in the colonial and post-colonial periods, commercial farmworkers lived and worked in appalling conditions, performing arduous agricultural tasks for very long hours only to earn parsimonious wages. Farmworkers had little or no access to decent social services, such as housing, healthcare, education, running water, or electricity. Attitudes of commercial farmers died hard as they neglected workers' rights as well as their material needs and, in extreme cases, even maltreated farm employees. The 1985 Labour Relations Act also discriminated against farmworkers in relation to wages, arbitration, and other rights.[18]

The above characterisation of white masters and black servants is important to understand how, at the time of independence, black and white elite farmers enjoyed more access to information about Makonde weather patterns. Their knowledge of climate change and environmental management in this tropical area had been passed down from earlier generations of farmers and was additionally acquired from agricultural colleges and schools within and outside Zimbabwe. The colonial state, meanwhile, had been pivotal in facilitating climatic knowledge among white farmers. This came as no surprise, because the majority of parliamentarians in the colonial parliament were CFU members. The CFU was very influential in pushing the state to provide farmers with climatic and agricultural knowledge. Over fifty percent of white farmers had

17 Interview with D. Matose, and G. Murembi, resettled commercial farmers, Chinhoyi, Makonde, May 22, 2013.

18 Zimbabwe Labour Relations Act, Act No. 16/1985, amended through Acts 12/1992, 20/1994.

a basic weather station on their properties.[19] From the 1980s, a typical weather station in Makonde had a rain gauge for measuring liquid precipitation, a thermometer for measuring air temperature, a barometer for atmospheric pressure, a hygrometer for humidity, an anemometer for wind speed, and a pyranometer for solar radiation. Richer farmers had sensors for identifying falling precipitation, a disdrometer for measuring drop-size distribution, a transmissometer for visibility, and a ceilometer to determine the height of the cloud ceiling.[20] A few farmers had acquired the scientific knowledge and the means to undertake cloud seeding to induce artificial rains. Their ideas of drought were not only passed down generations, but small farm archives and libraries contained farm histories and literature regarding the perils of deforestation, soil erosion, drought, desertification, and economic ruin suffered by previous farmers.[21] This body of knowledge was hugely important in informing farmers of lessons learnt from the past. Farmers not only worked closely with the centralised meteorological and weather observatory centres in Harare, but most importantly attended seminars, workshops, and conferences to share ideas and knowledge about the climate with neighbours especially their struggling white counterparts throughout the country. Other ideas exchanged at such fora included desirable sustainable agricultural practices for soil conservation measures, such as wind breaks, contour ridges, crop rotation, and the application of organic fertilisers. Overall, their advantage in terms of their knowledge of climate, farming, market, and labour gave CFU farmers a big head start over their black competitors in less productive areas of Zimbabwe.[22]

These inequalities in terms of environmental knowledge are mirrored in those pertaining to labour and land. The existing scholarly literature on Zimbabwe already provides nuanced understandings of the glaring inequalities that remain as a consequence of prolonged British imperialism in the country. The 1992 Land Acquisition Act (LAA) had given the state the power to conduct compulsory purchase, though landowners retained the right to challenge the price set and to receive prompt compensation.[23] Such an arrangement did

19 Interview with D. Marilier, and C. Stevenson, former commercial farmers, Chinhoyi, Makonde, May 20, 2013.

20 Ibid.

21 Ibid.

22 Interview with D. Matose, and G. Murembi, resettled commercial farmers, Chinhoyi, Makonde, May 22, 2013. The ideas in this section arose from this interview.

23 Section 6 of Constitution of Zimbabwe Amendment (No. 11) Act 1990 (Act 30/1990); Section 16(2) states: "No such law [authorising acquisition of land] shall be called into question by any court on the ground that the compensation provided is not fair." Tshuma, L., *A Matter of Injustice: Law, State and the Agrarian Question in Zimbabwe* (Harare: SAPES

not satisfy the majority of landless rural dwellers in Zimbabwe. Indeed, social realities in Zimbabwe continued to reflect the late-colonial period, with 8,000 white farmers and commercial institutions owning 15.5 million hectares of prime land, or 39 per cent of the total land in the country, while 4.5 million black farmers in communal and resettlement areas eked out a living on 16.4 million hectares of mostly unproductive land, to which they had been confined by ninety years of oppressive colonial rule.[24] Land inequality is the key explanation for the observable pent-up resource nationalism that arose from 2000 onwards, in which the ZANU PF government legislated against agrarian capitalism amending the LAA four times leading to land nationalisation.

There are various terms used to describe the coercive takeover of land in Zimbabwe that began in 2000. Terms applied by different stakeholders include occupation, land squatting, land invasions, trespassing, land demonstrations, and land grabbing, the different terms reflecting varying views and ideological standpoints. In the process, large-scale commercial farms were taken over, being held by the state either directly or through a state entity under a title deed (in which case it was freehold state land). The Fast Track Land Resettlement Programme (FTLRP) meant that 6,250 of the approximately 8,000 commercial farms that existed before 2000 were acquired from about 4,500 owners.[25] In 2005 the government passed an amendment to the LAA declaring all agricultural land to be state land. Together, approximately 100,000 black large-scale farmers received 2.5 million hectares, and close to 500,000 smallholders received 4.5 million hectares, including farmers in Makonde.[26]

Land resettlement was generally structured into subsistence and commercial farms, or Model A1 and Model A2, respectively. Model A1 was defined by government as the decongestion model for the majority of landless people. Beneficiaries have access to the following average allocations: agro-ecological region (AER) I: one to twelve hectares; AER IIa: 15 hectares; AER IIb: twenty hectares; AER III: thirty hectares; AER IV: fifty hectares; AER V: seventy hectares. Each household is allocated three hectares for arable land, with the rest being for grazing and other forest requirements, including timber, firewood, and water. Settlers are provided with basic social services through administrative and social management systems. In this model, twenty per cent of all resettlement land is reserved for war veterans, namely former combatants in the

Books, 1997), 129; Moyo, S., *The Land Question in Zimbabwe* (Harare: SAPES Books, 1995), 1.

24 Government of Zimbabwe, Ministry of Agriculture Report, 1999.

25 Government of Zimbabwe, Ministry of Agriculture Report, 2001.

26 Ibid.

liberation war.[27] The A2 model is administered under the Agricultural Land Settlement Act (Chapter 20:01) and is meant to increase the participation of black farmers in commercial farming through the provision of easier access to land and infrastructure on a full cost-recovery basis. The land is issued on a 99-year lease with the option to purchase.[28] However, agricultural endeavours by blacks are constrained by severe undercapitalisation due to their lack of financial inclusion in government land resettlement schemes. The buzzwords among most Makonde farmers concern the lack of inputs, among others, especially seed, fertiliser, and draught power. Moreover, hiring farmworkers is beyond the reach of many Makonde farmers. Congestion by overpopulation and overstocking, particularly in Makonde's communal areas, demonstrate the glaring economic inequalities between resettled and peasant farmers in the region, a demographic overload generally leading to ecological collapse on peasant farms and an increase in inequalities among farming households. Such influences have pushed government to enact the 2002 Environment Management Act (EMA) and the 2007 Indigenous Economic Empowerment Act (IEEA).[29] The EMA sought to manage and control an ecological crisis hitherto unknown in Zimbabwe, while the IEEA stipulated that foreign-owned companies worth over half a million dollars should cede 51 per cent of their shares to indigenous people.[30]

In contrast to material disparities, very little is known about the imbalance between blacks and whites in relation to their knowledge about weather and climate, knowledge that is essential not only for the 600,000 resettled farmers but also for communal peasants in rural Zimbabwe. Communal and resettled farmers, unlike white farmers, have been excluded from gaining formal climatic knowledge in schools, colleges, and universities. Blatant racial discrimination forced black farmers to rely on traditional ways of reading and understanding climate change. Mothers and fathers but especially grandparents were repositories of climatic and environmental knowledge. After independence, formal education in agriculture and climate was available to secondary school pupils but with a very weak curriculum in the area of climate and climate change. The status quo did not change at the tertiary level, where formal education still emphasised the learning of agricultural knowledge and practices more than understandings about the climate. Makonde farmers receive little knowledge

27 Government of Zimbabwe, Ministry of Agriculture Report, 2001.

28 Ibid.

29 Government of Zimbabwe, Environment Management Act, 2002, (Chapter 20:27).

30 Government of Zimbabwe, Indigenous Economic Empowerment Act, 2007, (Chapter 14:33).

about the climate from governmental agricultural extension services. Teaching farmers about the climate and climate change has never been government priority since independence.[31]

To fill this void, Makonde farmers generally draw on traditional climate information for farming and planning purposes. They have no weather stations, and those who access climate change ideas and knowledge from meteorological and weather observation centres via radio and newspapers find this information neither adequate nor meaningful. Media presence in rural Makonde is also very thin, and most farmers do not have access to television. Mobile phones are different as almost every household has one, and this technology might revolutionise knowledge about the climate in the future in the same way as it has done with mobile banking. In addition, farmers generally mistrust meteorological information. Although Makonde farmers listen to climate forecasts, they show a preference for traditional knowledge systems as a control. When scientific climate forecasting deviates from traditional forecasts, the farmers' inclination is towards indigenous information for reasons that it blends well with their culture, has been tried and tested over the years, and is transmitted in a language that they understand. There is often a striking similarity between indigenous and scientific climate indicators. Some indicators are the same in both systems, such as wind direction, clouds, and temperature. In addition, indigenous climate predictions are also based on plant and animal behaviour.[32]

For example, farmers associate the heavy production of tree leaves with a good season, while high fruit production is a sign of a poor rainfall season. The reasoning behind this observation is that high fruit production implies that people will subsist on fruits for lack of alternative foods. The production of white flowers by a tree species known as *mukuu* signals a dry season, while flower production on the upper branches of the *mukonde* tree indicates a good rainy season. Other indigenous signs of an imminent drought include a heavy infestation of most tree species by caterpillars during springtime, late bearing and lack of figs from July to September by the *mukute* tree, delayed maturing of acacia trees in the valleys, and the drying off of *chigamngacha* fruit from

31 Interview with Makonde District Chief Executive, Crynos Gandiwa, Mhangura administrative offices, July 25, 2014. "Indigenous" refers to cultural knowledge passed down through the generations, and "scientific" means climatic knowledge learnt formally in school, college, or university; focus group discussion with resettled farmers Amos Bwerinofa, Maria Tinarwo, and Anelli Dzapata, Ward 19, Makonde, July 22, 2014.

32 Interview with Makonde District Chief Executive, Crynos Gandiwa, Mhangura administrative offices, July 25, 2014.

September to early November.[33] These observations are discussed formally and informally at farmers' meetings and social gatherings without government involvement.

One of the most important animal indicators is the behaviour of spiders. When spiders close their nests, an early onset of rain is expected because they do not like any moisture in their nests. When a lot of crickets are observed on the ground, a poor rainy season is expected. The movement of elephants is associated with the occurrence of rainfall because they need a lot of water. A stork flying at very high altitude is associated with a good season. Observing certain types of birds singing while facing downwards from the top of a tree is a good indicator that it is about to rain, while seeing many birds is a sign of heavy rain. The wind blowing from west to east, and from north to south is assumed to bring with it a lot of moisture and a good rainy season. The prevalence of a strong wind from east to west during the day and at night from July to early November is an indicator of drought.[34] Generally, droughts are interpreted both in religious and scientific terms, and often perceived as punishment for sins like murder. Drought is also understood as part and parcel of the extreme climate variability caused by global warming. Specifically, climate change has muddled the ability of farmers to plan farming operations with any certainty. If they plant early, the rains may vanish at a moment critical for plant growth, and when they delay their planting, the rains might still not be assured of coming. Either way, the farmers find themselves gathering regularly to discuss new ideas on how to cope with fast-changing weather patterns.[35]

3 Climate Change Perspectives in Makonde

Debates on climate change in Makonde District typically revolve around shifting seasons in relation to drought, temperature, and rainfall patterns. Five former CFU members in Makonde who lost their properties during the land reform process reveal how the white community of farmers understood and dealt with climate variability and unpredictability. Dukus, Foriti, Rivonia, Tiripano, and Vrede agree that differences in understanding and managing climate change existed among white farmers. Some had better resourced weather stations, while others had none and were left to rely on

33 Focus group discussion with resettled farmers; Amos Bwerinofa, Maria Tinarwo, and Anelli Dzapata, Ward 19, Makonde, July 22, 2014.

34 Ibid.

35 Ibid.

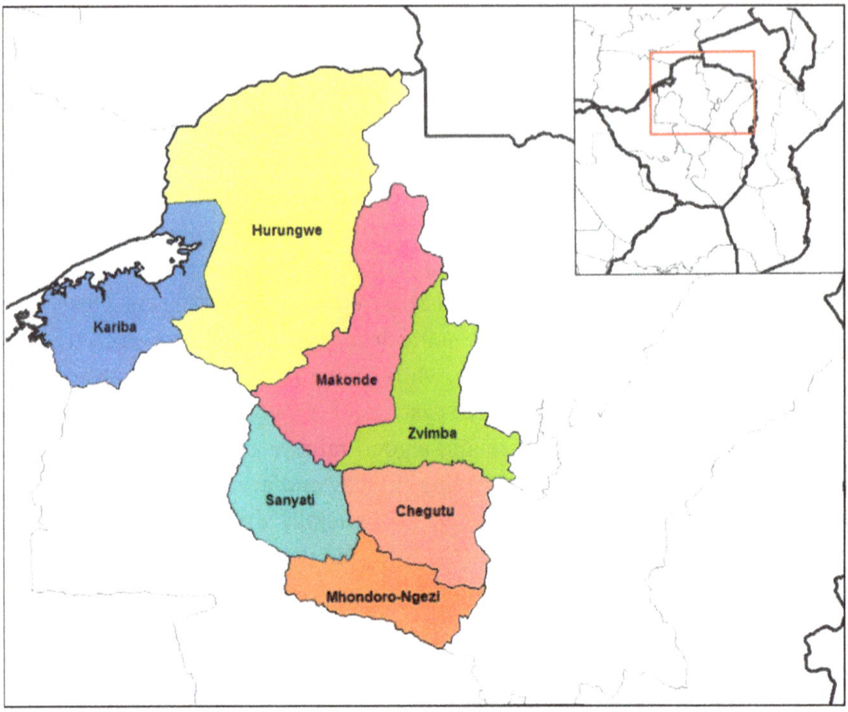

MAP 2.1 Makonde District
SOURCE: WIKIPEDIA, "DISTRICTS OF ZIMBABWE," LAST MODIFIED JUNE 8, 2017. HTTPS://EN.WIKIPEDIA.ORG/WIKI/DISTRICTS_OF_ZIMBABWE

their neighbours for climate information. Regardless of the privileges they had been accorded in colonial and post-colonial times, mastering ideas and knowledge about changes in tropical climate remained a major challenge for a number of commercial farmers. In particular, two climate-related issues preoccupied both black and white elite farmers: unpredictable weather and recurring drought. Foriti and Vrede note that during the 1970s rainfall patterns in Zimbabwe were predictable, as rainfall invariably came in October. Droughts too, they claim, could be forecast with some level of accuracy, having occurred in five-year cycles within a small margin of error.[36]

From the 1990s onwards, black and white Makonde farmers have been attending agricultural meetings, workshops, and conferences that confirmed

36 Interview with Foriti, and Vrede, former Makonde commercial farmers, Chinhoyi, Makonde, June 12, 2013.

many of their fears gained from their practical experience and knowledge that the rains had become more and more erratic and droughts more acute and frequent. Much of what farmers know about the climate has been learnt on their farms through parents, observations, recordings of weather elements, and knowledge shared by neighbours. However, huge gaps in information on climate change remained among Makonde farmers, a situation comparable to other farmers in Zimbabwe and throughout Africa. The lack of comprehensive climate information among Makonde farmers has meant that their farming operations before and after land reform have always been subject to uncertain planning and planting arrangements.[37] Not every elite black or white farmer was successful in balancing this agricultural and climatic mix for the purpose of maximum benefit in each farming season. While a few had become extremely wealthy through cultivation and animal husbandry, others became bankrupt due to climate change, drought, lack of inputs, and land reform. However, since commercial farmlands averaged 2,200 hectares, elite black and white farmers had been able over the years to practise relatively sustainable agriculture through crop and paddock rotation. Very low densities in human and livestock populations permitted such rotations—in sharp contrast to the demographic overload in peasant communal and resettlement areas.

There were parallel and competing structures for distributing pieces of land on properties seized from white Makonde farmers. District land committees included traditional authorities, village heads and chiefs, government administrative bodies, ministry of agriculture officials, local government councillors and civil servants, and government-leaning groups such as war veterans, the army, the police, the secret service, and ZANU–PF supporters, all under the patronage of the ZANU–PF party hierarchy. In Makonde, members of land committees prioritised land distribution, disregarding the implications of their behaviour and actions on climate change. For example, little attention was paid to the vandalism and burning of farm forests by 'armed frontier crack teams' spearheading commercial takeovers. Debates about the environmental impact of the widespread forest fires that were started to intimidate white landowners and on poaching activities during the violent farm confiscations were conspicuously absent. Traditional leaders, land committees, and government conservation officers from the EMA clashed over natural resource exploitation and conservation. During and after land reform, the state gave space to customary beliefs and

37 Interview with Foriti, and Vrede, former Makonde commercial farmers, Chinhoyi, Makonde, June 12, 2013.

traditional authorities in the usage of natural resources in addition to state-run and more 'scientific' institutions.[38]

Concerned with safeguarding natural resources, including land, forests, water, and wildlife, EMA officials engage with the issue of environmental degradation when interacting with land reform farmers in Makonde. However, neither the EMA nor any of the other actors involved in the land reform process—ZANU–PF, war veterans, peasants, or civil servants—prioritise climate change as a topic of discourse. ZANU–PF championed land redistribution for its own political survival and paid little, if any, attention to issues of climate change. For its part, the state emphasised food and cash crop production, but remarkably lacked robust measures and policies to mitigate the climatic impacts of farm takeovers, the clearing of land for farming, panning for minerals, and hunting. In addition, traditional leaders, land committees, farmers' clubs, and government conservation officers do not offer any educational programmes to enlighten farmers on climate change. Government policy on land usage prioritises food self-sufficiency in periods of both plentiful rain and drought and aims at producing surpluses for export and foreign exchange earnings.[39] Generally, Makonde farmers, like their counterparts across Zimbabwe, have embraced these government policy measures in their plans and farm operations in order to feed themselves and earn essential income for their family's needs. However, since farm offer letters, permits, and leases fail to provide security of tenure to farmers, landholders generally lack strong incentives and resources to invest in environmental conservation to mitigate climate change.

The ideas and knowledge on climate change of former commercial farmworkers have not been studied, although they have has been the backbone of commercial farming in Zimbabwe. Their knowledge of the weather does not appear to have been utilised by their employers. Much of what the farm employees knew about the weather would have been learnt through experience and would have been of little practical use to them as a landless group of people. Before and after land reform, they stood out as the most exploited class in post-colonial Zimbabwe. They were generally denied access to land because they were not only perceived as having voted against the 2000 draft constitution alongside their white employers, but were also discriminated against as *mabhurandaya* (read Blantyre)—'aliens', or descendants of the interwar migrant workers from Malawi, Mozambique, and Zambia. According to Juma Phiri, the majority of households of former commercial farmworkers in

38 Focus group discussion with resettled farmers Amos Bwerinofa, Maria Tinarwo, and Anelli Dzapata, Ward 19, Makonde, July 22, 2014.

39 Government of Zimbabwe, Ministry of Agriculture, 2002.

Makonde did not benefit from land reform, managing to keep only their homesteads on repossessed farms. Besides diversifying their economic activity into crafts, trade, fishing, and basic construction, former commercial farmworkers also became casual farm labourers for the largely Shona land beneficiaries.[40]

If understanding the climate appeared of little value to former commercial farmworkers, it was certainly important for the land reform farmers in Makonde, where agriculture continues to be the cornerstone of livelihoods. The Shona believe that humans are sons and daughters of the soil; they come from and return to the soil.[41] Land is venerated through many different practices, including the inhaling of fine tobacco snuffed by elders while at the same time sprinkling some of it on the ground to beseech ancestral spirits to bless the community with adequate rains, preserve the soil in a good state, and protect the community from harm. Elders also pour specially brewed beer from *rapoko* and sorghum millets on the soil in prayers for rains, calm weather, and bumper harvests. Elders of land beneficiaries in Makonde continue to perform watered-down versions of similar customary practices before the clearing of woodlands for farming purposes. Similar customary practices and rituals are still performed before burials. Land reform and Western concepts of agriculture learnt in schools have often challenged these beliefs and customs as obsolete, and many farmers' concerns appear to gravitate towards production rather than concerns about old traditions of environmental conservation. Nonetheless, Shona traditions in Makonde continue to be based on recognising the intimate bond between humans and their environment, according to which total realisation of the self is impossible without peaceful coexistence with the climate, soil, minerals, plants, and animals. In Makonde, society takes precedence over individuals because people are born and die while society lives on; hence there is a desire to preserve the basic tenets of customary practices with regards to climate and nature.[42]

However, state policies appear to push farmers away from long-standing cultural attitudes and towards scientific understandings of the climate, soil, forests, and game without necessarily providing the means with which to acquire that knowledge besides the basics taught in school. At the peak of land reform from 2000 to 2005, the environment played second fiddle to politics and economics, with the state turning the other cheek to rampant environmental

40 Focus group discussion with former commercial farmworkers, Juma Phiri, Phenius Dzore, Tipei Zinhu, and Jim Zhuwawo, Ward 16, Makonde, June 8, 2005.

41 Ibid.

42 Ibid.

destruction by settled Makonde farmers. Through the ministry of agriculture, the state often encouraged the use of, and regularly supplied fertilisers and chemicals regardless of their short and long-term environmental consequences, such as causing acidity in soils.[43] Resettled Makonde farmers 'mine' the soil to produce export-oriented tobacco (*fodya*), maize (*chibage*), and cotton (*donje*). The government and many farmers generally neglect the notion of climatic responsibility and environmental ethics, that is, the impact of current practices on future generations of farmers.

Human-habitat interrelationships in Makonde are largely a consequence of strategic livelihood-centred agricultural activities in response to a controversial land resettlement process undertaken in a hostile macro-economic environment. Makonde land reform farmers came from very diverse backgrounds. While some had no prior experience in farming, the majority had been working as communal peasant cultivators and livestock herders before they migrated to Makonde in search of adequate and productive land. Others had been resident in communal areas in Makonde and would be advantaged in knowing Makonde weather patterns that they learnt from elders, in contrast to the newcomers, whose knowledge is limited to what they learnt in agriculture and geography lessons at school as well as what they hear from long-standing Makonde farmers. Climate change knowledge among A1 and A2 Makonde farmers draws from formal and informal discussions mixing folktales, customs, traditions, superstitions, and science. Much folklore and many tales exist in Makonde that transmit climate knowledge to children at home and in schools. From the tender age of five, children are taught about daily weather—cloud cover, wind, sun, moon, rain, cold and heat, and temperature. All these factors of weather are captured in one famous tale recited to children by grandparents, parents, relatives, and teachers.

In this ecological story, an animal kingdom resolved impacts of a serious drought, which had left the animals with neither food nor water. Putting aside their differences, they began to dig a well. Months passed as they toiled under the scorching heat, the cloudless skies, and the windy hot air. The animals also prayed to their god for the rains to come. Hare, who had refused to participate in the digging, appeared vindicated when the animals almost gave up. Frog asked to do his bit. There was a raucous laughter that pierced the entire valley as the animals did not think much would come out of Frog's efforts. What miracle could frog perform? Zebra, however, asked that Frog be given a chance, tossing him down the dark pit. Frog burrowed into the soil

43 Focus group discussion with former commercial farmworkers, Juma Phiri, Phenius Dzore, Tipei Zinhu, and Jim Zhuwawo, Ward 16, Makonde, June 8, 2005.

and after what seemed an eternity a fountain of water burst from the large well pit. Frog became an instant hero. Hare, meanwhile, was barred from the pool and the pool was guarded each time the animals went foraging for food. On two occasions, however, Hare managed to fill his calabash with water and even bathed after tricking Monkey and Baboon by enticing them with honey. When Tortoise became guard, he hid under the water, tricking Hare into thinking that there was no sentinel. After filling his calabash with water, Hare proceeded with a bath when his leg was suddenly gripped from below. "It is a root you are holding!," Hare tries to trick Tortoise but to no avail.[44] The animals came back and found Hare captive punishing him for lacking a communal spirit in the face of an adverse climate. The moral of this story is that humans have battled drought and climate change before and have managed to put their heads together to mitigate the effects of an adverse climate. As future farmers, children hereby learn lessons about the benefits of collaboration and the risks of deviant behaviour.

Riddles teach children about the climate. In reference to changing and shifting winter seasons, Makonde elders remark that "*charova sei chando kukwidza hamba mumuti*" ["the biting cold has compelled tortoise up the tree"]. Another saying states that "*chamupupuri chine varoyi*" ["the whirlwind carrying witches"], which refers to an observed increase in dust winds in Zimbabwe. Yet another riddle claims that "*gehena harina moto*," or "hell has no fury," in reference to acute droughts that farmers must not fear.[45] Climate change ideas informed by indigenous knowledge are also enshrined in spirit guardians as a means through which ritually controlled ecosystems function. In pre-colonial Makonde, the ownership, allocation, and control of land, game, forests, and water resources all fell within the spiritual realm. Several weather and woodland phenomena, trees, rocks, mountains, and pools were understood and made sacred, consumed and conserved by cultural and spiritual design. In modern Makonde, the epistemology of climate and climate change continues to be generally conceived from these spiritual and religious perspectives, in which an omnipotent God (*Mwari*) superintends over rainfall and drought.[46]

Contemporary spirit mediums in Makonde are custodians of these traditions, leading and presiding over traditional rainmaking ceremonies known as *mupwerera* that take place either in sacred forests (*rambotemwa*) or on holy hills or mountains like Chirorodziva. Chiefs Chinhoyi and Makonde argue that

44 Focus group discussion with former commercial farmworkers, Juma Phiri, Phenius Dzore, Tipei Zinhu, and Jim Zhuwawo, Ward 16, Makonde, June 8, 2005.

45 Ibid.

46 Interview with Chiefs R. Chinhoyi, and M. Makonde, Chinhoyi, May 15, 2013.

these beliefs and customs have been heavily diluted and are less understood now due to influences from Western education that seek to explain the climate from a scientific perspective.[47] *Mupwerera* is supposed to be performed annually in times of both plentiful rain and drought as a way of cleansing society of its sins, such as murder and theft. Like many other cultivators, Tabitha Zengeni, a resettled farmer in Makonde, states that she learnt about the climate and customary cultivation strategies both from her elders and in secondary school. Indeed, while most farmers are thought to carry out unsustainable farming practices, there are nonetheless some cultivators who farm sustainably. Consequently, crop production on Zengeni's farm and those of others who employ sustainable methods comprises a diverse range of grains and other plants, such as maize, sorghum, millets, rice, beans, groundnuts, melons, sweet canes, vegetables, cotton, tobacco, and a bit of discreet cannabis or dagga. Furthermore, sustainable farmers also leave important standing trees on their fields for shade, fruit, and the benefit of birdlife. When struck by lightning, large trees are thought to capture and destroy thunderbolt eggs harmful to people, crops, and livestock.[48]

Makonde farmers do not usually evoke scientific concepts connected to climate change in their conversations; however, there are discourses that engage with the idea of environmental conservation, changing weather patterns, and harmful agricultural practices. For instance, while some practised cut-and-burn agriculture, Makonde farmers did not link these practices and the burning of vegetation generally to carbon emissions.[49] For many, environmental protection is a matter of mixed crop cultivation. Farmers Zengeni and Gatsi, for instance, argue that letting some of the land lie fallow was an age-old sustainable traditional agricultural technique that not only allowed fields to regenerate but also to regain their lost fertility.[50] Crop cultivation dominates most farmers' lives, and some believe that ploughing damages or scars the face of the earth and pollutes the atmosphere through widespread forest fires, especially towards the summer months. Farmers who farm sustainably in Makonde argue that land is supposed to be rested by letting it lie fallow as restitution for the wounds inflicted on the soil.[51]

47 Interview with Chiefs R. Chinhoyi, and M. Makonde, Chinhoyi, May 15, 2013.
48 Ibid.
49 Interview with Rueben Gatsi, resettled farmer, Ward 14, Makonde, June 12, 2013.
50 Interview with Tabitha Zengeni, resettled farmer, Ward 20, Makonde, June 27, 2013.
51 Focus group discussion with resettled Makonde farmers Betty Muchada, Moses Dzawo, Kunaka Imbayago, Denigo Dzvairo, and Mary Chaitezvi at Nkosana primary school, Makonde, March 10, 2012.

Makonde farmers also observe that, unlike in the 1970s when June was known to be the only cold month, since the 2000s winter has stretched from the end of May to August. August rains known as *gukurahundi* ("clean the chuff") have disappeared. In addition, October was generally known with some measure of certainty to be the month that saw the onset of summer and the rains. Since the beginning of the 21st century, however, Makonde farmers lament decreases in rainfall across the summer season, but especially during the early and late parts of the season. Increased incidences of drought and late heavy downpours have become a common feature. Other notable changes in the climate include increased extreme temperatures (especially in summer), localised floods, and decreased/varying river flow.[52] Farmers note that there are many insects and birds in Makonde whose presence and sounds are interpreted to signal extremely high temperatures (such as the edible nyenze insects) or heavy downpours (for example, the noisy horitoto birds). Makonde agricultural production processes, particularly plant growth, are sensitive to climatic conditions, especially for those farmers dependent upon rain-fed agriculture. The greatest challenge to government lies in creating more awareness among farmers on the impacts of climate change. This is because managing the environment sustainably becomes a tool for avoiding the excesses of climate change, notably through reducing forest fires that contribute to carbon emissions.[53]

Some Makonde farmers appear to use formally acquired scientific information to determine the predictability, occurrence, and severity of drought. However, even for climate scientists, the prediction of drought is both complicated and unreliable. In Southern Africa more generally, cycles of drier years—which occur as a consequence of the El Niño effect, the large-scale warming of the equatorial eastern and central Pacific Ocean—are followed by successive seasons with opposite conditions. However, after two dry years in a recognised drought cycle, there is no guarantee that the third year will also be a drought year. Actual climate conditions in recent years have, to a large extent, not corresponded with the predicted scientific outcomes. Until the mid-1990s, the general practice of declaring drought was based on the actual occurrence of drought. The severe drought of the 1991–1992 season in Zimbabwe was thus only recognised officially as late as in January 1992, well into the agricultural season. The first time that drought was forecast at a very early stage on the basis of global interpretations of the effects of El Niño was in June

52 Focus group discussion with resettled Makonde farmers Betty Muchada, Moses Dzawo, Kunaka Imbayago, Denigo Dzvairo, and Mary Chaitezvi at Nkosana primary school, Makonde, March 10, 2012.

53 Report of Governance and Social Development Centre, 2009.

1997, when severe drought was predicted for the 1997–1998 season.[54] This led to action by governments in the Southern African Development Community (SADC) region towards information dissemination and the providing of planting advice to farmers. Recommendations to farmers ranged from the planting of drought-tolerant and early-maturing crop varieties to destocking.[55] Even with improvements in the reliability of climate forecasts, the occurrence of recurrent drought and related risks have to be accepted and integrated into land use systems that are sustainable under climate variability. The prospect of accelerated global warming, and associated regional changes in climate, reinforces the need for the consideration of the longer-term constraints that future climate may place on Makonde farmers. Makonde farmers' clubs discuss and share knowledge on climate change and its implications for sustainable farming and livelihoods in the face of changes in natural vegetation, agriculture, range conditions, and water resources.[56]

Besides concerns about climate change, drought, and their impacts, most Makonde farmers try to mitigate the problem of soil erosion through knowledge acquired both from their elders and from formal education. Soil erosion is understood as an outcome of not only deforestation but also of unsustainable agricultural methods. For example, Makonde farmers express the view that the destruction of vegetation exposes soil to the elements of rain and wind, which erodes the fertile topsoils essential for agricultural production. They also link deforestation to climate change, noting that the loss of vegetation cover is directly connected to extremes in temperature and desertification. These issues are expressed in environmental terms governing land use, linking conservation ideas to government officials and the connections those officers have to international discourses about trees, erosion, water supply, and climate change.

Agricultural officials believe Makonde farmers threaten forests, the soil, and water sources because their settlements and economic activities have spread out beyond the floodplain, the result of a hurried and unplanned land allocation programme implemented by non-experts. Official thinking and state intervention are concerned about the impact of cutting down vegetation near

54 Scripps Institution of Oceanography, Experimental Climate Prediction Centre, "The El Niño of 1997/98," accessed September 20, 2018. http://meteora.ucsd.edu/~pierce/elnino/en97/en97.html.

55 Focus group discussion with resettled Makonde farmers Betty Muchada, Moses Dzawo, Kunaka Imbayago, Denigo Dzvairo, and Mary Chaitezvi at Nkosana primary school, Makonde, March 10, 2012; Report of Governance and Social Development Centre, 2009.

56 Focus group discussion with resettled Makonde farmers Betty Muchada, Moses Dzawo, Kunaka Imbayago, Denigo Dzvairo, and Mary Chaitezvi at Nkosana primary school, Makonde. March 10, 2012.

water sources such as rivers, streams, and wetlands. Agricultural officials argue that the clearance of bush and trees by farmers along river and stream banks increases the erosive powers of run-off water. However, soil erosion problems do not seem to concern other Makonde farmers and certain sections of the central government who believe that they are neither serious nor widespread enough to warrant attention.[57] Government intervention is constrained by a serious shortage of manpower and resources within the ministry of agriculture, responsible for various agricultural matters including educating farmers about the climate and climate change as well as forest, game, soil, and water conservation. There are no formal institutions and structures for inculcating climate knowledge among farmers. The Agricultural Extension Service, which used to provide such services to farmers before and after the 1980s, no longer has adequate funds nor sufficient personnel to visit farmers for educational purposes. According to Alois Mbedzi, a resettled farmer: "It is a very slow process teaching soil conservation to farmers as most of them are naturally conservative persons who, before spending any money, wish to see with their own eyes that the proposed remedy is a certain cure and, therefore, worthwhile."[58] Soil conservation regulations appear unpopular, since the EMA instructors try to enforce rules in arbitrary ways causing local resentment of conservation measures. The strict dismissal of indigenous knowledge and the policing function of EMA officials like donor-funded natural resource management programmes, fail to build upon, or even acknowledge, local practices and knowledge.[59]

The government has placed more emphasis on promoting food production to cut the food import bill as well as on raising the value of land. According to John Chipengo, during the first few years of land reform farmers took soil fertility for granted despite clear evidence of severe soil erosion continuing unabated on their farms. However, from 2010 many producers became aware of the impoverishment of their fields and reduced crop yields compelling them to use organic and artificial fertilisers in significant quantities due to the loss of soil fertility that had occurred mainly through soil erosion.[60] Poor crop husbandry practices, notably maize monoculture, exploitative cultivation methods, ploughing down slopes, and overcropping,

57 Focus group discussion with resettled Makonde farmers Betty Muchada, Moses Dzawo, Kunaka Imbayago, Denigo Dzvairo, and Mary Chaitezvi at Nkosana primary school, Makonde. March 10, 2012.

58 Interview with Alois Mbedzi, resettled farmer, Ward 2, Makonde, March 30, 2012.

59 Focus group discussion with resettled Makonde farmers Betty Muchada, Moses Dzawo, Kunaka Imbayago, Denigo Dzvairo, and Mary Chaitezvi at Nkosana primary school, Makonde, March 10, 2012.

60 Interview with John Chipengo, resettled farmer, Ward 2, Makonde, March 30, 2012.

have contributed to the environmental destruction that has been occurring on Makonde croplands. Simon Chamboko observes that the way soils were used until 2010 made little short-term economic sense, and that farmers lacked the willingness both to deal effectively with the problem of soil erosion and to concede the need for radical change in individual and collective attitudes towards natural resources. Jane Moyo also attributes rampant forest and soil destruction to ignorance caused by the "mining of the soil for profit." In particular, the practice of perennially cultivating tobacco, maize, and cotton on the same fields causes soil exhaustion, which has been a major concern among agricultural officials, such as Onesmo Zishiri, Mashonaland West's Director of Agriculture. A few Makonde farmers regard the subject of soil erosion as of more academic than practical importance. Others such as Chenai Chinhoyi are critical of monoculture agriculture. Chenai Chinhoyi observes that many streams and rivers in her area carry away fertile silt soils as a result of uninterrupted monocultural cultivation.[61]

In the Kenzamba, Hombwe, and Godzi areas of Makonde, farmers generally protect arable lands with contour ridges to guard against erosion. They also leave strips of unploughed land with vegetation to protect fields and keep wildlife like birds and bees. In the Chivende and Obva areas, EMA officials have experienced resistance from farmers who cut down forests to create more farmland and fail to practise sustainable agriculture that protects the soil through contour ridges (see Figure 2.1). Partly, this is because some remember the coercive measures used by colonial officials in the implementation of forest and soil conservation measures. Soil erosion is seen by government as the major threat to agricultural productivity, prosperity, and farm income. For instance, among other farmers, Julius Mafunga in Obva has suffered a reduction in maize crop yields. Mafunga's ten-year-old, thirty-acre farm in Chivende, which has no anti-erosion works, has become denuded in many parts. Having once produced twenty bags of maize per acre, that yield has since decreased to five per acre, while on certain plots the application of artificial fertilisers has had little impact.[62] On the other hand, farmers practising soil conservation have seen their yields rise on a yearly basis. For example, Never Chambati in Kenzamba harvested thirty bags of maize per acre in 2014.[63]

Soil erosion removes nutrients such as nitrogen, phosphorus, and organic carbon. The wider implications of high rates of soil loss from predominantly

61 Focus group discussion with resettled farmers Simon Chamboko, Jane Moyo, and Chenai Chinhoyi, Ward 14, Makonde, July 30, 2014.

62 Interview with Julius Mafunga, resettled farmer and Ward 15 councillor, March 23, 2012.

63 Interview with Never Chambati, resettled farmer and Ward 15 councillor, July 20, 2014.

FIGURE 2.1 Deforestation in the Obva area, Makonde; cutting down trees for crop fields and firewood
PHOTOGRAPH © VIMBAI KWASHIRAI

tobacco-, maize-, and cotton-growing farms in Makonde are yet to be measured and appreciated. The loss of organic matter suggests that the natural fertility of soils in Makonde has been seriously undermined. The Makonde District is by no means an exception in this regard and appears to be representative of what has been happening nationwide. Quite clearly, neither the district nor the country can sustain such high levels of soil erosion indefinitely. Farmers experiencing high soil losses have been squeezed out of the farming business, caught between the imperatives of increasing production costs and the reality of decreasing yields. At the same time, the growing extent of land that is already seeing depreciating production is gradually threatening the viability of the agricultural industry, particularly as farmers encroach more and more upon fertile lands set aside for pasturage and forestry.

In communal areas in Makonde, farmers follow the communal land tenure system, which also gives rise to a number of environmental problems similar to those experienced in resettlement areas, including on grazing land. Livestock owners lack incentives to improve pasturelands, forests, and freeways because benefits accrue to the community rather than to the individual investor. The full cost of holding livestock in excess of the carrying capacity of the rangelands is borne by the community at large rather than the individual. Individual farmers have incentive to keep as much livestock as possible with limited direct costs in veterinary services. The community ultimately suffers the consequence of degraded grazing resources, the conversion of grazing and forests into residential and crop land, and the cutting down of trees for fuel without any replacements. As demand for food and living space increases due to human pressure, a corresponding encroachment occurs through the cultivation of marginal lands as well as the settlement of hilly or fragile terrain. In addition, the fragmentation of land units encourages land degradation and erosion. Generally, there is a lack of coordination in policing mechanisms that could sustainably oversee sound management of grazing and forestry resources. Regardless, communal land ownership remains popular because it allows the poor free access to land. However, it discriminates against women because of the patriarchal customary laws that govern it, but married women, divorcees, and widows are usually assured of pieces of land.[64]

4 Conclusion

Climate change debates in Makonde and Zimbabwe do not preoccupy the lives of farmers, nor are they a priority area for government. Farmers and the government are more concerned about issues of agricultural productivity for the benefit of food security and foreign exchange earnings. As a result, they worry more about the loss of soil fertility than the effects of climate change, which they blame on industrial pollution. This does not suggest in any way that farmers and government do not agonise about climate variability as manifested in extremes of temperature, inadequate and erratic rains, as well as acute and more frequent drought. Ideas and knowledge about climate change are taught formally in school and informally in the home to equip young future farmers with knowledge about this important influence in agrarian society. There are no clear-cut or formal ways of gathering climate change information in Makonde. Climate knowledge is learnt from various sources like family and school and its applicability to agricultural planning and planting is also learnt along the way.

64 Focus group discussion with resettled Makonde farmers Betty Muchada, Moses Dzawo, Kunaka Imbayago, Denigo Dzvairo, and Mary Chaitezvi at Nkosana primary school, Makonde. March 10, 2012.

PART 2

Past

CHAPTER 3

Environmental and Climate Change in Africa: Global Drought and Local Environmental Infrastructure

Emmanuel Kreike

Abstract

This chapter discusses major approaches to the environmental history of Africa in terms of the drivers of environmental change, including both human and non-human factors. This allows for the contextualisation of what currently is considered perhaps the largest single environmental threat: global climate change. A key factor in identifying Africa as the most vulnerable continent is the perception that African societies are directly dependent upon their fickle environment. Africans are often seen as living virtually in and of nature. Societies elsewhere, by contrast, are considered to be shielded from nature's whims by a cultural environment created through modern technology and science. The differences in the relationship of Western and African societies to nature, however, have been vastly overstated. By way of critiquing the underlying nature (Africa) versus culture (West) dichotomy, this chapter investigates African environmental infrastructures, including land management systems, elaborate systems of water harvesting and food storage, and burning regimes, which cushion the impact of weather and environmental extremes. It substantiates its arguments by means of a case study from 1920s and early 1930s north-central Namibia. A global climate event (a severe drought), the global economic crisis, and regional political and demographic developments nearly led to a killer famine. The history of this drought demonstrates how environment and climate are embedded in and fractured through social, economic, and political factors.

1 Introduction

This chapter discusses some of the key approaches to understanding the dynamics of environmental change. It focuses on both human and non-human factors to contextualise the possible impact of global climate change on Africa. The global climate change argument emphasises environmental agency even as human atmospheric pollution is identified as the underlying cause. Global climate change is depicted as an unintended consequence of human activities,

 | DOI:10.1163/9789004410848_005

which set in motion processes of change with outcomes that are neither desired nor controlled by humanity. Global climate change is expressed in terms of nature's revenge and is in many ways reminiscent of the morality of the biblical flood narrative.[1] Thus, by and large, the global climate change debate highlights the effects of the powers of nature over human culture within an environmentally deterministic framework. This approach fits within a larger pattern of a post-modernisation theory paradigm of environmental change that no longer unambiguously sees humanity as having domesticated, conquered, or killed nature. An unbridled belief that Western science and technology would bring economic growth and prosperity through mastering the forces of nature and harnessing them for the good of humanity characterised the 1950s–1970s. Even the declinist paradigm, while lamenting the environmental cost of progress and questioning the morality of the domestication and death of nature, worked from the premise that humans were increasingly in control. Most approaches to environmental change in that period identified human actions as the critical engine of this process. The political ecology approach focused on power struggles over the environment or affecting the environment. In Africa, such studies highlighted how environmentalist agendas served as a legitimisation for or tool of imperial or colonial expansion, pitting the (colonial) state or settlers and their collaborators against local African communities. Thus African hunting, gathering, pastoralism, and crop cultivation were criminalised or severely constrained by the imposition of conservation schemes, such as the establishment of game reserves, national parks, and forests or the introduction of veterinary measures and anti-erosion projects.[2] The impact of

1 On the biblical flood and global climate change, see, for example, Keulemans, M., "Klimaatdebat is in Zekere Zin een Voortzetting van Bijbels Eindtijdverhaal," *Volkskrantkatern Vonk*, March 24, 2014.

2 On game conservation, see, for example, Anderson, D., and R. Grove, ed., *Conservation in Africa: People, Policies and Practice* (Cambridge: Cambridge University Press, 1987); MacKenzie, J.M., *The Empire of Nature: Hunting, Conservation and British Imperialism* (Manchester: Manchester University Press, 1988); Carruthers, J., *The Kruger Park: A Social and Political History* (Pietermaritzburg: University of Natal Press, 1995); Neumann, R.P., *Imposing Wilderness: Struggles over Livelihood and Nature Preservation in Africa* (Berkeley: University of California Press, 2000). On land use and conservation, see, for example, Beinart, W., and C. Bundy, *Hidden Struggles in Rural South Africa: Politics and Popular Movements in the Transkei and Eastern Cape, 1890–1930* (London: James Currey, 1987); Anderson, D., "Depression, Dust Bowl, Demography, and Drought: The Colonial State and Soil Conservation in East Africa during the 1930s," in *Colonialism and Nationalism in Africa, vol. 2: The Colonial Epoch in Africa*, ed., G. Maddox (New York: Garland, 1993), 209–231; McCann, J.C., *Green Land, Brown Land, Black Land: An Environmental History of Africa, 1800–1990* (Portsmouth: Heinemann, 1999); Jacobs, N.J., *Environment, Power, and Injustice: A South African History* (Cambridge: Cambridge

(Western) markets and commodification was another form of human agency that dramatically shaped the environment through overhunting and poaching. Commodification also changed the face of agriculture in Africa, for example through the introduction of large-scale commercial agriculture (plantations and settler agriculture) and subsequent land alienation.[3] Commodification and the imposition of colonial power also led to the colonial state's channelling of African rural labour from pastoralism and crop cultivation into the colonial economy through forced labour, forced cultivation, and migrant labour.[4] Often intertwined with the above approaches was a scientific/technological determinism model that highlighted the impact of modern 'Western' knowledge on Africans and Africa.[5] Modern Western medical knowledge and practices were identified as key factors in reducing mortality, for example through vaccinations and clinics. Rapid population growth after the Second World War was seen as the result of an unchanged 'culture' that emphasised large numbers of children to compensate for high infant mortality. During the 1970s and 1980s, the 'population bomb' was identified as the largest threat to Africa's future and development.[6]

University Press, 2003); Tropp, J.A., *Natures of Colonial Change: Environmental Relations in the Making of the Transkei* (Athens: Ohio University Press, 2006); Showers, K.B., *Imperial Gullies: Soil Erosion and Conservation in Lesotho* (Athens: Ohio University Press, 2005).

3 For example, Palmer, R., and N. Parsons, ed., *The Roots of Rural Poverty in Central and Southern Africa* (Berkeley: University of California Press, 1977); Bundy, C., *Rise and Fall of the South African Peasantry* (London: Heinemann, 1979); Beinart, W., P. Delius, and S. Trapido, ed., *Putting a Plough to the Ground: Accumulation and Dispossession in Rural South Africa, 1850–1930* (Johannesburg: Ravan Press, 1986); Anderson, D., *Eroding the Commons: The Politics of Ecology in Baringo, Kenya 1890–1963* (Oxford: James Currey, 2002).

4 On forced cultivation, see Isaacman, A., and R. Roberts, ed., *Cotton, Colonialism, and Social History in Sub-Saharan Africa* (Portsmouth: Heinemann, 1995).

5 On technology, see Goody, J., Technology, Tradition, and the State in Africa (London: Oxford University Press, 1971). On firearms, see Headrick, D.R., The Tools of Empire: Technology and European Imperialism in the Nineteenth Century (New York: Oxford University Press, 1981), 83–126. For a critique of linear models of technological determinism, see Kreike, E., Deforestation and Reforestation in Namibia: The Global Consequences of Local Contradictions (Leiden: Brill, 2010), 113–138. More recently, knowledge production has been depicted as a two-way process. See Grove, R.H., Green Imperialism: Colonial Expansion, Tropical Island Edens and the Origins of Environmentalism 1600–1860 (Cambridge: Cambridge University Press, 1997); Harries, P., Butterflies and Barbarians: Swiss Missionaries and Systems of Knowledge in South-East Africa (Oxford: James Currey, 2007); and Tilly, H., Africa as a Living Laboratory: Empire, Development and the Problem of Scientific Knowledge, 1870–1950 (Chicago: University of Chicago Press, 2011).

6 Ehrlich, P., *The Population Bomb* (New York: Ballantine, 1968). For a critique, see Boserup, E., *The Conditions of Agricultural Growth: The Economics of Agrarian Change under Population Pressure* (New York: Aldine Pub. Co., 1965); Tiffen, M., M. Mortimore, and F. Gichuki,

During the 1970s and 1980s, the era of the modernisation-cum-developmentalist paradigm, studies that emphasised environmental forces over human agency were relatively rare and mostly considered the impact of desertification and climate change in the Sahel region in the 1970s.[7] Early studies about the history of famines in Africa also interpreted famines as outcomes of the vagaries of climate and environment. In the 1980s, however, humans or humanity (including the impact of colonial political and conservation measures) and issues of entitlement (or rather the lack thereof) were increasingly identified as the deeper causes of famines.[8] Influenced by the declinist paradigm, which emphasised human-caused environmental destruction, the agency of culture over nature was probably exaggerated. Upon closer inspection, even the population bomb argument was ambiguous in terms of agency. Modern culture (through medicine and public health) was said to reduce mortality, privileging human agency, while continued high fertility rates were attributed to 'traditional African culture'. But, at the same time, African traditional culture was defined in terms of Africans living in or very close to and by nature: their proclivity towards large families was ascribed to the (natural) urge of (primitive) people to procreate without bounds. Alternatively, per a moral economy argument, it was claimed that Africans invested in family and community as insurance against old age, sickness, and famine.[9] In that sense, the population bomb argument is in many ways reminiscent of the nature's revenge component of the current debate on global climate change. Similarly, another approach to environmental change, which highlights biological invaders, also acknowledges nature's agency. This latter model is best known for its relevance to (North) America,

---

More People, Less Erosion: Environmental Recovery in Kenya (Chichester: John Wiley and Sons, 1994).

7 On climate change and African history, see Brooks, G.E., Landlords and Strangers: Ecology, Society, and Trade in West Africa, 1000–1630 (Boulder: Westview Press, 1993); Webb, J.L.A., Desert Frontier: Ecological and Economic Change along the Western Sahel, 1600–1850 (Madison: University of Wisconsin Press, 1995); McCann, J.C., "Climate and Causation in African History," International Journal of African Historical Studies 32.2–3 (1999): 261–280.

8 On famines in Africa, see Watts, M., Silent Violence: Food, Famine and Peasantry in Northern Nigeria (Berkeley: University of California Press, 1983); Vaughan, M., The Story of an African Famine: Gender and Famine in Twentieth-Century Malawi (Cambridge: Cambridge University Press, 1987); de Waal, A., Famine that Kills: Darfur, Sudan (Oxford: Oxford University Press, 2005).

9 On the moral economy concept, see Scott, J., *The Moral Economy of the Peasant: Rebellion and Subsistence in Southeast Asia* (New Haven: Yale University Press, 1976). The moral economy is upheld as a pre-modern era precursor to the modern rational economy in the developmentalist/evolutionary view of history.

where invasive species, in particular smallpox, but also horses, sheep, and a myriad of plants, caused dramatic environmental upheaval.[10] The literature has since expanded by not only emphasising European species in this ecological imperialism, but also African biological invaders (such as yellow fever).[11] In the context of Africa, invaders from outside the continent have received less attention (exceptions are the prickly pear and maize); instead, the emphasis has been on 'indigenous' species that have run amok because of disturbances to the continent's ecological equilibrium. The violence of colonialism, including conquest and population displacements, caused a number of endemic diseases, like sleeping sickness and malaria, to turn into epidemics.[12] These approaches effectively also fit into the nature's revenge pattern, since the disturbance of the ecological equilibrium is depicted as if awakening the dark forces of nature. In this context, the model of global climate change as a backlash from nature triggered by a long history of human abuse is thus not new. The nature's revenge language provides a powerful call for us to change our ways. However, despite the fact that humanity bears great responsibility in the squandering of environmental resources, the

10 Crosby, A.W., The Columbian Exchange: The Biological and Cultural Consequences of 1492 (Westport: Greenwood, 1972).

11 Kiple, K.F., *The Caribbean Slave: A Biological History* (Cambridge: Cambridge University Press, 2002); Curtin, P.D., *Disease and Empire: The Health of European Troops in the Conquest of Empire* (Cambridge: Cambridge University Press, 1998); Fenn, E.A., *Smallpox Americana: The Great Smallpox Epidemic of 1775–1782* (New York: Hill and Wang, 2002).

12 Law, R., The Horse in West African History: The Role of the Horse in the Societies of Pre-Colonial West Africa (Oxford: Oxford University Press, 1980); McCann, J.C., Maize and Grace: Africa's Encounter with a New World Crop, 1500–2000 (Cambridge, Mass.: Harvard University Press, 2005); van Sittert, L., " 'The Seed Blows About in Every Breeze': Noxious Weed Eradication in the Cape Colony, 1860–1909," Journal of Southern African Studies 26.4 (2000): 655-674; Kreike, *Deforestation*, 83–112. On invading microbes in Africa, e.g. rinderpest, see van Onselen, C., "Reactions to Rinderpest in Southern Africa, 1896–1897," Journal of Southern African Studies 13.3 (1972): 473–488; and Peires, J.P., The Dead will Arise: Nongqawuse and the Great Xhosa Cattle Killing Movement of 1856–7 (Johannesburg: Ravan Press, 1989). On malaria and environmental management, see Kjekhus, H., Ecology Control and Economic Development in East African History: The Case of Tanganyika, 1850–1950 (London: Heinemann, 1977); Giblin, J., The Politics of Environmental Control in Northeastern Tanzania, 1840–1940 (Philadelphia: University of Pennsylvania Press, 1992); Lyons, M., The Colonial Disease: A Social History of Sleeping Sickness in Northern Zaire, 1900–1940 (Cambridge: Cambridge University Press, 1992); Hartwig, G.W., and K.D. Patterson, ed., Disease in African History: An Introductory Survey and Case Studies (Durham: Duke University Press, 1978); Maddox, G., J. Giblin, and I.N. Kimambo, ed., Custodians of the Land: Ecology and Culture in the History of Tanzania (London: James Currey, 1996); Giles-Vernick, T., and J.L. Webb, ed., Global Health in Africa: Historical Perspectives on Disease Control (Athens: Ohio University Press, 2013).

nature's revenge language glosses over the intricate dynamics that mark environmental and climate change. Nature and culture need to be considered in unison because they are inextricably intertwined. The nature's revenge interpretation is derived from and embedded in the 20th-century nature-culture dichotomy and is conceptually inadequate to understand and diagnose the new challenges of the 21st century.

Global climate change is expected to further marginalise African livelihoods and rural economies, causing some parts of the continent to become drier and other parts wetter.[13] The most pernicious impact, however, may be that Africa's environment will become even more capricious. The African environment has often been portrayed as hostile to human existence (consider, for example, its historical reputation as the 'white man's grave').[14] It also harbours such dangerous diseases as malaria, bilharzia, AIDS, and Ebola. Africa's population is concentrated in semi-arid regions where rainfall is scarce and unreliable: droughts are frequent, and torrential rains cause floods. A key factor in identifying Africa as the most vulnerable continent is the very perception that African societies are directly dependent upon their fickle environment. Africans are often presented as living virtually in and by nature. Societies elsewhere, by contrast, are considered to be shielded from nature's whims (and therefore, by extension, from climate change) by a cultural environment created from and through modern technology and science.[15]

Yet, the inverse argument has also been made: that modern Western societies are more vulnerable to the vagaries of the environment (and therefore climate change) because commercial industrialised production systems are geared towards maximising production, are highly specialised, and rely on a narrow range of resources and technologies. Industrial agriculture, for example, is dependent on the monocropping of a few key cultivars with little or no genetic diversity, dramatically increasing crops' exposure to drought and disease. In contrast, African 'traditional' agriculture is less vulnerable because it is much more diversified and geared towards risk evasion rather than maximising

13 A 2012 study noted that extreme rainfall and flood events in Africa corresponded to El Niño occurrences (e.g. 1982–1983 and 2006–2007), see Kundzewicz, Zbigniew W. et al., "Flood Risk and Climate Change: Hydrological Perspectives," *Hydrological Sciences Journal* 59.1 (2012): 1–28.

14 For Africans as a source of contagion, see Farley, J., *Bilharzia: A History of Imperial Tropical Medicine* (Cambridge: Cambridge University Press, 1991), 13–20, 137–139. Iliffe frames African history as a long struggle to overcome a hostile environment, see Iliffe, J., *Africa: The History of a Continent* (Cambridge: Cambridge University Press, 2007).

15 Kreike, Deforestation.

production.[16] In fact, the main criticism of African 'traditional' household agricultural production is that risk-evasive strategies obstruct development. According to this argument, African farmers are hesitant to substitute low-productivity land races with high-yielding genetically improved cultivars and invest in social networks rather than in economic production.

The differences in the respective relationships of Western and African societies to nature, however, have been vastly overdrawn. The misconception is embedded in a paired dichotomy that subsumes the non-West under nature and the West under culture, thereby contrasting nature with culture and the non-West (and the pre-modern) with the modern West. As demonstrated by the dangerous dependency of modern industrial (and post-industrial) society on a single non-renewable, increasingly scarce, and highly polluting source of energy (hydrocarbons) as well as on monoculture food sources, and a vulnerability to 'wild' fires (from the Oakland Hills to the Cape Flats) and 'wild' waters (such as the New Orleans and Rhine floods), modern Western technological infrastructure is far from impervious to nature and global climate change. African societies have equally relied on an infrastructure, such as elaborate systems of water harvesting, food storage, and black soil creation, and land management structures and institutions to regulate use of and access to environmental resources. African environmental infrastructure also includes burning regimes and vegetation management (to control insects like the mosquito and the tsetse fly and to maintain pastures) and transhumance (to rotate pastures and ration water). Thus, although rural Africa is dramatically deficient in terms of such 'modern' technological infrastructure as all-weather roads and bridges, the presence of *environmental infrastructure* cushions the impact of weather and environmental extremes. Environmental infrastructure has to be constantly maintained, repaired, and re-invented in the face of new challenges: it is and has to be highly dynamic.[17] Furthermore, environmental infrastructure may fail, either as a result of repeated drought or as a result of war and population displacement. In turn, weak environmental infrastructure may worsen the impact of droughts and floods. This chapter focuses on late

16 This is a major argument of the indigenous knowledge literature, see, for example, Bassett, T.J., and D. Crummey, ed., *African Savannas: Global Narratives and Local Knowledge of Environmental Change* (Oxford: James Currey, 2003); Leach, M., and R. Mearns, *Beyond the Fuelwood Crisis: People, Land and Trees in Africa* (London: Earthscan, 1988); Fairhead, J., and M. Leach, *Misreading the African Landscape: Society and Ecology in a Forest-Savanna Mosaic* (Cambridge: Cambridge University Press, 1996).

17 On the environmental infrastructure concept, see Kreike, E., *Environmental Infrastructure in African History: Examining the Myth of Natural Resource Management in Namibia* (Cambridge: Cambridge University Press, 2013).

1920s and early 1930s north-central Namibia, when a global climate event (a drought that affected Southern Africa as severely as it did the North American Dust Bowl), a global economic crisis, and regional political and demographic events nearly led to a killer famine. The history of the drought demonstrates how the environment and climate are embedded in and fractured through social, economic, and political factors.

Climate and drought were key factors in north-central Namibia's 'Famine of the Dams' in the 1920s and 1930s. The drought was so severe and prolonged that the environmental infrastructure that served to store water and food was strained to breaking point. The collapse of the water and food storage infrastructure, however, was not only due to climatic or environmental factors, and neither the causes nor the impact of the famine were linear or homogenous. Refugees from war and famine with limited or no access to their own or others' environmental infrastructure proved most vulnerable, as is reflected in how the famine was depicted and explained: the famine, for example, is remembered as both the 'Famine of the Dams' and, paradoxically, 'the enriching famine'. Memories of the famine thus demonstrate that it was seen as both a natural and a social phenomenon.

2 The Famine of the Dams

In 1915, South African colonial forces invaded what was then German South West Africa and marched beyond the German colony's northern border to occupy what became known as Ovamboland. Two colonial officials administered Ovamboland through a system of indirect rule: the local chiefs and headmen were held responsible for order, justice, taxation, and labour recruitment. In an area that is semi-arid, millet was the staple crop and most households also kept livestock. Most men engaged in migrant labour on farms and mines in Namibia or South Africa. The area is highly semi-arid. The *Ondjala yOmatale*, the Famine of the Dams, affected much of colonial Namibia's Ovamboland, especially the two largest and most populous regions, Ondonga and Oukwanyama. Yet, in its early stages, the impact of the famine was very limited and localised. Independent environmental variables over which local communities had no influence, such as rainfall patterns, played an important role in the emergency, but the extent to which local ecosystems had been transformed by human activity was also critical. While another section of Oukwanyama that belonged to the Portuguese colony of Angola was filled with villages and was referred to by its inhabitants as *oshilongo* (the settled zone), Namibian Oukwanyama south of the border was mostly referred to as *ofuka* (wilderness). Most of the *ofuka* area had been settled in the decades before the famine by refugees and migrants from Angola. The colonial boundary between the two Oukwanyamas therefore largely coincided with an (environmental) infrastructural fault line that emerged during the early stages of the famine: Oukwanyama south of the border became a recipient of food aid, while Oukwanyama north of the border supplied most of the food aid.

During the early stages of the famine, mobility as well as social networks enabled the worst affected households to seek resources from other households, villages, and areas that controlled surpluses, either locally or further afield. Less fortunate households approached more privileged relatives or local headmen for the millet staple, or, alternatively, for livestock with which to purchase grain. Livestock owners exchanged their animals for food. Cattle herds from the drought-affected areas were led north into Angola to Oshimolo, where water and grazing remained available. A considerable number of people who recently had moved to Namibian Ovamboland chose to return to the Angolan side, where the food situation was much better. The latter strategy, however, led to the direct intervention of the colonial state, and transformed the drought from an agricultural and ecological crisis into a political and economic emergency.

3 The Ecology of a Famine

Environmental factors partially account for the uneven impact of the famine. Rainfall in the semi-arid environment was localised and highly erratic. Pastures declined rapidly under the severe drought conditions, and by September 1928 grazing on the Namibian side of the border was virtually depleted.[18] As water became scarcer, competition between herdsmen increased.[19] Herdsmen tried desperately to save their animals:

> all over the country one sees emaciated cattle, unable to move lying down. The owners [...] build shelters to protect the beasts from the heat. It all depends on the owner whether he is prepared to carry food and water, anything from five to seven miles, to these animals.[20]

---

18 National Archives of Namibia (henceforth NAN), Native Affairs Ovamboland (henceforth NAO) 18, *Monthly Reports for Ovamboland*, July–September 1928; and NAO 40, "Note of interview with his honour the administrator [of SWA] 9/2/29 in regard to prospective famine conditions in Ovamboland."

19 NAN, NAO 18, *Monthly Reports for Ovamboland*, July–August 1928.

20 NAN, NAO 18, *Monthly Report for Ovamboland*, January 1929.

By early 1929, over 5,000 head of cattle had perished in Namibian Oukwanyama alone, perhaps up to one-quarter of the total herd.[21] The severity of the drought prompted Namibian residents to seek out water sources and pastures in Angola. The villages on the Angolan side of the border were older villages with an established human-created water infrastructure. Considerable numbers of men, women, and children, sometimes led by prominent headmen, moved across the colonial border to the Portuguese-controlled side towards the end of the 1920s.[22] Migration to Angola continued throughout the first half of 1930.[23] Except for a strip of land directly north of the border, rainfall and harvests in Angolan Oukwanyama, in the heart of the old Oukwanyama *oshilongo*, had been much better.[24] Throughout the 1929–1931 famine years, Angolan Oukwanyama supplied both individual households and the colonial authorities in Namibian Ovamboland with large quantities of millet.[25]

During the January 1929 rainy season, patchy rains caused the situation in Namibian Oukwanyama to deteriorate further. By the end of January, many households had not yet planted any grain. Although some households survived on tree fruit, the fruit harvest was a total failure in the western half of Ovamboland. An invasion of grubs—so-called armyworms—resurrected terrifying images of the devastating 1915–1916 famine known as the Famine of the Insects,

21 NAN, NAO 18, *Monthly Report Oukwanyama*, January 1929; NAO 40, Officer Commanding (henceforth O/C) Oshikango to O/C NAO, Oshikango, March 11, 1929, "Stock Losses: Ukuanyama;" and "Particulars Obtained from Headmen in Western Ukuanyama Area re. Famine etc.;" "Particulars in Regard to Famine Ukuanyama Areas received from headmen east of the main road from Ondonga to Oshikango," and Appendices to O/C Oshikango from O/C NAO, Oshikango July 31 and August 8, 1929.

22 Arquivo Histórico Ultramarino, Lisbon, Portugal (henceforth AHU) 9, 590, 20, Fronteira Sul de Angola, Processo No. 265-C, C.R. Machado, Chefe da Delegação Portuguesa, Oluchanja, 23 Julho 1927, annex: considerações de carácter reservado, to Chefe da Delegação do Governo da União; NAN, NAO 17, O/C NAO, October 16, 1927, to Mr. Clarke; NAO 18, *Monthly Report Oukwanyama*, October 1927; ibid, *Monthly Report for Ovamboland*, September 1928. See also the following interviews by the author: Ester Nande, Onengali, May 20, 1993; Moses Kakoto, Okongo, February 16, 1993; and Petrus Mbubi, Onanime, February 26, 1993.

23 NAN, NAO 18, *Monthly Reports Oukwanyama*, June–August 1930.

24 Kalolina Naholo, interview by author, Ohamwaala, January 26, 1993.

25 NAN, NAO 41, Native Commissioner Ovamboland (henceforth NCO), *Famine Reports*, September-October and December 1930; Union (of South Africa) Government Representative Namacunde (henceforth UNG), *Union Administration 2 folder 1922–1946*, UGR Oshikango, *Monthly Report*, December 1929; NAN, NAO 40–41, *Reports on Famine Relief Works: Oukwanyama*, August 1930–February 1931; and Trust Fund Ovamboland, Statement for February 1930.

which had killed hundreds, perhaps even thousands, of people.[26] Fortunately, February rains heralded the return of migratory birds that curtailed the spread of the insect plague.[27] The environmental and social variables that contributed to the famine are closely intertwined. The old Oshikango district was the only district south of the modern border that had formed part of the settled core of the pre-colonial Oukwanyama kingdom in its entirety. Its landscape differed sharply from the other districts south of the border, as described in colonial archives and oral histories. The new farms and villages across most of the adjacent Omhedi and Onenghali districts and throughout the Ohaingu, Onahulu, Onamine, Onamunama, and Okalongo districts had been recently carved out of the *ofuka*, the wilderness. Although this area was abundant in game and wood, settlement in the *ofuka* was risky because it lacked the necessary environmental infrastructure (waterholes, fruit trees, raised fields, fenced farms, and food stores) to sustain its inhabitants in the face of drought, floods, and marauding wild animals and birds. Thousands of refugees who had fled violent Portuguese colonial exactions to settle on the South African-controlled side of the border during the 1910s and 1920s had only just begun to transform the *ofuka* into *oshilongo* by the late 1920s.[28] This unfolding yet incomplete process of creating environmental infrastructure explains to a significant extent why the drought conditions in most of Oukwanyama on the Portuguese side did not lead to a similar emergency and why even south of the border the impact of the famine was uneven.

Two of the four Namibian Oukwanyama districts in which senior headmen approached the colonial administration with requests for food aid were located entirely outside of what until 1915 had been the southernmost settled part of pre-colonial Oukwanyama: Ohaingu and Onanime. The remaining two, Onenghali and Omhedi, each contained only a very small slice of the old pre-colonial Oukwanyama *oshilongo*.[29] These districts lacked the infrastructure

26 On the previous famines, see Kreike, E., *Re-Creating Eden: Land Use, Environment, and Society in Southern Angola and Northern Namibia* (Portsmouth: Heinemann, 2004).

27 NAN, NAO 18, *Monthly Reports Oukwanyama*, January–February 1929. On the failure of the fruit harvest, see NAO 40, O/C NAO to Secretary South West Africa (henceforth SWA), *Ondangwa*, January 15, 1929, and Assistant Priest St. Mary's Mission to O/C NAO, [Odibo], January 10, 1929.

28 Kreike, Re-Creating Eden.

29 NAN, NAO 40, O/C NAO to Secretary SWA, *Ondangwa*, March 18, 1929; O/C Oshikango Station to O/C NAO, *Oshikango*, March 11, 1929 (both letters of that date); O/C Oshikango Station to O/C NAO, Oshikango, April 18, 1929; O/C Oshikango Station to O/C NAO, Oshikango, May 8 and 15, 1929; and O/C NAO to O/C Oshikango Station, Ondangwa, May 12, 1929.

(especially waterholes and water reservoirs) that marked the old Oukwanyama across the border in Angola.

Moreover, Namibian Oukwanyama was flooded by refugees from the Ondonga area, which was even more seriously affected by the drought. With the failure of the 1929 harvest, many Ondongas abandoned their farms for neighbouring communities. At Ombalantu, to the southwest of Oukwanyama in Ovamboland, they were turned away. "Large numbers" of Ondonga refugees migrated to Omulunga, east of Oshiede in Angolan Oukwanyama, where land and water were abundant. Others went as far as Ombadja and Humbe, deeper into southern Angola. Hundreds of Ondonga stayed closer to home and 'squatted' in Oukwanyama villages south of the border. The influx of so many refugees was an additional burden on already scarce food and water supplies in Namibian Oukwanyama.[30]

Detailed insights into the impact of the famine in Namibian Oukwanyama are offered by data taken from reports about the drought conditions in July 1929. The data was gathered from 135 Oukwanyama village headmen and probably included all the villages in Oukwanyama south of the border. The village locations were indicated according to their position east or west of the old Ondangwa–Oshikango road (a rather arbitrary distinction), which ran approximately one to two miles to the west of the modern tarred road.[31] Of the 135 villages, 132 reported a poor harvest, the remaining three a fair harvest. All but one headman reported having a famine-affected village. Cattle had been sent to remote cattle posts: 94 village headmen stated that the cattle from their villages were in Angola. Grazing conditions were different east and west of the road for the livestock remaining in the villages (mainly goats). In Oukwanyama, east of the Oshikango–Ondangwa road, 31 of the 42 villages (74 per cent) reported grazing in or near the village as "nil" and nine (21 per cent) as "good". West of the road, where the majority of villages were located, grazing was slightly better: only five of the 93 villages (five per cent) reported grazing conditions as "nil", 84 (90 per cent) as "poor", and only four (four per cent) as "fair" or "good".[32]

30 NAO 18, *Monthly Reports Oukwanyama*, May–July 1929.

31 The old road was known as Shongola's road (*shongola* was the nickname for Native Commissioner C.H.L. Hahn). The old road cut the Oshikango district in two, but Onenghali, Onanime, and Onamunama districts were located east of the road and Omhedi, Ohaingu, Onahulu, and Okalongo to its west.

32 NAN, NAO 40, "Particulars Obtained from Headmen in Western Ukuanyama Area re. Famine etc.;" and "Particulars in Regard to Famine Ukuanyama Areas Received from Headmen East of the Main Road from Ondonga to Oshikango;" and appendices from O/C Oshikango to O/C NAO, Oshikango July 31 and August 8, 1929.

Extreme water scarcity also illustrates the severity of the drought. West of the Ondangwa–Oshikango road, 18 per cent of the villages relied on water located at least 1.6 kilometres away, while ten per cent of the villages relied on water sources at a distance of three to eight kilometres away. In Oukwanyama, east of the road, 16 per cent of the villages relied on water at least 1.6 kilometres away, while 33 per cent relied on water sources situated three to eight kilometres away.[33] All water sources were man-made; during the dry season villagers had no access to any natural sources of water in the region.

Finally, the survey demonstrated that the drought led to substantial population displacement. 42 per cent of all Oukwanyama villages hosted Ondonga-based households, and thirty per cent reported Ondonga refugee households squatting in their villages. In addition, 24 per cent of all Oukwanyama's villages reported losing households that had fled to Angola.[34]

In 1929 Ovamboland's colonial administration initiated a single food for work project in Oukwanyama in the village of Etale. Local enthusiasm for contributing labour towards the construction of a dam in exchange for food aid, however, proved limited. Alternative sources of food—such as patronage networks—prevailed, while October rains relieved water shortages and allowed for field preparation and seeding. Locals participated in the food for work program only in December when drought conditions persisted and the plantings were lost. The Etale dam, a water storage reservoir meant to alleviate watershortages during the dry season, was completed in mid-December.[35]

After losing their first batches of millet seed during the normal planting period (October–December) because of the continued drought conditions, people were loath to commit their dwindling seed reserves to parched fields, although the longer they waited, the less time remained in the cropping season. Most of the cattle herds remained at remote cattle posts north of the border in Oshimolo because of the shortage of water and grazing in the Namibian villages. Optimistic herdsmen who had driven herds in their care back to Namibia after the first promising rains had returned to Angola by December.

---

33 NAN, NAO 40, "Particulars [...] Western Ukuanyama Area re. Famine;" "Particulars [...] East of the Main Road from Ondonga to Oshikango;" and appendices to O/C Oshikango Station to O/C NAO, Oshikango, July 31 and August 8, 1929.

34 Ibid.

35 NAN, NAO 18, *Monthly Reports Oukwanyama*, September–October 1929; and NAO 41, Famine Relief Ukuanyama, *Works Issue Schedules*, October–December 1929.

Several of the most recently established Angolan refugee villages in Oukwanyama were virtually abandoned because they lacked reliable sources of water. In the far eastern Namibian Oukwanyama village of Eenhana, drinking water had to be fetched in drums from Angola (requiring a trip of ten kilometres to the border alone) or Oshikango (forty kilometres). In the most recently settled areas in far western Namibian Oukwanyama—Okalongo, the situation was equally serious: only a single well dug by a missionary remained as a source of drinking water for all the villages in the area. By the end of 1929, therefore, large areas of Namibian Oukwanyama were scarcely better off than Ondonga. In December 1929, the headmen of the villages of Onekwaya and Ohalushu reported the first cases of starvation in Oukwanyama.[36] The onset of the Marula season in January 1930, when tree fruit became available, reversed this trend. Onekwaya, located within the confines of the pre-1915 Oukwanyama *oshilongo*, was blessed with an established environmental infrastructure, including fruit trees. Many of the newer villages established by Angolan refugees before the famine, however, had few or no fruit trees, since the most common ones, such as the Marula (*Sclerocarya birrea*) and Birdplum (*Berchemia discolor*), do not bear fruit until ten to fifteen years after planting.[37]

In May 1930, colonial Namibia's Director of Works conducted a sample survey of 172 randomly selected households in Ovamboland. The surveyors concluded that only the three consecutive years of drought could explain why Ovamboland's "excellent system of grain storage" could no longer keep "an acute general famine" at bay. Despite the owners' "considerable resentment", the surveyors opened the clay-sealed grain storage baskets in the sampled households and measured their contents. They found "[m]ore than one kraal […] where the people had no *mahangu* [millet] left, and had substituted watermelon pits as part of their diet". Almost half of the sampled households in Ondonga had depleted their millet stores. The same was true for one-quarter of the sampled households in Oukwanyama. Most of the livestock of the surveyed households had been herded to remote cattle posts in Angola, depriving households of access to dairy products and compounding their nutritional stress. The smaller communities in western Ovamboland generally fared better.[38]

36 NAN, NAO 18, *Monthly Reports Oukwanyama*, September–December 1929.

37 On the history of the fruit trees, see Kreike, *Environmental Infrastructure*, 36–62.

38 NAN, NAO 40, Director of Works to Secretary SWA, Windhoek, June 3, 1930.

4 Ondjala Yokoyolangudja or "The Enriching Famine": the Moral Economy of a Famine

During the early phase of the drought, social networks such as clans and patron–client networks mitigated the effects of the food crisis, but social networks for the redistribution of food soon collapsed.[39] During the first phase of the Famine of the Dams, social networks in colonial Namibian Oukwanyama were a critical component of the arsenal of drought-coping mechanisms. The colonial official in Oshikango in the heart of Oukwanyama noted the large-scale destruction of palm bushes that took place during the peak of the 1929 famine. The palm bushes were tapped to make palm wine, which not only provided nutrition but was also exchanged for grain with the "principal headmen" and important "small headmen", since "they will not give away any grain except for some sort of payment".[40]

In essence, the 'exchange' of palm wine for grain may be compared to the giving of Marula wine (made from the fruit of the Marula tree) to village headmen, district headmen, and kings, which is strongly symbolic as recognition of a tributary or patron–client relationship.[41] It thus suggests that people sought to widen or strengthen existing social networks around patrons in order to gain access to additional sources of food, a moral economy transaction.[42] Offering a small share of one's own home-made Marula wine or palm wine to a social superior established a patron–client relationship, 'obligating' the patron to support the client in times of need. Patrons often took their obligations seriously. For example, Elisabeth Ndemutela could barely recall the Famine of the Dams because during emergencies her household had been supplied with food by her grandmother or her 'uncle', the powerful senior headman Nuyoma Moshipandeka. Mwulifundja Linekela Haiyaka had just become a widow during the Famine of the Dams and, together with her mother and grandmother, survived on food aid supplied by senior headman Jikuma.[43] In recently settled villages, however, social networks proved too frail. During the 1920s, Twemuna Shifedi's family had moved from the village of Edundja to settle in Ondaanya,

39 Hayes. P., "The 'Famine of the Dams': Gender, Labour and Politics in Colonial Ovamboland, 1929–1930," in *Namibia under South African Rule: Mobility and Containment, 1915–1945*, P. Hayes et al., ed. (Oxford: James Currey, 1998), 117–146.

40 NAN, NAO 18, *Monthly Report Oukwanyama*, December 1929.

41 Loeb, E.M., *In Feudal Africa* (Bloomington: Indiana University, 1962), 177.

42 Scott, The Moral Economy of the Peasant.

43 Interviews by author: Elisabeth Ndemutela, Okongo, February 16–17, 1993; and Mwulifundja Linekela Haiyaka, Omhedi, February 8, 1993.

in the depths of what was still considered *ofuka*. The isolated villagers could not call on wider social networks during the famine because other settlements were simply too far away. All but two households subsequently abandoned the village.[44]

Although the Famine of the Dams exposed the weaknesses of local patronage networks, paradoxically, government food aid in Oukwanyama during the famine not only resurrected, but even strengthened selected local patronage networks. The colonial government favoured a small number of headmen to assist with managing food aid. In 1931, after the famine in Namibian Oukwanyama had passed, the colonial administration distributed food to thirty Oukwanyama headmen who had been especially helpful during the famine because "they had great difficulty in meeting demands for food, which their status [as patrons] obliged them to supply and at the same time they did not want to ask for free food".[45] For some, the famine proved to be a windfall, as is suggested by an alternative name for the Famine of the Dams: *Ondjala yokoyolangudja*, or the 'Enriching Famine'. The 'enriching' sentiment filters through in the monthly reports for Oukwanyama for late 1929, when the food situation for many became increasingly difficult. Palm bushes were cut by the thousands for use as food and to placate those who still had grain supplies. Food theft rose sharply as social networks faltered:

> the natives who own a small supply of corn or dried melon pits are afraid to sleep at nights and spend their time patrolling their kraals. The principal headmen will on no account assist their subjects and remind me of a lot of ghouls in the way they guard their grain baskets.[46]

People with access to money or cattle, either directly or through patronage networks, suffered less as a result of the famine. Handsome profits could be made by selling grain at high prices and/or extending client networks through food gifts. Some headmen who assisted colonial officials with the distribution of food aid were remembered to have abused their positions.[47]

44 Interview by author: Twemuna Shifidi, Ondaanya, January 28, 1993.

45 See NAN, NAO 41, Famine Relief Report Oukwanyama, February 1931, Special Issues; and Famine Relief Reports Oukwanyama, January–February 1931; and Famine Report Ovamboland, January 1931.

46 NAN, NAO 18, Monthly Report Oukwanyama, December 1929.

47 Interview by author: Helena Nailonga, Ekoka laKula, February 23, 1993; and Nahango Hailonga, Onamahoka, February 4, 1993.

5 The Poverty of Patronage: the Political Economy of a Famine

Although the threat of a serious famine in Ovamboland loomed large by early 1929, Namibia's highest colonial official was reluctant to intervene; the cost of an emergency food aid programme in Ovamboland was estimated at 5,000 pounds sterling. The administrator for colonial Namibia relented only after being assured that taxation would be subsequently introduced in Ovamboland. These taxes served to establish 'Tribal Trusts Funds' to finance any future emergency relief, all local development projects, and the introduction of salaries for Ovamboland chiefs and principal headmen designed to make them more accountable to the colonial authorities.[48]

Against the spectre of starvation, the introduction of a five shilling tax on all adult men in Ovamboland met with little opposition from the headmen and chiefs, yet the project collapsed on the ground. In Ondonga, tax registration and collection started in October, although little was accomplished. In Oukwanyama, registration seemed promising at first, but collapsed due to a tax revolt among Oukwanyama migrant labourers in Luderitz in southern Namibia and the fear of massive tax flight to Angola.[49] To the chagrin of the South African colonial officials, who generally looked down upon their Portuguese counterparts across the border, the Portuguese administration in Angola managed to collect its taxes as usual during 1930, highlighting the uneven impact of the drought across the colonial border. The South African official in Oshikango snidely noted that "much comment is made amongst the natives on the attitude of the two governments in respect of tax during the famine period".[50]

The fear of drought-induced massive migrations from colonial Namibian Oukwanyama to southern Angola—which would prejudice South Africa's labour supply—figured prominently in the decision to cancel tax collection and extend food aid to Oukwanyama. Although local officials downplayed the refugee dislocation caused by the drought, the above-mentioned July/August 1929

48 NAN, NAO 18, *Monthly Reports Oukwanyama*, January–February 1929. See also NAO 40, O/C NAO to Sec. SWA, Ondangwa, January 15, 1929; Assistant Priest St. Mary's Mission to O/C NAO, [Odibo], January 10, 1929; and "Note of interview with the administrator 9/2/29 in regard to prospective famine conditions in Ovamboland."

49 NAN, NAO 18, *Monthly Report Oukwanyama*, January 1930; *Accession* 450, 7, Administrator SWA to Prime Minister Pretoria, Windhoek, September 6, 1932; NAO 40, "Note of interview with his honour the administrator 9/2/29 in regard to prospective famine conditions in Ovamboland." See also NAO 42, Courtney-Clarke to Secretary SWA, Windhoek, May 6, 1929; Ibid, document with heading "Tribal Trust Funds" with nine points regarding taxation [1929]; O/C NAO to Secretary SWA, Ondangwa, October 13, 1929.

50 NAN, NAO 18, *Monthly Report Oukwanyama*, October 1930.

survey revealed that one out of every four villages in Namibian Oukwanyama had lost one or more households to Angola.[51]

For financial reasons, the Ovamboland authorities ceased to accept cattle in lieu of cash for the food aid it supplied at subsidised prices. Households without cash were thus forced to take their cattle across the border to Angolan Oukwanyama to purchase grain.[52] Meanwhile, the Portuguese authorities used this dependence on Angolan food supplies to coerce Oukwanyama refugee households in colonial Namibia into returning to Angola. The colonial Namibian administration responded by increasing its food aid programmes in the form of food for work projects and free rations.[53] The colonial government began food distribution in Ondonga at the end of January 1929 in order to bridge the period until the May harvest.[54] Throughout the colonial era, the South African and Portuguese authorities competed to attract migrants from the well-populated border region. During the 1910s and 1920s, violent Portuguese rule had caused migration and flight from southern Angola into Ovamboland. The Famine of the Dams, however, threatened to reverse this population flow, causing much anxiety among South African officials in Namibia because Ovamboland was the territory's main source of migrant labour.

The Namibian administration first introduced food for work programmes in hard-hit Ondonga in July 1929 to construct so-called dams, water reservoirs to sustain large numbers of people and livestock during the dry season.[55] The dams consisted of relatively shallow excavations in seasonal watercourses or in pans in which water accumulated during the rainy season. The excavated earth was used to raise the banks of the dams to increase their storage capacity. The excavations could not be too deep, since salty water lay beneath the layer

51 NAN, NAO 18, *Monthly Reports Oukwanyama*, June–August 1930; NAO 40, "Particulars [...] Western Ukuanyama"; "Particulars [...] East of the Main Road"; and appendices from O/C Oshikango to O/C NAO, Oshikango, July 31, and August 8, 1929.

52 NAN, NAO 40, O/C NAO to Secretary SWA, Ondangwa, May 25 and 30, 1929; Acting Secretary SWA to O/C NAO, Windhoek, May 18, 1929; O/C NAO to Secretary SWA, Ondangwa, May 25, and June 6, 1929; and V. Alho (Finnish Mission Society) to O/C NAO, Olukonda, April 17, 1929.

53 NAN, NAO 9, O/C NAO, September 6, 1929, to Secretary SWA and NAO 18, *Monthly Report Oukwanyama*, September 1929.

54 NAN, NAO 41, Statement of First Consignment of 948 bags of maize meal received at Ondangua from 28/1/29 to 24/4/29 for Famine Relief Ovamboland. In addition to the food bought from the colonial administration, Ondonga households purchased millet in Oukwanyama and western Ovamboland from 1928 to 1930, NAO 11, Annual Report, Finnish Mission Society, 1930.

55 NAN, NAO 40, Director of Works to O/C NAO, Windhoek, June 17, 1929, and O/C NAO to Secretary SWA, Ondangwa, July 17, and August 10, 1929.

of impregnable subsoil.[56] The name of the 1929–1931 "Famine of the Dams" derived from the fact that most food for work projects consisted of the manual construction of large dams/water reservoirs.

In Namibian Oukwanyama, relief work began only in October, after the Portuguese administration had clamped down on the transborder grain-for-cattle trade and grain shortages increased. The Portuguese authorities halted the grain trade to force Oukwanyama households that had fled to the Namibian side of the border in previous years to return to Angola.[57] Construction of the Etale dam in Oukwanyama, as for that of the two dams in Ondonga, relied on manual labour. Officials supplied picks and shovels and a few spades and wheelbarrows at each dam site.[58] The bulk of the labour was performed by women, girls, boys, and older men, with individuals from each of these categories receiving different amounts of food aid. Children received 0.5 lb; girls 1 lb; women and boys 1.5 lb; and men, "police boys", and headmen 2.25 lb.[59] Recruiting workers for the Etale dam proved to be difficult. The Oshikango-based official, H.L.P. Eedes, despaired that "apparently" people were unwilling to work for food alone.[60] Most adult men had left their villages to trade for food, herd, or seek work in central and southern Namibia.[61]

The food for work programme in Namibian Oukwanyama ended with the onset of the rainy season in December 1929, although food aid in neighbouring Ondonga continued.[62] By January 1930, however, an increasing number of people who had exhausted their patronage and family sources appeared on the doorstep of the local colonial government representative in Namibian Oukwanyama, Eedes. Eedes urgently requested food supplies from his superior, Hahn, but the latter refused to accept his subaltern's assessment of the situation leading to a bitter conflict between the two men. Food for work

---

56 NAN, NAO 40, Director of Works to O/C NAO, Windhoek, June 17, 1929.

57 NAN, NAO 9, O/C NAO to Secretary SWA, September 6, 1929; NAO 18, *Monthly Report Oukwanyama*, September 1929.

58 NAN, NAO 40, O/C Oshikango to O/C NAO, *Oshikango*, July 8, 1929.

59 NAN, NAO 41, Famine Relief Ukuanyama, Works & Issue Schedules, October–December 1929.

60 NAN, NAO 18, *Monthly Reports Oukwanyama*, September and October 1929; and NAO 41, *Famine Relief Ukuanyama, Works & Issue Schedules*, October–December 1929.

61 NAN, NAO 40, O/C NAO to Secretary SWA, Ondangwa, May 25, June 6, July 17, and August 10, 1929; ibid. O/C Oshikango to O/C NAO, Oshikango, May 8 and 15, August 8, 1929 and O/C NAO to O/C Oshikango, Ondangwa, May 12, 1929; NAO 18, *Monthly Report for Ovamboland*, January 1929, and *Monthly Report Oukwanyama*, September 1929; Pauline, interview by author, Onengali, December 15, 1992.

62 On food aid, see NAN, NAO 40, O/C NAO to Secretary SWA, Ondangwa, September 5, 1929 and NAO 41, *Famine Reports Ovamboland*, October–December 1930.

programmes and free food aid were only resumed months later after Eedes had been relieved of his duties and replaced at Hahn's urging.[63]

Eedes' successor was either sufficiently intimidated by the fate of his predecessor or too inexperienced to be able to assess the full extent of the famine. His first report—for July 1930—stated that there were "no real cases of hardship up to the present. There are undoubtably [sic] a number of families who have very little food, for whom some relief must be afforded in the coming months. I am of [the] opinion that the majority are not so bad off as might be expected."[64] A month later, after travelling through Oukwanyama to meet headmen anxious about the food situation and being confronted with mounting evidence that households were fleeing to Angola in increasing numbers, he sounded much less confident. In July, he reported that "numbers of the poorer people are at the end of their resources and will expect food assistance very soon". At the same time, to avoid disagreement with his superior, he added: "[it is] difficult [...] to gauge the food supply, natives naturally make the position much worse than it really is [...] Mr. Hahn informs me that he has the question of corn and mealie [maize] meal well in hand and as any sudden emergency can be met there is no cause for anxiety."[65] Evidence of his growing anxiety, however, is strongly suggested by his obvious relief at the "timely" arrival of emergency food supplies, which in his words "released the tension and re-established confidence in the Administration".[66]

As had been the case during the previous year, most government food aid was distributed through food for work programmes aimed at improving the water infrastructure. Free distribution was kept to a minimum; able-bodied women and children constructed the dams in exchange for food rations. To prevent competition with the migrant labour supply, only adult men who were not targeted for migrant labour recruitment—heads of households, old men, and those put in charge of homesteads and livestock—were employed at the

63 NAN, NAN, NAO 18, *Monthly Reports Oukwanyama*, December 1929 and January 1930; NAO 40, O/C NAO to Secretary SWA, Ondangwa, March 11, 1930; Weekly Drought Reports, weeks ending March 22 and 29, and May 5, 1930. Sold as famine aid for forty shillings per bag, grain prices were subsidised since one bag cost fifty shillings. The Portuguese traders in southern Angola sold grain at forty shillings per bag, see NAO 40, Secretary SWA to O/C NAO, Windhoek September 28, 1929.

64 NAN, NAO 18, *Monthly Reports Oukwanyama*, June and July 1930 and September–November 1930.

65 NAN, NAO 18, *Monthly Reports Oukwanyama*, June, July, and September–November 1930.

66 NAN, NAO 18, *Monthly Report Oukwanyama*, August 1930.

dams.[67] Whereas during the previous year labour recruitment for the single dam project in Oukwanyama had been challenging, during September 1930 many more men were engaged in the projects than the administration deemed advisable. The projects primarily aimed to employ women.[68] Migrant labour opportunities for men, however, declined steeply due to the impact of the global economic depression.[69] Since food was distributed in a limited number of locations, people often walked for long distances on a regular, if not daily, basis. Although her relatives provided her with some food, Pauline worked at Onenghali dam, nine kilometres from her village. Salome Tushimbeni Haihambo, a girl from a single-parent household that had recently migrated from Angola to Omataku, twenty kilometres east of Onenghali, also worked at Onenghali Dam.[70]

Native Commissioner Hahn remained extremely reluctant to acknowledge the extent of the famine in Oukwanyama district, but the threat of large-scale population flight from Ovamboland to the Portuguese colony forced him to introduce famine aid in the district. However, he limited any food aid to a minimum in a district that was largely populated by recent refugees and migrants from southern Angola. Before the famine, Hahn had actively encouraged migration from Angola to Namibian Oukwanyama to enlarge the colony's black labour supply. The Famine of the Dams coalesced local and state patronage networks and blended the spheres of moral economy and political economy. When local patronage networks collapsed, the colonial state stepped in, becoming the supreme patron: the state supplied much of the emergency food during the height of the Famine of the Dams. Government food assistance brought the colonial state into much closer contact with the population of Ovamboland, especially women, suggesting that the reach of the colonial state had expanded.[71] In this sense, food aid serves as an indicator of the strength and penetration of the colonial state. Paradoxically, however, because the colonial administration in Ovamboland did not have the staff to supervise food distribution, emergency supplies were redistributed through local 'big men', which consequently strengthened and widened the range of select local patron–client relationships.

---

67 NAN, NAO 40, Office of the Administrator SWA to O/C NAO, Windhoek [August or September] 1930, and NAO 41, Famine Relief Report Oukwanyama, *Works & Issue Schedules*, September–November 1930.

68 NAN, NAO 41, Report on Famine Relief Works: Oukwanyama, September 1930.

69 NAN, NAO 41, *Famine Relief Report Oukwanyama*, December 1930.

70 Pauline, interview by author, Onengali, December 15, 1992, and Salome Tushimbeni Haihambo, interview by author, Oipya, June 19, 1993.

71 Hayes, "A History of the Ovambo," 330–331.

Distributing food aid through local 'big men' solidified a layer of 'traditional leaders' known as 'principal headmen'. Hahn relied significantly on Ovamboland's kings (renamed chiefs) and headmen for his system of indirect rule. By relying on the headmen for much of the day-to-day administration, South Africa could rule the large population of Ovamboland through a handful of white officials headed by Hahn and, later, by his successor Eedes. The two-layered structure of principal or senior headmen and junior headmen or sub-headmen is comparable to the structure of village headmen and district or provincial headmen in the pre-1915 Oukwanyama kingdom. The difference between the pre-colonial and colonial structures was that, rather than being the king's district headmen, the South African colonial officials in Ovamboland selected their own favourites. 'Indirect rule', which in the case of Ovamboland also functioned as a discourse to camouflage the inherent weakness of the colonial overlords, facilitated the consolidation of these revamped patron–client relationships, allowing them to be institutionalised in wider territorially based units and offices rather than in individuals. However, the senior headmen often also exploited their key positions in the system of indirect rule to expand their personal positions of power vis-à-vis the majority of other headmen.[72]

Ironically, food aid in northern Namibia exported the famine to southern Angola. A large percentage of the food distributed through the government centres was purchased in Angola. In addition, many households located near the Angolan border bought grain directly from producers in Angolan Oukwanyama or from Portuguese traders. Although the Portuguese administration kept the border formally closed for the trade in grain and cattle to all but Portuguese traders throughout most of the famine, by September 1930 a large number of traders from Namibian Oukwanyama, especially from the "wealthier class", were buying food from producers and traders across the border. In fact, supplying the high demand for grain in northern Namibia seriously depleted food stores in southern Angola, and by September 1930 food shortages just across the border from Namibia were as acute as in Ovamboland. Deeper into southern Angola, in Mupa (Evale) and Mulondo "fair" quantities of food were still available. In October, sales by Portuguese traders dropped because the Portuguese administration restricted the purchase of grain from producers in districts where supplies were running low.[73]

The Portuguese administration in southern Angola, however, lacked both the means and the will to enforce its policies. By December 1930, Portuguese

72 Kreike, *Deforestation*.

73 NAN, NAO 41, NCO, *Famine Reports*, September-October and December 1930; UNG, Union Administration 2 folder 1922–1946, UGR Oshikango, *Monthly Report*, December 1929.

traders were again conducting a "fairly big trade", although they only accepted payment in the form of breeding stock. The increase in the price of grain—in September 1929 a heifer bought two bags of millet (40 lb), whereas in December 1930 it bought only 1.25 bags (25 lb)—should therefore be seen as a reflection of both an increased demand in Oukwanyama south of the border and a decreased supply in Oukwanyama north of the border and in southern Angola in general.[74] Taxation and grain sales to buyers from Ovamboland and Portuguese traders depleted reserves in southern Angola, rendering households north of the border less resistant to drought. Already in August 1930, food shortages immediately to the north of the border were "every bit as bad" as they were to the south. Despite famine conditions in Ombadja (west of Angolan Oukwanyama) since June, the Portuguese administration only introduced very limited emergency aid in September. Moreover, the Portuguese authorities never relaxed their forced labour policies. In March 1931, Portuguese officials arrested eight Ombadja principal headmen "for failing to provide free labour for road making. The headmen and their followers were forced to work for two weeks, when they were released."[75] The famine conditions spreading into southern Angola while the Portuguese officials refused to relax their heavy taxation and forced labour recruitment, extinguished South African fears for a mass exodus from their territory into the Portuguese colony.

6 The Return of the Rains

Good rains in December 1930 and January 1931 radically improved food availability throughout Ovamboland, especially of highly nutritional vegetables and tree fruit. In Namibian Oukwanyama, rains allowed millet to be sown "in the sandy areas", the softer soils that required a less abundant soaking before they could be tilled, in as early as October 1930. However, seed shortages prompted the Oshikango-based officer to supply small quantities of seed to the "poorest", but a dry spell from mid-January to mid-February, with only very "patchy and irregular" rainfall, damaged the emerging crops. In Ondonga, food for work programmes continued until early March. The remaining famine supplies were sold or issued to the "very old and indigent natives who will have no means of subsistence until they reap their own crops". By the end of March, abundant rains had not only filled waterholes and pans, but also the newly constructed

74 NAN, NAO 41, NCO, *Famine Reports*, September-October and December 1930; UNG, Union Administration 2 folder 1922–1946, UGR Oshikango, *Monthly Report*, December 1929.

75 NAN, NAO 18–19, *Monthly Reports Oukwanyama*, June–September 1930 and March 1931.

dams in Ondonga and Oukwanyama. In early April, the semi-annual flood or *Efundja*, fed by rains further upstream in Angola, submerged the lower-lying areas in the Ovambo floodplain on either side of the colonial boundary.[76]

Conditions improved more drastically in Namibian Oukwanyama than elsewhere in Ovamboland: by January all fields had been planted. Cabbages, melons, and tree fruit became available in January and were plentiful in February. Abundant rains, however, also increased weed growth. With many people physically weakened because of the famine, weed control was difficult. The colonial official instructed the headmen "to keep a watchful eye on fieldowners who may neglect to keep the fields free from weeds". In February, remaining free food rations were halved.[77] The 1931 harvest in Namibian Oukwanyama was above average; the headmen estimated that the millet supplies would last for two years.[78] For the first time in several years, the cattle herds were brought back from the remote cattle posts to the villages and paraded at the customary cattle fests. In September, conditions had improved to the extent that clan elders and headmen allowed women's initiation (*Efundula*) ceremonies to take place, paving the way for bridewealth exchange and marriage, demonstrating that the crisis was over and normal life could be resumed.[79]

7 Conclusion

The Famine of the Dams was not merely about entitlement, nor was it solely the result of an episode of climatic drought episode. 'Nature's revenge' narratives are equally as one-dimensional as explanations that privilege social factors or emphasise the political ecology. The late 1920s drought in Southern Africa was in many ways caused by a global climatic phenomenon. However, the extent to which drought conditions morphed into a famine was the product of a constellation of environmental, political, and social factors. Climate fluctuations intersected with recent massive population displacements in a politically unstable border region. Refugee households proved highly vulnerable to the vagaries of the climate. In Namibian Oukwanyama, recent migrants

76 NAN, NAO 41, *Famine Relief Reports Oukwanyama*, January-March 1931; and NAO 18–19, *Monthly Reports Oukwanyama*, October and December 1930, and January–February 1931.

77 NAN, NAO 41, *Famine Relief Report Oukwanyama*, January-February 1931; *Famine Relief Reports Ovamboland*, January 1931; and NAO 18–19, *Monthly Reports Oukwanyama*, December 1930, January and February 1931.

78 NAN, NAO 19, *Monthly Reports Oukwanyama*, June/July 1931.

79 NAN, NAO 19, *Monthly Reports Oukwanyama*, March and September 1931.

and refugees from north of the border had not yet managed to construct the fully developed environmental infrastructure of villages, farms, fields, water sources, food stores, and fruit trees that had marked the southern Angolan homes that they had abandoned during the first two decades of the 20th century. Moreover, the drought turned into a famine not only due to environmental factors, but also as a consequence of the colonial Namibian political economy. The timing and protocol of emergency government intervention in Oukwanyama clearly show how political considerations dominated the food aid agenda. The colonial administration was reluctant to declare a famine emergency given the expected cost of food aid in the context of a global economic recession. Only concern that the early 1900s refugee flow from southern Angola to north-central Namibia might be reversed convinced the colonial authorities of the need to introduce food aid, which was purchased from food stores in southern Angolan villages. This food aid, as well as water and grazing in southern Angola, proved crucial in staving off famine in Namibian Oukwanyama. Yet, ultimately this strategy drained food and water supplies in southern Angola, leaving the spectre of famine to loom large across the border until the drought finally broke in 1931.

A new look at African famines may offer avenues for reassessing how climate and society interacted in the past, which is suggestive of how environment and society in Africa may fare under the impact of global climate change in the present and future. The example of the Famine of the Dams suggests that the impact of global climate change cannot be understood in terms of a linear cause-and-effect relationship, with climate (nature) causing direct environmental change that in turn shapes human societies. The current debate about climate change is constrained by operating very much within a nature–culture paradigm that identifies forces of nature and forces of culture (human agency) as discrete analytical and narrative categories. The inhabitants of north-central Namibia in the 1920s and 1930s, however, did not live in and by nature, nor had they harnessed nature and replaced it with culture. Instead, they created environmental infrastructure to interpret, divert, or reshape climatic effects, in the process diluting or exacerbating their impact through religion, migration, and social, economic, cultural, political, and environmental reinvention and reimagining.

CHAPTER 4

Shamba Forestry in Colonial Kenya: Colonial Dominance or African Opportunity?

Ben Fanstone

Abstract

Shamba forestry is an agroforestry system widely used in Africa that incorporates food production within plantation forestry. The highly cost-effective nature of *shamba* meant that it was the only economically viable method of plantation establishment that could be practised in Kenya under British rule. It provided a reliable source of wood to the colony and allowed landless Kikuyu people to farm successfully on fertile forest soils even during times of drought, but it also institutionalised corruption and fostered crime because the system was designed to meet the needs of the government rather than those of the farmers it employed. *Shamba* relied heavily on a patron–client relationship of strict social and economic control of forest plantations. Despite its draconian administration, *shamba* was extremely popular because it allowed for the continuation of pre-colonial patterns of Kikuyu expansion into forest areas, providing a release from the overcrowded native reserves of the colony. *Shamba* continues in Kenya and across Africa today partly because it is a viable adaptation strategy to climate change-induced extreme weather events (drought, floods) and the Kenya Forest Service is attempting to transform it into a more egalitarian plantation system. However, activities classified as 'illegal' have long been used as a survival strategy by *shamba* farmers, highlighting the need to develop alternative revenue sources within the modern system.

1 Introduction

The *shamba* system is a forestry scheme first adopted in colonial Kenya in 1910[1] that allocates farmers small plots of cropland (in Swahili: *shamba*) among the seedlings of a forest plantation. The farmers are responsible for clearing the

1 Kenya Forest Department, "Taungya in Kenya: The 'Shamba System' " (paper presented at the "World Symposium on Man-Made Forests and their Industrial Importance," Canberra: Food and Agriculture Organization of the United Nations, April 14–25, 1967), 2.

 | DOI:10.1163/9789004410848_006

land, planting tree seedlings, and sowing their own crops, and then tending to both for one to five years (until tree cover makes crop cultivation untenable). After this period is over, the farmer will be allocated another plot of land of approximately half a hectare in a new plantation, and the process repeats itself.[2] This system is fundamentally identical to the *taungya* system developed in British-controlled Burma in the 1850s[3] but that most likely evolved from a 16th century plantation system developed in southern China.[4] The name "taungya" is commonly applied to this system where it is used elsewhere in the world today. At its peak in the 1960s, approximately 8,500 families, representing between 42,000 and 50,000 individuals,[5] were engaged in Kenya's *shamba* system. Despite constituting less than one per cent of Kenya's population, these farmers produced between six and ten per cent of the country's maize in the 1960s.[6] For farmers, the system gives access to very productive yet otherwise alienated land. For the forester, the system has been proven to benefit the establishment of commonly used cypress and pine tree plantations by reducing weed growth; moreover, in comparison to other trialled methods of plantation establishment and maintenance, such as ploughing and herbicides, *taungya* is extremely cheap.[7]

The productivity of the *shamba* system hints at its continued relevance today. Across the developing world, this agroforestry method is lauded for its combination of plantation forestry, whether this be for commercial timber

2 Oduol, Peter A., "The Shamba System: An Indigenous System of Food Production from Forest Areas in Kenya," *Agroforestry Systems* 4.4 (1986): 365, accessed October 10, 2014. doi:10.1007/BF00048108; Holman, Dennis, *Inside Safari Hunting, with Eric Rundgren* (London: W.H. Allen, 1969), 61–62; Kenya Forest Department, *Plantation Management Costs & Practical Hints*, Bulletin No. 34 (September 28, 1939), 9. The length of food crop cultivation was also dependent on tree species, with a shorter cultivation period for crops planted among fast-growing trees.

3 Bryant, R.L., "The Rise and Fall of Taungya Forestry: Social Forestry in Defence of the Empire," *The Ecologist* 24.1 (1994): 21–26; Evans, Julian, *Plantation Forestry in the Tropics* (Oxford: Clarendon Press, 1992), 290–292.

4 Menzies, Nicholas, "Three Hundred Years of Taungya: A Sustainable System of Forestry in South China," *Human Ecology* 16.4 (1988): 361–376.

5 Kenya Forest Department, "Annual Report of the Forest Department, 1963" (Nairobi Government Printer, 1963), The National Archives of the UK (TNA), CO 544/106; Oduol, "The Shamba System," 367. Official sources only indicate the number of families employed; therefore, the total population figures come from an estimation of family size at between five and six individuals, which is based on figures from Odoul and estimates presented in the Annual Reports of the Kenya Forest Department (1920–1963).

6 Kenya Forest Department, "Taungya in Kenya," 4.

7 Imo, Moses, "Interactions amongst Trees and Crops in Taungya Systems of Western Kenya," *Agroforestry Systems* 76.2 (2009): 265–273, accessed October 3, 2014. doi:10.1007/s10457-008-9164-z.

production or for local fuelwood needs, and agriculture that can increase food security and raise the families involved above the subsistence level. Already recognised as a form of drought-resistant agriculture, *taungya* and agroforestry in general hold promise as robust adaptation strategies to the climatic changes that are predicted to affect sub-Saharan Africa particularly harshly.[8] The rain-fed nature of the majority of agriculture in Africa marks it as particularly vulnerable to the variations in the amount of rainfall and the changes to rainfall patterns that are projected to occur over the next century. Concurrently, farmers in Kenya have indicated a willingness to engage with agroforestry because of its ability to enhance well-being through increased food security and the creation of additional sources of income.[9] A recent survey conducted in Kenya found that 87 per cent of respondents viewed the *shamba* system as "very good" and 95 per cent supported its reintroduction.[10] Within this context of increased interest in *shamba*, this chapter will explore the development and operation of the *shamba* system in colonial Kenya. I argue that, based on historical analysis, the celebration of *taungya* as a viable climate change-resistant farming and forestry strategy is warranted, but only to a limited extent. Successful forest management requires an adequate understanding of the historical development and management of those forests and their economic, social, and cultural roles in a given society, a point supported by recent research.[11]

In opposition to the current celebration of *taungya* is the assertion by Bryant, based on evidence from colonial Burma, that *taungya* contains a

8 Mbow, Cheikh et al., "Agroforestry Solutions to Address Food Security and Climate Change Challenges in Africa," *Current Opinion in Environmental Sustainability, Sustainability Challenges* 6 (February 2014): 61–67, accessed October 3, 2014. doi:10.1016/j.cosust.2013.10.014; Lasco, Rodel D. et al., "Climate Risk Adaptation by Smallholder Farmers: The Roles of Trees and Agroforestry," *Current Opinion in Environmental Sustainability* 6 (2014): 83–88; Kalame, Fobissie B. et al., "Modified Taungya System in Ghana: A Win–Win Practice for Forestry and Adaptation to Climate Change?" *Environmental Science & Policy* 14.5 (2011): 519–530, accessed October 3, 2014. doi:10.1016/j.envsci.2011.03.011.

9 Thorlakson, Tannis, Henry Neufeldt, and François Collart Dutilleul, "Reducing Subsistence Farmers' Vulnerability to Climate Change: Evaluating the Potential Contributions of Agroforestry in Western Kenya," *Agric Food Security* 1.15 (2012): 6–7.

10 Witcomb, Mark, and Peter Dorward, "An Assessment of the Benefits and Limitations of the Shamba Agroforestry System in Kenya and of Management and Policy Requirements for Its Successful and Sustainable Reintroduction," *Agroforestry Systems* 75.3 (2009): 264, 267, accessed October 3, 2014. doi:10.1007/s10457-008-9200-z.

11 Imo, Moses et al., "Professional and Societal Mismatch in Kenyan Forestry: Is There a Right Way to Manage Our Forests?" in Forest Landscape and Kenya's Vision 2030: Proceedings of the 3rd Annual Forestry Society of Kenya (FSK) Conference and Annual General Meeting Held at the Sunset Hotel, Kisumu. 30th September–3rd October, 2008, ed. D.O. Ogweno, P.S. Opanga, and A.O. Obara (Kisumu: Forestry Society of Kenya, 2009), 74.

fundamental rift between forestry and farmer. While farmers are permitted to plant crops, Bryant argues that they remain disconnected from the benefits of the system that the government enjoys: the output of the trees that they also tend.[12] Essentially, a conflict exists between the needs of the national economy and the local needs of the farmers employed within the system. In relation to this, the chapter will show how *shamba* in Kenya was dependent on its farmers also being able to access income sources that were not sanctioned by government, precisely because of this rift between the objectives of the government and those of the farmers.

Discussion will first focus on the rationale behind the establishment of *shamba* forestry in colonial Kenya and then move on to consider the social and economic effects on the farmers involved. The final section engages with the ongoing debate concerning *taungya*, highlighting the historical evidence both for the system's effectiveness and for its negative consequences.

2 The Colonial Shamba System in Kenya

2.1 *Establishment of the System*

For much of Kenya's colonial period (1895–1963), the Kenya Forest Department's (KFD) primary goal was to supply the timber and fuelwood needs of the colony. The Uganda railway, which ran through Kenya to Lake Victoria, was almost totally dependent on wood for its fuel requirements prior to the Second World War, while the colonial economy hinged on the Uganda railway for transportation.[13] However, Kenya was not and is not a country with extensive forests. At the time of independence, forests only covered approximately 2.7 per cent of Kenya's total land area.[14] This low level of forest cover, together with difficulties in accessing the even smaller percentage of exploitable native trees, meant that the establishment of plantations, typically of exotic eucalypts, became the only way to guarantee the colony's wood supply. Early proposals for a labour force to clear land and to plant and tend trees called for the recruitment of white South African forest labourers.[15] This, however, would have been incompatible with the colony's policy to allow only gentry, officers, and the

12 Bryant, "The Rise and Fall of Taungya Forestry," 25.

13 Logie, J.P. and W.G. Dyson, *Forestry in Kenya: A Historical Account of the Development of Forest Management in the Colony* (Nairobi: Colony and Protectorate of Kenya, 1962), 8.

14 Ibid., 2.

15 East Africa Protectorate, *Report on the Forests of British East Africa by D. E. Hutchins. With a map and 25 photographs*, published 1909, Cd. 4723. LX. 1, 77.

otherwise wealthy to settle in the territory;[16] instead, Africans would have to provide labour for the KFD. Private contractors holding KFD-issued extraction and processing licences employed wage labourers, but the only cost-effective method to establish and maintain plantations would be to use workers who were themselves resident in the forests. This was a point repeatedly made by the KFD:

> [...] this Department is only able to plant the large acreage that it does owing to the availability of Kikuyu squatters who clear, cultivate and maintain the land in which we plant [...]. Without this system our cost would go up by 400% or 500% and acreages planted could decrease by a similar figure.[17]

Despite this colonial necessity, the wider forestry community was sceptical of the system. At the Second Empire Forestry Conference in 1923, heated debate developed over the issue of *taungya* or *shamba* because it attempted to control shifting cultivation, an agricultural practice demonised by foresters because of its apparent forest destruction. It was further feared that it would encourage "nomadic habits" in its practitioners and thus retard their agricultural development, required as they were periodically to move their farming plots. However, the economic arguments of the KFD representatives prevailed, and it was decreed that the *shamba/taungya* system would be used in those colonies where there was no alternative for the economic development of plantations.[18] Such was its acceptance that the Empire Forestry Journal printed a short guide to using the system, written by a forester from Kenya Colony.[19] Following this, the system was also used in Ghana, while in Nigeria trials began in 1927 after a KFD forester was transferred there.[20] The success of *shamba* in Kenya can therefore be seen as directly leading to its uptake in other territories in the empire.

16 Kennedy, Dane K., *Islands of White: Settler Society and Culture in Kenya and Southern Rhodesia, 1890–1939* (Durham: Duke University Press, 1987), 44.

17 Senior Assistant Conservator of Forests, Londiani, to Secretary, Molo Farmers' Association, Molo, March 10, 1930, Kenya National Archives (KNA), FOR/1/210.

18 Rajan, S. Ravi, *Modernizing Nature: Forestry and Imperial Eco-Development 1800–1950* (Oxford: Oxford University Press, 2006), 171–178.

19 Gardner, H.M., "Re-Afforestation in Kenya Colony by Means of Shifting Cultivation," *Empire Forestry Journal* 2.1 (1923): 61–64.

20 Kalame et al., "Modified Taungya System in Ghana," 520; von Hellermann, Pauline, "Things Fall Apart? Management, Environment and Taungya Farming in Edo State, Southern Nigeria," *Africa* 77.03 (2007): 375, accessed October 3, 2014. doi:10.3366/afr.2007.0052.

2.2 *Opposition, Oppression, and Opportunity*

At the heart of the *shamba* system was the contracting of so-called "squatters",[21] officially 'resident labourers', to reside within a forest reserve where a plantation was being established through conversion of native forest or the planting of trees on previously unforested land. In return for cultivation rights within the forest, farmers were expected to provide the KFD with 180 days of paid labour per year, largely used to tend to the plantation and fire prevention measures.[22]

The system faced significant settler opposition throughout the colonial period, largely because the KFD squatters had to reside in the forests. The settlers also employed their own squatters on their farms, although settler attitudes towards them and KFD squatters hardened after the 1920s. By the 1940s, many settlers believed that squatting in general led to "wastage of labour & from the wrong use of land, and should be abolished as soon as possible".[23]

The settlers engaged in programmes to reduce the number of squatters on their land and to reduce or eliminate squatter livestock, which was competing with their own and was seen to carry disease. The settlers believed such policies should also be applied to KFD squatters who resided in the forests. The precarious finances of the settlers and support for their situation from other branches of government increased the anti-livestock pressure on the KFD. As early as 1918, farmers' associations argued in letters to government officials that KFD squatter livestock were "practically under no control, and owing to the many diseases, they are a source of great danger to the cattle industry".[24] In the 1920s such pressure led the KFD to prohibit its farmers, except for those in Nairobi division who did not threaten settler livestock, from keeping cattle.

At the beginning of the 1930s, when their declining economic position was at its lowest point, the settlers sought the elimination of all forest squatter

21 Although used in the pejorative and well-known sense of "illegal occupier of property" in modern Kenya, "squatter" was a neutral term during the colonial period used to describe Africans who legally resided and laboured on settler or crown land. Official documents from the period also use the descriptive term "resident labourer".

22 Kenya Forest Department, "Annual Report of the Forest Department, 1954–1955" (Nairobi Government Printer, 1955), TNA, CO 544/80, 31–32. Forestry in colonial Kenya also depended on squatters employed by private sawmilling companies. While these sawmill squatters also cultivated in the forests, they were not a part of the *shamba* system and were under only very limited control of the KFD. This study is thus concerned only with the squatters engaged in *shamba* itself.

23 Extract from a letter from Mr H.G. Prettijohn, quoted in Conservator of Forests to the Chief Secretary, Nairobi, April 4, 1944, KNA, VF/1/11.

24 Roberts, Arthur, Secretary of the Southern Uasin Gishu Farmers' Association, to Director of Agriculture, July 20, 1918, KNA, QB/1/209.

livestock. The KFD response was initially strong, with the Conservator of Forests declaring to his subordinates that the KFD:

> [...] has always helped the Stock Industry by refusing to allow squatter cattle. No proof or evidence of any sort has been produced to show that the sheep and goats in the Forest Reserves have caused any harm or loss to the farms [...].[25]

Due to the settlers' influence in government, however, the KFD sought a compromise in 1932 by implementing a programme to "get rid of goats", allow only sheep, and "obtain squatters with no stock or with sheep only".[26] Ridding the forests of goats was a goal with which many foresters agreed, arguing that, as browsers, goats had the potential to damage the saplings in plantations.[27] One forester protested that this policy would limit the ability of forest farmers to adapt to changing environmental conditions, such as in years of drought, floods, crop failure, or locust plagues. Sheep were stated to be far more susceptible to environmental changes and more limited in their ecological range as they were dependent on good grazing land.[28] As farmers in Kenya sold livestock during times of food shortage,[29] the restrictions on the number of livestock that could be kept by families within the *shamba* system was likely to have decreased their food security. Moreover, sheep were less able to survive through periods of environmental stress and held a significantly lower monetary value than cattle.[30] Squatters protested, arguing that the removal of goats was causing them great hardship and that "if our goats are removed from us what our [sic] children will be helped with?"[31] Despite these appeals, by 1935

25 Gardner, H.M., Conservator of Forests to Senior Assistant Conservator of Forests, Londiani, April 11, 1930, KNA, FOR/1/210.

26 Gardner, H.M., Conservator of Forests to the District Commissioner, Kiambu, January 5, 1934, KNA, FOR/1/210 (quotation); Cooper, A.M., Forester Elburgon, to the Assistant Conservator of Forests, Londiani, February 9, 1932, KNA, FOR/1/210.

27 Eliot, C.F., Assistant Conservator of Forests, Nyeri, to the Conservator of Forests, March 1, 1932, KNA, FOR/1/210.

28 Graham, R.M., Assistant Conservator of Forests to the Conservator of Forests, January 27, 1932, KNA, FOR/1/210.

29 Thorlakson, Neufeldt, and Dutilleul, "Reducing Subsistence Farmers' Vulnerability to Climate Change," 4. "Food shortage" is defined as difficulty in feeding every member of the family.

30 Kenyatta, Jomo, *Facing Mount Kenya: The Tribal Life of the Gikuyu* (London: Secker & Warburg, 1938), 64–66.

31 Samuel Ngichu, representing the Mbari ya Mwenda Achera clan, to District Commissioner, Kiambu, care of the Kikuyu Land Hoard Association (translated), December 25, 1933, KNA, FOR/1/210.

all the goats had been removed; no goats were legally kept by squatters in the forests thereafter.

By limiting the economic opportunities of the forest squatters, the KFD ensured that the *shamba* system prioritised departmental needs—the production of timber—rather than those of the squatters. Moreover, the inherent racial discrimination contained within the power dynamics of the colonial state meant that the KFD placed the needs of the white settlers ahead of those of its own workforce. The case of livestock regulation reflects more generally that the tension between the needs of the farmers within the *shamba* system and the interests of the department held considerable potential for conflict.

Livestock restrictions within the *shamba* system also had consequences for squatters that went beyond the nutritional into the social. For Kikuyu men—and the Kikuyu ethnic group represented the vast majority of those engaged in the system—possessing livestock was essential for social advancement. Forest squatters outside of Nairobi division were prohibited from owning cattle in the 1920s, a ban that was extended to all forest squatters after 1939. This ban constituted a serious constraint on the accumulation of wealth and prestige. While sheep and goat ownership was common among the Kikuyu, only a minority could afford to keep cattle and this ownership was a key symbol of their high social status.[32]

Wealth in the form of livestock was and is particularly important because of its use in bridewealth. While cash could constitute a part of the transfer to the bride's family from the groom's family, as it certainly does today, livestock, especially cattle, is valued highly as a part of bridewealth. The ability to pay bridewealth was essential to Kikuyu farmers because it allowed for the creation of a large labour pool through having multiple wives and children.[33] Unless forest squatters could successfully keep cattle illegally in the forest, which based on conviction rates for illegal grazing many tried to do, the majority were hindered in terms of marriage and social advancement if they remained within the *shamba* system.

These social and economic consequences also affected the *shamba* system's demographics. For already wealthy households, the precondition of

32 Kenyatta, *Facing Mount Kenya*, 64; Stitcher, Sharon, *Migrant Labour in Kenya: Capitalism and African Response 1895–1975* (Harlow: Longman, 1982), 32–38; Ambler, Charles H., *Kenyan Communities in the Age of Imperialism: The Central Region in the Late Nineteenth Century* (New Haven: Yale University Press, 1988), 25–29.

33 Adams, Bert N., and Edward Mburugu, "Kikuyu Bridewealth and Polygyny Today," *Journal of Comparative Family Studies* 25.2 (1994): 159–166.

abandoning cattle and goats was a deterrent to accepting forest squatter contracts. During the phase of goat eradication in the early 1930s, a forester expected that:

> [...] some will prefer to leave rather than lose their goats, but there should be sufficient goatless Kikuyu about to fill their places. [...] The new squatters taken on recently agree quite readily to the ban on goats. I propose, in future, to discharge any squatter, whose goats are caught in plantations or parts of the forest in which they are prohibited.[34]

The reference to "goatless Kikuyu" indicates a class of farmer who was willing to give up the right to possess the two most-valuable livestock species—cattle and goats—for the chance to cultivate in the forests. It seems likely that concurrent with the livestock eradication programme was a gradual shift in forest squatter demographics and social structure away from those with livestock savings to the growing class of young and landless Kikuyu. Comments made by foresters further support this theory; for example, the squatters inhabiting Ngong forest near Nairobi were described in 1934 as being "nearly all [...] of a very poor class among the wakikuyu".[35] These poorer Kikuyu were most likely *ahoi* (singular form: *muhoi*, in Kikuyu), the term used to describe those who lived as tenants on the *githaka* (clan land) of others.

Prior to the establishment of British control over Kenya, landless *ahoi* were able to strike out into the forest. The creation of native reserves organised on ethnic lines as well as the alienation of land for forest reserves, settler estates, and national parks essentially prohibited the pre-colonial Kikuyu method of relieving land pressure.[36] While wealthy Kikuyu laid claim to *githaka* within the native reserves, *ahoi* were often forced to leave, seeking work on settler farms and in the rapidly expanding capital of Nairobi or taking up squatter contracts offered by the KFD. Moreover, *ahoi* could also accept that trees grown on the land that they farmed were not their property, as this distinction

34 Cooper, A.M., Forester Elburgon, to the Assistant Conservator of Forests, Londiani, February 9, 1932. KNA.

35 Forester, Assistant, Ngong Forest Station, to Assistant Conservator of Forests, Nairobi, July 20, 1934.

36 Middleton, John, *The Central Tribes of the North-Eastern Bantu: The Kikuyu, Including Embu, Meru, Mbere, Chuka, Mwimbi, Tharaka, and the Kamba of Kenya, Ethnographic Survey of Africa—East Central Africa 5* (London: International African Institute, 1953), 52–53; Muriuki, Godfrey, *A History of the Kikuyu, 1500–1900* (London: Oxford University Press, 1974), 74–81.

also existed in relation to trees grown on *githaka* in the native reserves and had occurred before the arrival of colonialism.[37]

The system also opened up areas to Kikuyu cultivation that otherwise would have been used by other ethnic groups. In 1932, for example, squatters were introduced into the moorlands of Mount Elgon; yet, rather than being drawn from the local population of predominantly Ogiek peoples, the squatters were Kikuyu. As Mount Elgon lies far from the Kikuyu-inhabited areas of Kenya, the *shamba* system was their only way to access the area.[38]

The *shamba* system therefore allowed the temporal continuation and spatial extension of the pre-colonial pattern of Kikuyu expansion into forests, while livestock restrictions ensured that forest cultivation was attractive only to the poor. The imposed colonial *shamba* system usurped the pre-colonial power structures of which *ahoi* Kikuyu were part, establishing the KFD as landlord and patron to the squatters it employed. Such squatters had no control over where or for how long they farmed, and they could be summarily evicted for violation of KFD rules.[39] The *shamba* system, then, provided no means for its African farmers to secure ownership of land, instead formalising their Kikuyu status of *ahoi* into that of a tenant under the colonial regime. While the presence of secure land tenure is a key reason why similar agroforestry regimes persisted for so long in China,[40] its absence within Kenyan *shamba* is identified as a key structural weakness that would take effect in the 1970s (see Section 3). However, while creating a structural weakness, this usurpation also represented an opportunity for *ahoi* to farm in fertile forests, providing a means for them to escape land shortages in the native reserves.

2.3 *Mau Mau and Villagisation*

The Mau Mau period in the 1950s and its aftermath marked the completion of the colonial development of *shamba*, exposing the latent oppression within the system and bringing in changes that would continue to characterise it until its structural overhaul in 1976.

---

37 Dewees, Peter A., "Trees and Farm Boundaries: Farm Forestry, Land Tenure and Reform in Kenya," *Africa: Journal of the International African Institute* 65.2 (1995): 19, accessed October 3, 2014. doi:10.2307/1161191; Leakey, L.S.B., *Mau Mau and the Kikuyu* (London: Methuen & Co. Ltd., 1952), 12.

38 Mason, District Commissioner, Kitale, to PC Rift Valley Province, December 10, 1952, KNA, NK/2/17/18; Kenya Forest Department, "Annual Report of the Forest Department, 1932" (Nairobi Government Printer, 1932), 20, TNA, CO 544/38.

39 Kenya Forest Department, *Squatter and Shamba Management*, Bulletin No. 33, published May 20, 1932.

40 Menzies, "Three Hundred Years of Taungya," 371–374.

When the predominantly Kikuyu-led Mau Mau rebellion against British rule broke out in 1952, Mau Mau fighters sought refuge in the forests, turning them into a warzone. The government closed access to the forests and viewed all Kikuyu, including the majority of the squatters and their families employed in the *shamba* system, as potential collaborators. By December 1952, the KFD had begun forcibly removing Kikuyu from its forests and transporting them to the already overcrowded native reserves. Initially, squatters were removed "following conviction for being actively connected with any subversive activities".[41] This pretence was soon dropped, at least in private, and it was decided that:

> The numbers to be discharged would be a percentage of the total squatter force [...] The percentage is 20% to 25% and the individuals to go would be the less satisfactory workers and those upon whom any suspicion has fallen.[42]

Rather than simply a measure to get rid of potential Mau Mau fighters and supporters, many squatters were removed for failing to meet the KFD's expectations of labour productivity. Powers of martial law in force during the Mau Mau emergency were used to circumvent the legal rights of forest squatters. The livestock of these squatters, which amounted to their life savings and most-prized possessions, were forfeited and sold to the Kenya Meat Commission.[43] In pursuing this policy of evictions, the KFD was undermining a central pillar of the *shamba* system: the security of having a written contract that provided signatories with the right to farm forest plots for a defined period of time and that protected them from the intense competition for land within the native reserves from which they came.[44] This would be a key issue in the eventual collapse of the *shamba* system after the end of colonialism, which is explored in more detail below.

41 Honoré, E.J., Conservator of Forests, West of Rift, to Provincial Commissioner, Rift Valley Province, December 16, 1952, KNA, NK/2/17/18.

42 Johnston, C.M., Provincial Commissioner, Rift Valley Province, to Provincial Commissioner, Central Province, December 23, 1952, KNA, NK/2/17/18.

43 C.M. Johnston, Provincial Commissioner Rift Valley Province to Chief Native Commissioner, December 23, 1952, KNA, NK/2/17/18.

44 In the predominantly Kikuyu district of Kiambu, for example, there was a 313 per cent increase in population density per square mile between 1921 and 1951 (from 52 to 215 inhabitants per square mile); Mosley, Paul, *The Settler Economies: Studies in the Economic History of Kenya and Southern Rhodesia, 1900–1963* (Cambridge: Cambridge University Press, 1983), 78.

Mau Mau laid bare the priority of the *shamba* system: the maintenance of the forest always trumped the needs of its workers. With all Kikuyu deemed suspect, the KFD quickly began to search for potential squatters among "Kipsigis, Jaluo and South Nyanza tribes, Wanderobo, Tugen, Elgeyo, and North Nyanza tribes".[45] Most of these other ethnic groups, however, did not possess a ready-made landless class eager to take up forest squatting. In a bid to make the system more attractive, the KFD backed down on its cattle restrictions for non-Kikuyu squatters, allowing approximately five cattle per squatting family. The results, however, were reported to "have been disappointing". Apart from two groups of Elgeyo and Tiriki "of appreciable size resident labourers of non-Kikuyu tribes still number a few handfuls only".[46]

By 1957, many Kikuyu squatters who had been 'displaced' by the events of Mau Mau were being accepted back onto the plantations, albeit with the cattle restrictions back in place, signalling the failure of the KFD to attract non-Kikuyu into the system. However, these returnees now found themselves within a system that had been altered through the process of villagisation.

Prior to Mau Mau, forest squatter villages were dispersed, with the location of huts chosen according to their proximity to the primary cultivation area. The farmers could keep close watch over their crops, livestock, and allocated tree saplings, curbing wildlife and fire damage, while journey times for farming or firewood collection were short. This arrangement was, however, difficult to administer and police. The role of the forests in the Mau Mau struggle meant that dispersed villages would not be tolerated; in as early as December 1952 the KFD had plans in place to begin a process of villagisation, whereby the squatters would be concentrated into dense villages that would greatly ease their supervision.[47]

Prior to Mau Mau, the KFD regarded its squatters as a source of pride, yet the rapid drafting and enacting of villagisation plans exposed the latent mistrust that the KFD had for Africans as well as its underlying conflict of interest. For the forest squatters, villagisation meant longer journeys to cropland and trees alike; thus, while supervision of the squatters increased, supervision of the plantations decreased, thereby reducing the efficiency and cost effectiveness

45 Honoré, E.J., Conservator of Forest, West of Rift, to Provincial Commissioner, Rift Valley Province, December 15, 1952, KNA, NK/2/17/18.

46 Kenya Forest Department, "Annual Report of the Forest Department, 1951–1953" (Nairobi Government Printer, 1953), TNA, CO 544/78.

47 Johnston, C.M., Provincial Commissioner Rift Valley Province to District Commissioners Nakuru, Eldoret, Kitale, & Thompson's Falls, District Officer Naivasha, December 5, 1952, KNA, NK/2/17/18.

of the *shamba* system. The emphasis on security meant that "little thought was given to such essential factors as water supplies, health, sanitation, range from fire, and so on".[48] The consequence was a greater input, and associated expense, for the KFD into matters of people's welfare.

2.4 *Popularity and Success*

Research on the development of *taungya* forestry in Burma, the progenitor of the *shamba* system, mirrors the analysis hitherto presented: this was a system that usurped the needs of the local population. Protest in Burma occurred on a large scale: forest fires were deliberately set, and seedlings were often destroyed.[49] By contrast, in Kenya the KFD maintained throughout the whole colonial period that "[t]his form of employment is very popular with the labourers",[50] a statement that is indeed supported by evidence.

While for the majority of squatters the restrictions imposed by the *shamba* system curtailed social advancement, for a minority the system also represented an avenue into government service. Squatter workgroups were led by African headmen drawn from their ranks, and headmen who showed exceptional ability could be promoted into the KFD itself to become forest guards, rangers, seedsmen, or assistant foresters, of which there were a combined total of about 130 until the late 1940s when recruitment doubled. Those who sought work in the KFD but were not part of the *shamba* system were advised to first become squatters, where their practical abilities could be evaluated. Some African employees had spent their childhoods within the *shamba* system, increasing their affinity to it and having the monetary and prestige benefits of working in the KFD impressed upon them. Until the 1950s this was the only way for an African to join the KFD, which was dominated by European personnel in its higher ranks and Indian staff in lower positions.[51]

For those Africans who did attain KFD employment, the *shamba* system represented a transformative process. Lacking the means to establish their own wealth and prestige in native reserves already dominated by powerful African landowners, poorer farmers profited from the opportunity not only to

48 Kenya Forest Department, "Annual Report of the Forest Department, 1954–1955."

49 Bryant, "The Rise and Fall of Taungya Forestry," 21–23.

50 Kenya Forest Department, "KFD Annual Report," published 1955.

51 Rammell, J.C., Acting Conservator of Forests, to the Director of Education, Nairobi, December 17, 1931, KNA, FOR/1/166; Senior Assistant Conservator of Forests to the Forester, Kerita, December 15, 1938, KNA, FOR/1/166; Kicheru, Willis K. to Forest Department, July 26, 1938, KNA, FOR/1/166; Graham, R.M., for Acting Conservator of Forests, to Paul K. Ngugi, February 11, 1928, KNA, FOR/1/166.

cultivate land but also to obtain reliable employment. Furthermore, this work would have compensated for their lack of livestock by providing a cash income to supplement farming in times of environmental stress.

By the 1950s, the KFD was also implementing social policies for its squatters and those living near forests that increased the attractiveness of the system and heightened its support among existing squatters. Forest officers were specifically assigned to the task of overseeing African welfare, and the department began building and operating facilities as diverse as community halls (10 in operation), hide-drying sheds, dispensaries (12 in operation), schools (32 in operation), and water systems by 1953. Squatter children were not required to work within the *shamba* system and could attend schools built by the KFD, which made the system more attractive than squatting on settler farms, where access to education was more limited.[52] While the *shamba* system restricted the ability of its squatters to accrue wealth in terms of livestock accumulation, it did offer new opportunities in accordance with the interests of the colonial regime.

The attractiveness of these opportunities is reflected in the number of squatter families engaged in the *shamba* system, a figure that showed an upward trend across the colonial period. The earliest records, from 1910, indicate that the system immediately attracted 1,000 families. Twenty-two years later, the squatter population reached a pre-Second World War high of 2,466 families. That number continued to rise during the war, and by the time of Kenya's independence the *shamba* system was engaging 8,474 families.[53] However, this rise was not steady, with dips in the squatter population in the mid-1930s due to the KFD's financial constraints caused by the Great Depression as well as during the Mau Mau rebellion.[54] The expulsion of squatters during Mau Mau was rapidly reversed as soon as the threat of rebellion in the forests was considered to be over, and by 1958 the number of squatters exceeded the pre-Mau Mau figure by almost one thousand. These statistics show that as long as the KFD was open to it, more squatters would be forthcoming to join the system.

Records also indicate that drought, famine, and locust plagues were key drivers of *shamba* system uptake in Kenya. Assessments of the motivations

---

52 Kenya Forest Department, "KFD Annual Report," published 1953, 40–43; Kanogo, Tabitha, *Squatters and the Roots of Mau Mau, 1905–63* (London: James Currey, 1987), 83–89.

53 Kenya Forest Department, "Taungya in Kenya"; Kenya Forest Department, "Annual Reports of the Forest Department, 1920–1963" (Nairobi Government Printer, 1920–1963), TNA, CO 544/12–CO 544/106.

54 Kenya Forest Department, "Annual Report of the Forest Department, 1933" (Nairobi Government Printer, 1933), 26, TNA, CO 544/40.

behind squatter migration to settler farms have found that ninety per cent cited landlessness within their respective native reserves.[55] This corresponds with the increase in Africans who engaged in waged employment from the late 1920s onwards as a response to declining productivity in the native reserves due to land shortages and climatic conditions.[56] Landlessness was exacerbated by the drought-induced famines that affected many areas of highland Kenya from the mid-1920s until the mid-1930s and then again at the beginning of the 1940s, precisely the time periods that saw large rises in the number of *shamba* farmers.[57] Those farming within the *shamba* system fared better than those on settler farms or in native reserves, as the KFD argued:

> Normally, on the fresh forest soil they obtain better crops than they would elsewhere and though during the year locust damage was serious in many districts, the squatters were certainly as well or better off than they would have been anywhere else.[58]

Farms in the forests appear to have been at least as productive as those outside the forests and, crucially, were able to maintain production levels through periods of drought when crops in native reserves were failing. Alternatively, crop yields remained high enough to produce a surplus that allowed farmers to subsist on stored grain or sell this for food. This lends credence to the argument that *shamba* and *taungya*-like systems may have utility in a future Africa beset by the increased occurrence of drought.[59] Caution should be taken with this claim, however, as *shamba* can be overwhelmed. The drought of 1934 was so severe that many crops failed in the forest plantations, and with the KFD's inability to supply more work squatter numbers remained lower than they had been in 1932. Yet, the KFD still reported that: "The squatters were contented and gave no trouble, although their crops were in many districts sadly

55 Furedi, "Kikuyu Squatters," 181.

56 Kitching, Gavin, Class and Economic Change in Kenya: The Making of an African Petite Bourgeoisie 1905–1970 (London: Yale University Press, 1980), 34, 59; Stitcher, Sharon, Migrant Labour in Kenya, 75–79.

57 Anderson, David, *Eroding the Commons: The Politics of Ecology in Baringo, Kenya, 1890s–1963* (London: James Currey, 2002); Anderson, David, and David Throup, "Africans and Agricultural Production in Colonial Kenya: The Myth of the War as a Watershed," *The Journal of African History* 26.4 (1985): 327–345.

58 Kenya Forest Department, "Annual Report of the Forest Department, 1932" (Nairobi Government Printer, 1932), TNA, CO 544/38, 20.

59 Kalame, Fobissie B., "Modified Taungya System in Ghana."

damaged by drought."[60] An explanation for this apparent contentment can perhaps be found in the crimes reported to have occurred in the forests during that time. The limits placed on sheep numbers and the total ban on goats and cattle combined with the drought and economic slump of the 1930s to severely limit the forest squatters' food security. In response, as KFD statistics suggest, squatters employed natural resource utilisation strategies that were prohibited under colonial law to secure extra income or food.

Between 1929 and 1939, the squatter population increased by 18 per cent, while "illegal grazing" in the forests increased by 185 per cent, and "theft of forest produce" (mainly fuelwood) rose by 119 per cent. These huge surges in the number of acts deemed "criminal" by the KFD can be partly attributed to increased policing efforts by forest guards and partly to the activities of those residing outside of the forests. In particular, the Kikuyu Forest Division bordered several native reserves and in 1938, theft of fuelwood from those forests constituted half of the total recorded crimes.[61] Yet a significant proportion of the remaining crimes must have been committed by those who resided permanently in the forests: the squatters. The taking of wood, which could be sold, and the grazing of animals within forest reserves represented vital survival strategies at a time when the colonial economy could not supply sufficient work and crops were failing. This informal African economy existing outside of the colonial economic infrastructure was fully established by and utilised during the Second World War to offset crop failures.[62] Frequent cases of squatters accommodating and supporting Africans with no legal right to be in the forests highlight the role of the former in this economy.

Forest guards, who were direct KFD employees, were also implicated in such activities, with several being convicted for bribery, corruption, and aiding criminals during the 1940s.[63] During the 1950s, European KFD officers were routinely and frequently transferred between forest stations out of fear that remaining at the same station for too long would see them foster "corrupt" relationships with local Africans.[64] Such a tactic suggests that the KFD had experienced this

---

60 Kenya Forest Department, "Annual Report of the Forest Department, 1934" (Nairobi Government Printer, 1934), TNA, CO 544/44, 26.

61 Kenya Forest Department, "Annual Report of the Forest Department, 1938" (Nairobi Government Printer, 1938), TNA, CO 544/55, 13.

62 Berman, Bruce, Control and Crisis in Colonial Kenya: The Dialectic of Domination (London: James Currey, 1990), 267.

63 Assistant Conservator of Forests, Nyeri, to Conservator of Forests, January 2, 1941, KNA, FOR/1/244; R.M. Graham, Assistant Conservator of Forests, to Conservator of Forests, November 25, 1940, KNA, FOR/1/244.

64 Anonymous, family member of a KFD officer, email message to author, July 4, 2014.

challenge before, although official records are silent on the topic. One of the main attractions of the *shamba* system beyond its usual provision of fertile land may have been the opportunities it presented to successfully engage in economic activities that were not sanctioned by the government.

The forest crime statistics also reveal another significant aspect of the *shamba* system: the absence of large-scale resistance to it. In 19th-century Burma, fire was used as a form of protest against the imposed *taungya* system,[65] but colonial records in Kenya do not reveal any such use until reports of arson in 1960 and 1961. No official explanation is given for these cases, but the falling conviction rates for illicit grazing and theft of forest produce in those years indicate a more effective crime prevention strategy was being employed by the KFD. These were years of severe drought when squatters may well have also been prevented from turning to their usual fall-back strategies.

The KFD laid blame for the arson attacks on the "[Kenya] Land Freedom Army" (KLFA),[66] a militant group that continued the Mau Mau struggle for Kikuyu land until 1965. While the KLFA did operate in the forests, there is no evidence besides the KFD accusation to suggest that it was responsible for arson.[67] This accusation may also have been an attempt to disguise discontent at KFD policy and exploit the continued fear of Mau Mau. In saying that "without the co-operation of the local population no fire protection programme, no matter how costly, is likely to be effective",[68] the KFD was suggesting it had lost the cooperation of some of its squatters. This comment is in marked contrast to the KFD's lauding of the squatters' involvement in firefighting activities across the rest of the colonial period. Indeed, apart from these instances and excluding arson used as a tactic during the Mau Mau conflict, there are no obvious connections between forest fires and protest in colonial Kenya.

3 The Future of Shamba in Kenya

The *shamba* agroforestry system is also in use in Kenya today. The Forest Act of 2005 rebranded *shamba* as the Plantation Establishment and Livelihood

65 Bryant, "The Rise and Fall of Taungya Forestry," 23.

66 Kenya Forest Department, "Annual Report of the Forest Department, 1960" (Nairobi Government Printer, 1960), TNA, CO 544/97, 6; Kenya Forest Department, "Annual Report of the Forest Department, 1961" (Nairobi Government Printer, 1961), TNA, CO 544/100, 6; Kenya Forest Department, "KFD Annual Report," published 1963, 6.

67 Kanogo, Squatters and the Roots of Mau Mau, 162–175.

68 Kenya Forest Department, "Annual Report of the Forest Department, 1960," 6.

Improvement Scheme (PELIS) and made it a fundamental part of the policy of the Kenya Forest Service, the successor to the KFD, in response to declining tree cover and in an effort to mitigate rural poverty. This commitment represents an about-turn from the government position of the 1980s and 1990s.

Following Kenya's independence from British rule, the *shamba* system was maintained in much the same way it had been in the 1950s until 1976, when the resident labourers engaged within the system were made full-time employees of the forest department, which effectively turned them into civil servants, and cultivation rights in the forest were offered to any willing to pay the rent. This separated forest workers and cultivators, meaning that there was no longer any compulsion for workers to farm the forest. The cost of plantation establishment and maintenance increased dramatically as cultivators often ignored less fertile or remote areas while illicitly obtaining rights to cultivate over protected areas, and many plots were subleased. This disconnection of the roles of forest worker and forest farmer caused the virtual collapse of the *shamba* system and the plantation system it supported.[69] Initially designed to engage *shamba* farmers in forestry, this scheme of separating them from the land further highlighted the competing objectives of the KFD and its farmers.

The current director of the Kenya Forest Service, David Mbugua, has argued that what he identifies as corruption within the old Forest Department and among the *shamba* squatters developed after 1962 and led to the collapse of the system.[70] However, as seen above, actions in opposition to government regulations were very much a part of the *shamba* system during the colonial era. As such, post-1962 'corruption' was not new, but a continuation of practices begun in response to a colonial framework that denied the *shamba* farmer any say in the system or the ability to subsist during times of acute environmental stress (as in the mid-1930s).

In the mid-1980s, the *shamba* system—by then completely ineffective—officially came to an end. Following this, cases of land grabbing by elites, apparently with the complicity of government employees, increased tremendously and forced *shamba* farmers out of the forests.[71] When *shamba* was partially

---

69 Kiriinya, C.K., "The Rise and Fall of Taungya: Lessons from Kenya," *Agroforestry Today (ICRAF)* 6.3 (1994): 3–4; Oduol, "The Shamba System," 365.

70 Mbugua, D.K., "Deepening Forestry Sector Reforms in Kenya," in Forest Landscape and Kenya's Vision 2030: Proceedings of the 3rd Annual Forestry Society of Kenya (FSK) Conference and Annual General Meeting Held at the Sunset Hotel, Kisumu. 30th September–3rd October, 2008, ed. D.O. Ogweno, P.S. Opanga, and A.O. Obara (Kisumu: Forestry Society of Kenya, 2009), 18–22.

71 Geteria, Wamugunda, "Management of the Mau Ecosystem," in *Forest Landscape and Kenya's Vision 2030: Proceedings of the 3rd Annual Forestry Society of Kenya (FSK)*

reintroduced in 1994, the system was still structurally weak and marked by the claiming of forested areas for cultivation by elites without tree establishment. Environmental groups became concerned over the level of forest exploitation occurring at the time, particularly in the form of charcoal production, without any attempt to re-establish tree plantations, and brought pressure to bear on the government, which resulted in *shamba* again being banned in 2002.[72] In 2005, the successor programme to *shamba*, PELIS, was officially launched. It began operating in 2008 with the aim of re-establishing government control of *shamba* forests and reintroducing forest cultivation. By 2012, PELIS had made 2,049.6 hectares of land available to previously landless people across five counties, producing approximately three million bags of maize. This is heralded as a great success by the Kenya Forest Service, which now hopes that PELIS will continue to develop into an integral part of its forestry strategy while remaining free of the problems that plagued *shamba* after the 1970s.[73]

Understanding the failure of the *shamba* system is vital to ensuring the continuation of PELIS in the future. The breakdown of *shamba* represented a usurpation of a system that had favoured the poor throughout the colonial period. It also suggests that the livestock restrictions in place from the late 1920s onwards were a key feature in sustaining the success of the system, since they made forest cultivation less attractive to wealthier farmers with large herds. At the same time, *shamba* became a target for acquisition by elites as it developed to a state where it was extremely successful, as the high maize production levels recorded in the 1960s indicate it was. From this perspective, the future success of PELIS largely falls on the ability of the Kenya Forest Service to protect it from these dynamics.

In providing such protection during the colonial period, the KFD was ensuring that the right of its squatters to farm assigned plots was guaranteed by law. Even though this security was undermined during the KFD's purging of 'undesirables' in the 1950s, the importance of protection against land grabbing by African or settler elites is highlighted here. Similarly, Mbow et al. argue that farmers will only fully commit to agroforestry systems when they themselves,

Conference and Annual General Meeting Held at the Sunset Hotel, Kisumu. 30th September–3rd October, 2008, ed. D.O. Ogweno, P.S. Opanga, and A.O. Obara (Kisumu: Forestry Society of Kenya, 2009), 57.

72 Witcomb, and Dorward, "An Assessment of the Benefits and Limitations of Shamba," 262.

73 Odwori, Paul Okelo, Philip M. Nyangweso, and Mark O. Odhiambo, "Alleviating Food Insecurity and Landlessness Through Pelis in Kenya" (paper presented at the "4th International Conference of the African Association of Agricultural Economists," Hammamet, Tunisia, September 22–25, 2013), 1–3.

and not their successors, will reap the rewards.[74] Menzies' analysis of three hundred years of *taungya* in China also emphasises maintaining land security for the farmer as the key factor in the longevity of the system there.[75] Evictions that sow mistrust between farmers and the governing authority, such as those during the 1950s, must therefore be avoided if PELIS is to be a success.

Furthermore, Witcomb and Dorward present "two main factors" accounting for the failure of *shamba*: "poor and often corrupt management of the system and high levels of poverty in forest adjacent communities which compelled locals to exploit the system".[76]

As the historical analysis has shown, the deterioration in the management of *shamba* began during the colonial period. This 'corruption' stemmed from the wider trend, clearly evident by the 1950s, of a growing wealth gap between rich and poor Africans. This situation was fostered by a colonial regime that restricted African enterprise and economic development to those locked into a patronage relationship with the government.[77] While the effects of the corruption of the *shamba* system only became evident during the land grabs of the 1970s and beyond, they were born in the colonial era. Avoiding a similar breakdown of the system in the future is, again, dependent on ensuring that it is adequately insulated from forces more powerful than those that can be mustered by the *shamba* farmers.

Witcomb and Dorward's second factor concerning the high levels of poverty in forest communities adjacent to the *shamba* system can also be related back to the system's colonial foundations. While *shamba* proved effective in maintaining harvest yields in times of mild to moderate drought, forest-cultivated crops did fail in the mid-1930s, and the resistance of the KFD to livestock in its forests meant that it is extremely likely that *shamba* farmers turned to survival strategies that included generating income by means not sanctioned by the government. Colonial regulations meant that such survival strategies became categorised as 'illegal'. Grazing livestock in areas prohibited by the government and the taking of forest produce to sell via the informal market represented effective ways to generate cash and purchase food from areas less affected by famine, thereby compensating for failing crops. If *shamba* farmers had maintained a tree crop of value to them, for example fruit trees, and if they had the necessary utilisation rights to support themselves through periods of drought and engage in trade with neighbouring communities, this would have

74 Mbow et al., "Agroforestry Solutions," 63.
75 Menzies, "Three Hundred Years of Taungya."
76 Ibid., 263.
77 Kitching, Gavin, *Class and Economic Change in Kenya*, 315–316, 372.

alleviated the pressures pushing these adjacent communities to 'exploit the system'.

This is a point that is particularly pertinent to the future of *shamba* within Kenya in the context of increasing incidences of extreme weather events. Some projections indicate that sub-Sahelian Africa may experience a twenty per cent reduction in the production of staple crops by 2050 as a result of climate change and associated increases in the frequency of drought and flooding, highlighting the urgent need to develop responses that can safeguard agricultural production. Agroforestry has been championed in this regard because of the multiple benefits that trees and forests can bring to farmers, including protection of water catchments, shade and shelter, diversification of income and crops where fuel or fruit trees are used, protection of soil, and the enhancement of soil fertility and crop yields by the planting of fertiliser tree species.[78]

If current and future *shamba* systems are to be a success under conditions of climate change, there must be a realisation that farmers need additional revenue sources, which have often been deemed illegal in the past, especially in times of economic stress. Trees-on-farms-based agroforestry embraces this approach to multiple revenue sources, for example by growing trees for timber and fruit. Agroforestry involving fruit trees in western Kenya has shown that farmers were able to subsist on fruit from those trees when floods had wiped out all other crops.[79] However, the primary purpose of the *shamba* system, which is a farms-in-forest approach to agroforestry, throughout the colonial period and beyond was the production of timber and fuelwood through fast-growing eucalypts and pines. Integrating tree species that are of more use to the farmers and giving farmers rights over these trees is thus a radical alteration to the system. Considering the fact that over 75 per cent of the energy utilised in Kenya is derived from wood,[80] on which the country seems dependent for the foreseeable future, the central question remains whom the *shamba*

78 Lasco et al., "Climate Risk Adaptation by Smallholder Farmers," 83–86.

79 Thorlakson, Neufeldt, and Dutilleul, "Reducing Subsistence Farmers' Vulnerability to Climate Change," 7.

80 Malo, Meshack et al., "Visioning the Energy Dimension of Kenya: Changing Forest Landscape under Vision 2030," in *Forest Landscape and Kenya's Vision 2030: Proceedings of the 3rd Annual Forestry Society of Kenya (FSK) Conference and Annual General Meeting Held at the Sunset Hotel, Kisumu. 30th September–3rd October, 2008*, ed. D.O. Ogweno, P.S. Opanga, and A.O. Obara (Kisumu: Forestry Society of Kenya, 2009), 41; Koech, Eric, Eliud Kireger, and Joel Laigong, "Forest Resources as a Source of Energy and Food Security: Two Contradictory Issues or Two Sides of the Same Challenge?" in *Forest Landscape and Kenya's Vision 2030*, 77–80.

system is really designed to benefit. Whether the interests of the farmers are subordinate to those of the state, as they were during the colonial period, remains a key issue to be resolved if *shamba* is to be successful as an adaptation strategy to climate change.

The response within the international forestry community to this question has been the steady development of community-based forest schemes since the 1980s. In Kenya, these take the form of Community Forest Associations (CFAs), an initiative that has been described by the UN Development Programme as "the only true hope for transforming the management of forests and embrac[ing] democratisation of forest management".[81] Within PELIS, the role of CFAs is to allocate half-acre plots within plantation developments, with local people, specifically elders, playing a very active role in this process.[82] PELIS has so far proved successful,[83] but it remains to be seen whether CFAs are sufficient to bridge the gap between the traditional institutions of forestry and the people engaged in the *shamba* system, and thus avoid the conflict between these actors that was inherent in the historical precedent.

4 Conclusion

The title of this chapter asked whether the *shamba* system as practised in colonial Kenya presented the colony's African population with opportunities or whether it was an example of colonial domination. The answer, of course, is both. The conflict in aims between the forest authority and the farmers employed in *shamba* is evident from the system's inception to its collapse. *Shamba* was established not to provide land and employment to African farmers but to meet the timber and fuel needs of the colony. At each stage, this overriding of local interests was evident. At the expense of their adaptability to extreme

81 Nyandiga, Charles O., "Strengthening Forest Governance: How Far to Go in Decentralization and Participation?" in *Forest Landscape and Kenya's Vision 2030: Proceedings of the 3rd Annual Forestry Society of Kenya (FSK) Conference and Annual General Meeting Held at the Sunset Hotel, Kisumu. 30th September–3rd October, 2008,* ed. D.O. Ogweno, P.S. Opanga, and A.O. Obara (Kisumu: Forestry Society of Kenya, 2009), 112.

82 Kenya Forest Service, *PELIS Now Halted in Kimondi 2c Forest*, August 2011, accessed October 3, 2014. http://www.kenyaforestservice.org/index.php?option=com_content&view=category&layout=blog&id=228&Itemid=168&limitstart=5; Nyandiga, "Strengthening Forest Governance," 105–106.

83 Mathu, Winston, "Forest Plantations and Woodlots in Kenya," *African Forest Forum Working Paper Series* 1.13 (2011): 17.

weather events, the KFD banned its farmers from keeping livestock. When security needs required it, contracts with Kikuyu squatters were torn up and replacement farmers from other ethnic groups were sought, although never found. Yet, the Kikuyu readily took up contracts to be *shamba* farmers throughout the period. The system allowed poorer *ahoi* Kikuyu the chance to farm and gain access to colonial education and employment that they otherwise might not have had, while also legally protecting their right to farm from the land competition that existed in the native reserves.

The dichotomy within the *shamba* system is particularly exposed when it is examined as a response to climate change. Various agroforestry approaches have been heralded as viable responses to mitigate reductions in agricultural production induced by climate change.[84] In the colonial period, however, the capacity of *shamba* to increase food security was overwhelmed, providing only unsanctioned means to secure additional income. The central characteristic of the historical system—the provision of tree products to the benefit of the state economy and not to that of the farmers who tended the trees—continues to undermine the modern successor to that system, but does not completely negate its usefulness as a response to climate change. Unless *shamba*, now PELIS, is changed to provide tree products that can be exploited by the system's farmers, it too faces the possibility of being weakened by corruption, the unsanctioned taking of forest products, and pressure from wealthier farmers outside the system seeking to acquire forest land, all threats that are exacerbated during years of drought, floods, and locust plagues. Government control of the forests also represents the withholding of control from direct forest users, especially *shamba* farmers, and risks a replication of the colonial patronage system. The Mau Mau period is an extreme example of how this form of environmental management translated into an extension of state power and control. In view of the slide into corruption that has marred the system almost since its inception, the success of *shamba* as a way to mitigate climate change-induced drought and flooding will depend on its ability to engage with its users on an equal footing.

84 Mbow et al., "Agroforestry Solutions to Address Food Security and Climate Change Challenges in Africa"; Kalame et al., "Modified Taungya System in Ghana."

PART 3

Present

CHAPTER 5

I'm Staying! Climate Variability and Circular Migration in Burkina Faso

Jonas Østergaard Nielsen

Abstract

In the Sudano-Sahelian region of West Africa, the need for human adaptation to climate change and variability is not a new phenomenon. Since the 1970s, the Sudano-Sahelian region has been through a sustained climate crisis, with prolonged drought periods replaced in the early 1990s by significant annual precipitation variability. In response to these climatic trends, rural populations have diversified their livelihoods. One of the most used and described strategies is circular labour migration. This implies that rural dwellers migrate to areas with more widespread employment options such as major cities in search of wage labour, but return after some months or years to their villages of origin. In recent literature, the continued predominance of this migration pattern has been questioned. A major reason for this is that climate change might be encouraging more permanent forms of migration due to increasingly problematic environmental conditions in rural areas. This chapter explores contemporary migration patterns among residents of Ouagadougou, Burkina Faso by way of ethnographic research. The chapter shows that since the early 2000s rural in-migrants have been increasingly choosing (or planning) to stay permanently in Ouagadougou. Work opportunities, population growth, lack of land, and the Ivorian crisis are given as reasons for leaving villages as well as for staying in Ouagadougou, but the findings highlight that continued climate variability plays a major role when decisions regarding migration are made.

1 Introduction

Climate change and variability is a defining feature of the Sudano-Sahelian region of West Africa. Since the 1950s, this region has been through "the most dramatic example worldwide of climate variability that has been directly and quantitatively measured."[1] Very wet 1950s and 1960s were replaced by very dry

1 Hulme, M., "Climatic Perspectives on Sahelian Desiccation: 1973–1998," *Global Environmental Change* 11 (2001): 20.

 | DOI:10.1163/9789004410848_007

1970s and 1980s, while the 1990s and 2000s have witnessed major annual precipitation variability. While predictions of future climate in the region remain highly contested, inter-annual and inter-seasonal precipitation variability is likely to increase, and both prolonged droughts and extreme rainfall events will probably become more frequent in African drylands like the Sudano-Sahelian climate zone.[2]

In the Sudano-Sahelian region, climatic conditions are critical for rural households, for whom rain-fed subsistence agriculture represents the main livelihood.[3] The impact of climate variability on human populations in the region has consequently attracted immense international political and scholarly interest. A large part of this research has focused on how the rural populations of the Sudano-Sahelian region have adapted to the climate crisis experienced since the early 1970s by diversifying their agricultural and non-agricultural livelihood strategies.[4]

Of these, arguably, the most common and well-described is circular labour migration. This concept describes a process by which rural dwellers migrate to areas with employment opportunities to engage in wage labour but stay only temporarily before returning to their villages of origin.[5] In the West African Sudano-Sahelian region, close correlations between this strategy and historical as well as recent precipitation patterns have been identified since the early 1970s.[6] For a large part, rural dwellers engage in this strategy to counter the negative impact of drought and precipitation variability on rain-fed agriculture. In Burkina Faso, circular labour migration has been widely practised since the

2 Christensen, J. H. et al., "Regional Climate Projections," in *Climate Change 2007: The Physical Science Basis. Contribution of Working Group 1 to the Fourth Assessment Report of the Intergovernmental Panel on Climate Change, ed. S. Solomon* (Cambridge: Cambridge University Press, 2007).

3 Niemeijer, D., and V. Mazzucato, "Soil Degradation in the West African Sahel: How Serious Is it?" *Environment: Science and Policy for Sustainable Development* 44 (2002): 20–31.

4 Mortimore, M., and W. M. Adams, "Farmer Adaptation, Change and 'Crisis' in the Sahel," *Global Environmental Change* 11 (2001): 49–57.

5 Gugler, J., "The Son of the Hawk Does Not Remain Abroad: The Urban-Rural Connection in Africa," *African Studies Review* 45 (2002): 21–41.

6 Cordell, D. D. et al., *Hoe and Wage: A Social History of Circular Migration System in West Africa* (Boulder: Westview Press, 1996); Rain, D., *Eaters of the Dry Season: Circular Labor Migration in the West African Sahel* (Boulder: Westview Press, 1999); Henry, S. et al., "The Impact of Rainfall on the First Out-Migration: A Multi-level Event-History Analysis in Burkina Faso," *Population and Environment* 25 (2004); Nielsen, J. Ø., and A. Reenberg, "Temporality and the Problem with Singling Out Climate as a Current Driver of Change in a Small West African Village," *Journal of Arid Environments* 74 (2010); D'haen, S. et al., "Beyond Local Climate: Rainfall Variability as a Determinant of Household Nonfarm Activities in Contemporary Rural Burkina Faso," *Climate and Development* 6.2 (2014).

early 1970s and is similarly associated with climate variability and declining agricultural production following the wet 1950s and 1960s.[7] For the critical climatic period from the early 1970s to the late 1990s, it has been estimated that as many as sixty per cent of Burkinabe men and 15 per cent of women who left their villages after the age of 15 returned within ten years.[8]

Based upon ethnographic research conducted in the capital Ouagadougou in 2013, this chapter explores the contemporary practice of circular labour migration in Burkina Faso. By focusing on explanations given by rural migrants residing in Ouagadougou regarding the practice of circular labour migration, this chapter provides a qualitative insight for the period from the early 1950s until the present into the question of why rural in-migrants have increasingly chosen to stay in Ouagadougou rather than return to their villages of origin. A growing rural population, climate variability, and lack of land and economic opportunities are among the most important reasons given for leaving the village. These factors also play a role when a choice regarding return migration is made, but climate-related environmental impacts on agriculture and, in particular, more but not better rains are shown to have a crucial negative impact on decisions about return migration among migrants in Ouagadougou.

This finding has implications for our understanding of circular labour migration in this region. In recent literature, calls have been made for empirical research exploring the impact of climate change and variability on circular labour migration in sub-Saharan Africa (SSA).[9] Much of this literature comes from research on urbanisation. Hitherto, it has been difficult to link rural-to-urban migration to urban growth in SSA because of the high rates of return migration.[10] If, as is hypothesised,[11] climate change and variability across the

7 Bolwig, S., "Livelihood Practices and Land Use in the Sahel: Labour Allocation and Adaptive Capability among the Fulani Rimaybe in Northern Burkina Faso," (PhD diss., University of Copenhagen, 1999); Breusers, M., "Responses to Climate Variability in the Kaya Region, Burkina Faso," in *The Impact of Climate Change on Drylands: With a Focus on West Africa*, ed. A. Dietz et al. (Dordrecht: Kluwer Academic Publishers, 2004), 207–241; Hampshire, K., "Flexibility in Domestic Organization and Seasonal Migration Among the Fulani of Northern Burkina Faso," *Africa* 3 (2006); Nielsen and Reenberg, "Temporality and the Problem with Singling Out Climate."

8 Henry et al., "The Impact of Rainfall on the First Out-Migration."

9 McGranahan, C. et al., Africa's *Urban Transition and the Role of Regional Collaboration* (London: IIED, 2009); Parnell, S., and Walawege, R., "Sub-Saharan African Urbanisation and Global Environmental Change," *Global Environmental Change* 21 (2011).

10 Potts, D., "Challenging the Myths of Urban Dynamics in Sub-Saharan Africa: The Evidence from Nigeria," *World Development* 40 (2012); Henry et al., "The Impact of Rainfall on the First Out-Migration."

11 Parnell and Walawege, "Sub-Saharan African Urbanisation," (New York: UNDP, 2013).

continent make return migration less attractive due to corresponding environmental damage, it is likely that rural-to-urban migration will become more permanent.[12] Such a change will compound urban growth rates[13] but also influence circular labour migration patterns.

This chapter begins with an introduction to the research setting, focusing first on climatic conditions in Burkina Faso since the 1950s, second on circular labour migration, and third on Ouagadougou. I will then discuss my methodology as well as the results of my study. The latter will show that contemporary circular migration in Burkina Faso is a complex process influenced by economic, social, political, and environmental events of recent occurrence. However, a major finding is that return migration is, according to the informants, increasingly disregarded as an option due to continued precipitation variability and its impact on agricultural production across Burkina Faso. In the subsequent discussion, the results are embedded in recent literature on climate change and variability and (circular) labour migration in sub-Saharan Africa.

2 Research Setting

Burkina Faso is a poor, landlocked country located in the Sudano-Sahelian region of West Africa (Map 5.1). It is ranked 183 out of 187 nations on the latest Human Development Index, has an annual GDP per capita of USD 1,149, an under-five mortality rate of 176 per 1,000 live births, and an adult literacy rate of 28.7 per cent.[14] Burkina Faso has approximately 14 million inhabitants, a number expected to reach 19 million by 2020.[15] Subsistence agricultural production is the mainstay of most households in the country.[16]

2.1 *Climate Variability in Burkina Faso, 1950–2000s*

The most recent period of recurring drought to hit the Sudano-Sahelian region, including Burkina Faso, commenced in the early 1970s (Figure 5.1).[17] Following

12 McGranahan et al., Africa's *Urban Transition.*

13 Parnell and Walawege, "Sub-Saharan African Urbanisation."

14 UNDP, *Human Development Report 2013. The Rise of the South: Human Progress in a Diverse World* (New York: United Nations Development Programme, 2013).

15 Africapolis. *Urbanization Trends 1950–2020: A Geo-Statistical Approach, West Africa* (Africapolis, 2013).

16 Niemeijer, and Mazzucato, "Soil Degradation in the West African Sahel."

17 Brook, G., *Landlords and Strangers: Ecology, Society, and Trade in West Africa, 1000–1630* (Boulder: Westview Press, 1993); Nicholson, S., "Climatic Variation in the Sahel and Other African Regions during the Past Five Centuries," *Journal of Arid Environments* 1 (1978).

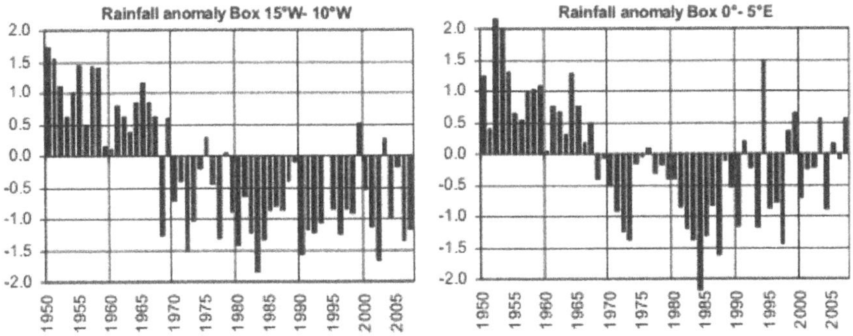

FIGURE 5.1 Rainfall anomalies in the Sahel. The figure captures the wet 1950s and 1960s and the dry 1970s and 1980s and the increased variability in precipitation since the early 1990s
SOURCE: LEBEL AND ALI (2009)

very wet years in the 1950s and 1960s, annual rainfall fell by twenty to thirty per cent in the following three decades, which included major drought periods in the early 1970s and in 1983–1984.[18] Since the late 1990s, a recovery from dry conditions has been observed as annual rainfall totals have increased in comparison to the previous three decades.[19] This is particularly the case for the eastern part of the West African Sudano-Sahelian zone, where Burkina Faso is located.[20] However, there are indications that the West African climate might only be changing slowly and that the inter-annual and inter-seasonal precipitation variability recorded since the early 1990s remains high.[21] During the 2000s, for example, both prolonged dry and wet periods during the rainy seasons were observed across all three climatic zones of Burkina Faso, as were large inter-annual fluctuations, as exemplified by the contrast in annual rainfall between 2003 and 2004 (Figure 5.1).[22] These events illustrate that rainfall continues to fluctuate across the region and over time, "leaving some areas in some years well supplied, yet other regions and other years dry and parched."[23] What has

18 Hulme, "Climatic Perspectives on Sahelian Desiccation."

19 Lebel, T., and A. Ali, "Recent Trends in the Central and Western Sahel Rainfall Regime (1990–2007)," *Journal of Hydrology* 375 (2009); Nicholson, S., "On the Question of the 'Recovery' of the Rain in the West African Sahel," *Journal of Arid Environment* 63 (2005): 615–641.

20 Lebel, and Ali, "Recent Trends."

21 Ibid.

22 D'haen et al., "Beyond Local Climate;" Author's adaptation of data from: Direction de la Meteorologie, Burkina Faso.

23 Hulme, "Climatic Perspectives on Sahelian Desiccation," 19.

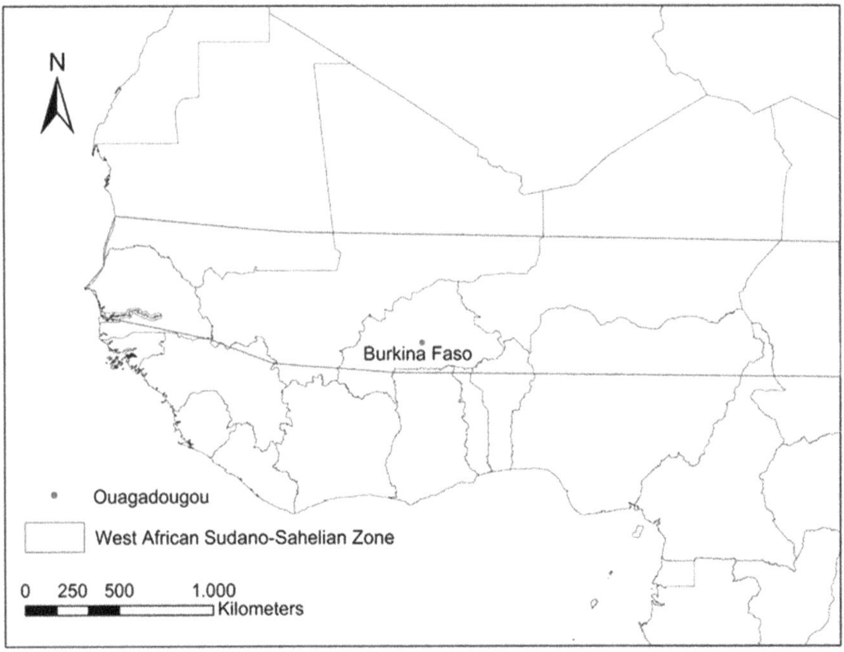

MAP 5.1 The Sudano-Sahelian region of West Africa, showing the location of Burkina Faso and Ouagadougou

characterised the climate in the region, including Burkina Faso, since the early 1990s is thus not a consistent annual rainfall total average, but a high degree of spatial and temporal variability in precipitation. It is to these dynamic climatic conditions that most environmental and social systems have to adapt.

2.2 *Circular Labour Migration*

A commonly used strategy to deal with fluctuating precipitation that causes either drought or flooding and results in insufficient agricultural production[24] is circular labour migration. This is not a new phenomenon in Burkina Faso,[25] but the drought at the beginning of the 1970s and its prolonged aftermath—including, since the early 1990s, increased annual and seasonal precipitation variability—have played a significant role in its development across the Sudano-Sahelian region in the last thirty to forty years, including in Burkina Faso.[26]

24 Nielsen, and Reenberg, "Temporality and the Problem with Singling Out Climate."

25 Breusers, "Responses to Climate Variability."

26 Cordell et al., *Hoe and Wage*; Hampshire, K., and S. C. Randall, "Seasonal Labour Migration Strategies in the Sahel: Coping with Poverty or Optimizing Security?" *International*

Burkinabe migration in the 1950s and 1960s was mainly orientated towards Côte d'Ivoire and Ghana. In the 1950s and earlier, labour migration had been closely linked to French colonial structures of production. Burkina Faso, especially the more densely inhabited central and southern regions of the country, was largely used by the French colonial administration as a labour pool for plantation work in Côte d'Ivoire.[27] In the post-colonial 1960s, this system was abolished, and restrictions on movement were lifted. As a result, voluntary labour migration, particularly to Ghana, began to be practised, mainly by younger men looking for cash—often for the purpose of marriage.[28] During the 1970s, Côte d'Ivoire substituted Ghana as the major migration destination, and circular labour migration became much more common, mainly due to the climatic crisis experienced throughout the country.[29] Starting from that time, money for food—no longer available from the predominately rainfed agricultural sector—needed to be made elsewhere. Civil unrest in Côte d'Ivoire during the late 1990s and 2000s, however, has resulted in more migrants going to Ouagadougou and Bobo-Dioulasso.[30] Typically, migrants leave at the end of the crop harvest (November–December) to earn money, primarily for food.[31] They commonly stay between six months and ten years, after which they return home to live and participate in agricultural activities.[32] A large section of the migrants continue with circular migration, going back and forth over a number of years until they become too old to endure the often very demanding journeys and physical labour.[33]

Journal of Population Geography 5 (1999); Hamshire, "Flexibility"; Henry et al., "The Impact of Rainfall on the First Out-Migration"; Mortimore, M., and W. M. Adams, "Farmer Adaptation, Change and 'Crisis' in the Sahel," *Global Environmental Change* 11 (2001); Rain, *Eaters of the Dry Season.*

27 Cordell et al., *Hoe and Wage.*

28 Ibid.

29 Ibid.; Henry et al., "The Impact of Rainfall on the First Out-Migration."

30 Bjarnesen, J., "Diaspora at Home? Wartime Mobilities in the Burkina Faso-Côte d'Ivoire Transnational Space" (PhD diss., University of Uppsala, Sweden, 2013); Boswell, K. V., "Migration, War, and Repatriation from Côte d'Ivoire" (PhD diss., Indiana University, 2010).

31 See, however, Nielsen, J. Ø., "Drought and Marriage: Exploring the Interconnection between Climate Variability and Social Change through a Livelihood Perspective," in *The Question of Resilience: Social Responses to Climate Change*, ed. K. Hastrup (Copenhagen: The Royal Danish Society of Arts and Letters, 2009).

32 Henry et al., "The Impact of Rainfall on the First Out-Migration."

33 Nielsen, J. Ø., and Reenberg, A., "Cultural Barriers to Climate Change Adaptation: A Case Study from Northern Burkina Faso," *Global Environmental Change* 20 (2010).

2.3 *Ouagadougou*

Ouagadougou is a rapidly growing city: in 1950, there were 35,000 inhabitants, in 1970 126,000, in 1990 562,000, and in 2000 900,000; currently, approximately 1.4 million inhabitants live in the city, while by 2020 the population is expected to reach 1.8 million.[34] Ouagadougou's urban area has expanded accordingly from around 25 square kilometres in 1956 to approximately 372 square kilometres in 2012 (see Map 5.2).

Ouagadougou is a hot, bustling place. Located on a flat plain and built with largely similar buildings, it has a fairly homogeneous appearance. Only in a few of the districts, particularly in the central business district, are concrete buildings—usually around four storeys high—and paved streets to be found. By contrast, most of the surrounding residential areas consist of a mix of modern constructions and more traditional clay huts, and almost all secondary streets are unpaved. The constructions found in a particular district depend to a great extent on whether the district is *lotis* or *non lotis* (see Map 5.2). The former is a district in which an official subdivision has taken place, implying that houses and plots are formally registered to individual owners. In these districts, houses are often constructed with concrete and organised in a grid structure along large, wide roads. Public amenities, such as sanitation facilities, access to running water, and electricity, are accessible (see Figure 5.3). In the latter, *non lotis* districts, usually found on the city margins, development occurs without any central planning. Houses are small and constructed with clay, with little space in between each other. Electricity is not available, roads and sanitation are non-existent, and water has to be bought from a common source, often a pump (see Figure 5.2).

The inhabitants of the *non lotis* districts are often engaged in the informal sector.[35] Working on construction sites or as maids, guards, waiters, and mechanics, selling food, beer, coal, or other products, often along the major roads, and making bricks are some of the most common occupations. Both the *lotis* and *non lotis* districts are ethnically very diverse, but the majority of residents are Mossi, the traditional inhabitants of the central plateau of Burkina Faso on which Ouagadougou is located. The spoken language in the city is accordingly their language, Mòoré.

---

34 Africapolis, *Urbanization Trends 1950–2020*.

35 Thorsen, D., "Weaving In and Out of Employment and Self-Employment: Young Rural Migrants in the Informal Economy of Ouagadougou," *International Development Planning Review* 35 (2013).

MAP 5.2 Satellite images taken in 2012 showing the difference between *lotis* and *non lotis* as seen from above
SOURCE: THORSEN, D., "WEAVING IN AND OUT OF EMPLOYMENT AND SELF-EMPLOYMENT"

3 Methodology

This study is based on qualitative ethnographic research conducted in September and October 2013 in Ouagadougou, Burkina Faso. It explores the period from the early 1950s to 2013. As such, it covers (1) the dramatic changes in climatic conditions experienced in the Sudano-Sahelian region; (2) the ascendance of circular labour migration in Burkina Faso; and (3) living memory. It incorporates two of the main sources of data collection and data recognised in

FIGURE 5.2 Typical houses found in *non lotis* areas of Ouagadougou
SOURCE: GOOGLE EARTH

FIGURE 5.3 Houses in *lotis* areas of Ouagadougou. Single-family homes consist of one or two rooms and are always mud-brick constructions

qualitative research: participant observation and interviews.[36] These methods were supplemented by transect walks, informal interviews, and GPS measurements. During my research, I lived at a private home in Tampouy, a district in northern Ouagadougou.

Participant observation was chosen for the research because it places the researcher at the heart of the action and allows for the collection of stories, figures and observations about daily life and its routines. To be of value, participant observation must lead to insights, "the noticing of apparently insignificant points, the making of connections."[37] Often this happens because details, however small they might at first appear, add up over time, pointing the way to other aspects and details that might otherwise go unobserved. Among such details, the disregard of return migration was significant for this research.

The basic insights gained via participant observation were explored further in 38 semi-structured interviews with rural in-migrants that lasted between one and two hours. Age, gender, and period of residence in Ouagadougou were important interviewee selection criteria (Table 5.1). Moreover, nearly all peripheral districts of the city were covered, as it is generally there where migrants settle and continue to reside.[38] The interviewees were further selected according to socio-economic status, place of origin, and ethnicity, thus covering many of the major differentiations within Ouagadougou and Burkina Faso as well as a variety of migration histories. Each semi-structured interview site was registered on a hand-held GPS (see Map 5.2 and Map 5.3).

The interviews were directed towards understanding the informants' perspectives and were thus relatively non-standardised and open-ended.[39] Nonetheless, all respondents were asked about: (1) their life history (e.g. where they were born; how long they had been in Ouagadougou); (2) their current situation (e.g. household and neighbourhood composition; main income sources, expenditures, concerns, and wishes); (3) housing (e.g. how long they had lived on a particular plot; how they attained the plot; from where they had moved to the

36 Bernard, H.R., *Research Methods in Anthropology: Qualitative and Quantitative Approaches* (Walnut Creek: Altamira Press, 2002); Denzin, K.N., and Y.S. Lincoln, *The Sage Handbook of Qualitative Research* (Los Angeles: Sage, 2011).

37 Cohen, A. P., "Participant Observation," in *Ethnographic Research*, ed. R. Ellen (London: Academic Press, 1984), 220.

38 Beauchemin, C., and P. Bocquier, "Migration and Urbanization in Francophone West Africa: An Overview of the Recent Empirical Evidence," *Urban Studies* 41 (2004).

39 Kvale, S., *InterViews: An Introduction to Qualitative Research Interviewing* (Thousand Oaks: Sage Publications, 1996); Spradley, J. P., *The Ethnographic Interview* (Fort Worth: Harcourt Brace Jovanovich College Publishers, 1979).

TABLE 5.1 Semi-structured interview distribution (N = 38)

Social distinction	Number of interviewees
Gender	
Women	20
Men	18
Age	
20–30	14
30–45	14
45<	10
Length of residence in Ouagadougou in years	
0–5	14
5–10	12
10–20	6
20<	6

plot); (4) Ouagadougou (e.g. how well they knew the city; changes to the city and neighbourhood since they had moved there and what the reasons behind these might be); (5) rural-urban relations (e.g. relations with their village of origin; the presence of family or village members in Ouagadougou and why they had come to Ouagadougou); and (6) return migration (whether they had thought of returning to their village; reasons for wanting or not wanting to return; whether they thought other people were currently returning and why/why not). The two latter categories of questions in particular were followed up with questions concerning changes in these parameters over the lifetime of the interviewee.

All semi-structured interviews were conducted in Mòoré, with the help of an interpreter, or in French, if the interviewee/s was/were sufficiently fluent. All interviews were digitally recorded, transcribed, coded, and analysed along with any fieldnotes using Nvivo10 software.

In most of the districts (7) in which interviews were conducted, I also arranged a transect walk with a local resident whom I had identified during the interviews. In all but one case, the selected residents were male. The walks consisted of a tour through the district, during which different developments, events, and challenges as well as changes over time were described. During these transect walks, informal conversations around issues of pertinence to the district were also held with people who joined the walk. Many of these conversations revolved around rural-urban relations and migration in contemporary Burkina Faso.

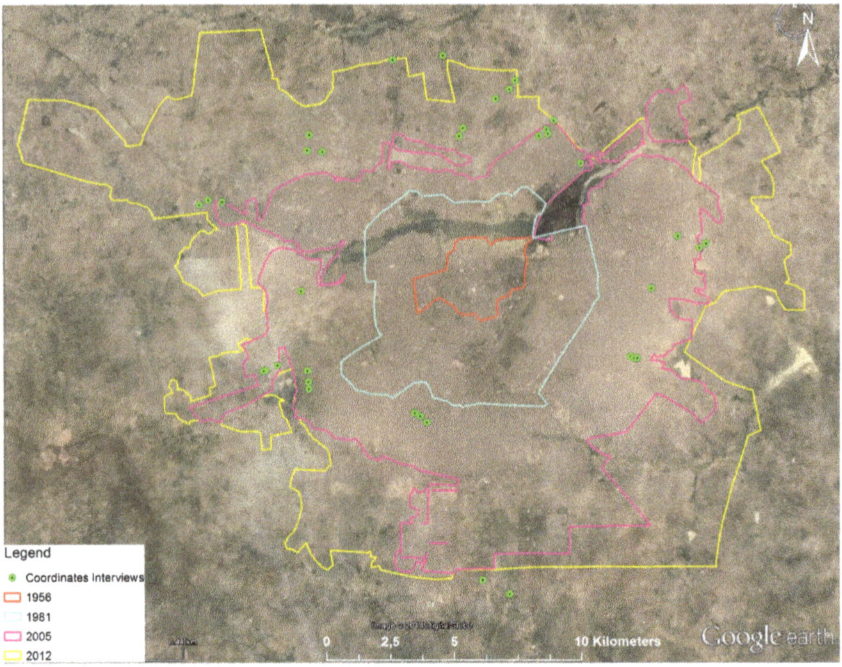

MAP 5.3 Interview sites and the spatial expansion of Ouagadougou in the period 1956–2012
SOURCE: IGB, INSTITUT GÉOGRAPHIE DU BURKINA AND GOOGLE EARTH. OWN ADAPTATION

4 Results

Circular labour migration is a complex process in present-day Burkina Faso, influenced by economic, cultural, political, and environmental factors alike. Of these, climate variability was the second-most important push factor and the most important reason for informants not wanting to return to their village of origin (see Table 5.2 and Table 5.3).

4.1 *Leaving the Village: Unemployment, Land, and Climate Variability*

Securing a living is extremely hard in many villages across Burkina Faso.[40] A lack of salaried work, a growing rural population, and little or no available land, machines, seeds, and fertiliser for agricultural production as well as continued problems with climate variability, especially in relation to rainfall, were among the most prevalent reasons for this situation given by informants (Table 5.2).

40 D'haen et al., "Beyond Local Climate."

TABLE 5.2 Reasons mentioned by interviewees (N = 38) for moving to Ouagadougou since the year 2000

Reason for migration	Number of interviewees mentioning this	Percentages of interviewees mentioning this	Number of interviewees mentioning this as most important reason	Percentages of interviewees mentioning this as most important reason
Growing population	32	84%	23	61%
No land	28	73%	16	42%
Divorce/ marriage	20	61%	18	47%
Climate variability	34	89%	25	66%
Lack of work	36	95%	27	71%
Other	6	16%	1	2%

Note that many interviewees mentioned several reasons as being equally important.

Regarding wage labour, all informants found it impossible to engage in the nonfarm economy in and around their home village. It is generally argued that participation levels in nonfarm economic activities in SSA are influenced by the economic opportunities and development characteristics of villages. Villages with markets, good road access and quality, and access to telecommunications and electricity are expected to be able to offer better opportunities for engagement in nonfarm activities than those without.[41] While mentioned as factors influencing the number of salaried jobs in their villages, the presence of these features was of little help according to the interviewees. A major reason behind this was, in the words of a 45-year-old man from the south of the country, "that more and more people are looking for the few jobs available."

41 Lanjouw, P., "Nonfarm Employment and Poverty in Rural El Salvador," *World Development* 29 (2001); see also D'haen, Nielsen, and Lambin, "Beyond Local Climate" for an overview.

TABLE 5.3 Reasons mentioned by interviewees (N = 38) for not returning to origin village

Reason for not returning to origin village	Number of interviewees mentioning this	Percentages of interviewees mentioning this	Number of interviewees mentioning this as most important reason	Percentages of interviewees mentioning this as most important reason
Better life in Ouaga	21	55%	4	10%
Climate variability	34	89%	31	81%
No land in village	14	37%	11	29%
Work opportunities in Ouaga	33	87%	30	79%
Other	3	8%	1	3%

Note that many interviewees mentioned several reasons as being equally important.

The interviewees linked the population growth experienced in villages across Burkina Faso to higher birth rates, but also to the crisis in Côte d'Ivoire. This broke out in the late 1990s and resulted in, among other developments, a targeting of Burkinabe living and working there.[42] All Burkinabe whom I interviewed or spoke with during fieldwork who had returned from Côte d'Ivoire mentioned that they had left the country because of attacks, rumours of imminent attacks, social tensions, deteriorating living conditions, and/or a desire to "live peacefully," as it was often put. Researchers have estimated that about 500,000 to one million people were forced to

42 Zongo, M. "Introduction," in *Les Enjeux Autour de la Disapora Burkinabé: Burkinabé á l'Étrangers au Burkina Faso*, ed. by M. Zongo (Paris: L'Harmattan, 2010), 11–14; Hagberg, S., and J. Bjarnesen, " 'Good Guys' and 'Bad Guys': The Burkinabe Public Debate on the Ivorian Crisis," in *Une Anthropologie entre Pouvoirs et Histoire: Conversations Autour de l'Œuvre de Jean-Pierre Chauveau*, ed. E. Jul-Larsen et al. (Paris: Karthala, 2011), 509–534; Bjarnesen, "Diaspora at Home?"

return to Burkina Faso from Côte d'Ivoire during the period from 2002 to 2006 alone.[43] A large proportion of these returnees have settled in Ouagadougou, but many also went back to their village of origin or to other urban areas in Burkina Faso.[44] "[T]his," as a young man from the central plateau explained in an interview, "caused increased competition for the few jobs available in my village."

Another effect of a growing rural population has been increased competition for land. According to my interviewees and other people with whom I spoken during fieldwork, the lack of access to agricultural and residential land in their villages was a major reason for leaving. A fifty-year-old man from the west of the country explained in an interview how he—upon his return from Côte d'Ivoire after ten years there—was "offered a very small piece of land so far from the village that it was not worth my effort [to farm]." According to several interviewees, many migrants returning from Côte d'Ivoire were offered no land on which to live and farm. Only interviewees from the very north of the country did not mention land as a reason to migrate.[45] Interviewees also often stated that the arrival of more people had resulted in parcels of land having to be shared among increasing numbers of relatives. A lack of access to machinery such as tractors and a steep increase in price for seeds and fertilisers further dampened agricultural prospects. Leaving for Ouagadougou or other cities and towns thus represented "the only real alternative," as was often stated.

Interviewees strongly associated the dismal state of agricultural production with precipitation variability. After the lack of work, this was mentioned as the second-most important reason for migrating to Ouagadougou (Table 5.2). As agricultural production is largely rain-fed in Burkina Faso, precipitation is critical. Considering the precipitation variability experienced in Burkina Faso since the early 1970s,[46] it is not surprising that informants considered this to be a major problem. Interviewees from all three major rainfall regions in the country explained how agricultural production remained difficult due to poor and unpredictable rains. Too little rain, rain falling at the wrong time, prolonged drought periods during the rainy season, heavy rain, flooding, run-off and erosion, and so-called false starts to the rainy season were commonly cited

43 Zongo, "Introduction"; Boswell, "Migration, War, and Repatriation."

44 Bjarnesen, "Diaspora at Home?"

45 Nielsen, and Reenberg, "Cultural Barriers to Climate Change Adaptation."

46 Lebel and Ali, "Recent Trends"; Epule, E. T. et al., "The Causes, Effects and Challenges of Sahelian Droughts: A Critical Review," *Regional Environmental Change* 14 (2014): 145–156.

problems in relation to rainfall.[47] For the people interviewed on this matter, however, the major challenge was that rain had not improved since the early 1970s. Older interviewees (35<) explained how in the 1970s, 1980s, and 1990s farmers in their villages had still hoped for a return to the rainfall regime of the 1950s and 1960s. Now, however, this hope was gone; "therefore we leave and stay away," a middle-aged woman explained.

4.2 *Staying in Ouagadougou: Work, Aspirations, and Climate Variability*

Among the migrants interviewed, remaining in Ouagadougou was a major issue. Only in four out of the 38 interviews conducted with rural in-migrants did the interviewee express a wish or plan to return to their respective village of origin (Table 5.3). Moreover, all but the four thinking about returning had left for Ouagadougou with the aim of settling there permanently, a trend that was mirrored in informal conversations conducted during fieldwork. The major reasons for this were increasing job opportunities in Ouagadougou, hopes and aspirations for a better life, and climate variability (see Table 5.3).

African migration is often explained in terms of finding employment, and most literature on migration posits that "[e]mployment is central to all rural-to-urban migration theories because the search for a job is seen as the primary, if not the only, motivation for migration."[48] As seen in Section 4.1 above, economic issues clearly have been a significant push factor, but increasing work opportunities in Ouagadougou coupled with a general economic upturn in Burkina Faso since the early 2000s was also cited during fieldwork as a reason for staying. Stories about the opening of new clothes shops, supermarkets, pharmacies, markets selling imported second-hand goods from Europe, banks, and housing developments were frequent. According to the informants, most people in Ouagadougou now also own a motorbike, and an increasing number own a car.

A better educated population, the return of better-off migrants from Côte d'Ivoire investing in Burkina Faso, a growing population generating more competition and activity, and, in particular, the growth of gold mining were

47 Mertz, O. et al., "Climate Factors Play a Limited Role for Past Adaptation Strategies in West Africa," *Ecology and Society* 15 (2010); Mertz, O. et al., "Adaptation Strategies and Climate Vulnerability in the Sudano-Sahelian Region of West Africa," *Atmospheric Science Letters* 12 (2011): 104–108; D'haen, Nielsen, and Lambin, "Beyond Local Climate."

48 Beauchemin, and Bocquier, "Migration and Urbanization in Francophone West Africa," 2257.

given as explanations. One middle-aged man explained how people working in the mines "invest the money they make here in Ouagadougou." This, in turn, created jobs for people in the city. Women worked as maids, but also in small-scale commerce, as hairdressers, as shop assistants, or as waitresses, while men found work in the construction sector or as mechanics, vendors, and guards.[49]

Plans to stay permanently in Ouagadougou also reflect that for many migrants the city represents a chance to live a different kind of life than that on offer in the village. Many of the women moving to Ouagadougou do so, for example, to get away from arranged marriages, hostile in-laws, threatening or violent husbands and ex-husbands, and a village environment in which there is little security in terms of policing (Table 5.2). According to older women with whom I spoke during fieldwork, another reason for female migration and permanent settlement in Ouagadougou is that young women today have much more freedom than their older relatives. They do not "make do," as it was often expressed, and instead choose to come to Ouagadougou in order to obtain the personal and financial freedom that is not available to them in their villages. For women, the appeal of Ouagadougou is also related to the fact that the city offers recently divorced women a chance to start anew. This idea is very closely linked to the aim of finding a new husband in Ouagadougou, something that all of the divorced women whom I interviewed spoke of with desire. "So why," asked a young, recently divorced woman of about thirty years of age rhetorically, "would I go back to a village with hostile former in-laws, no jobs, and the prospect of working all day in fields producing no food because of bad rain?"

Climate variability is a major deterrent to return migration. According to the interviewees, many of the people who move to Ouagadougou do not plan to return—and in fact do not return—to their village because of climate variability (Table 5.3). This included most of the interviewees, as 34 out of the 38 people interviewed had left their village without a plan to return. Especially in the period from the early 1970s until the early 2000s, as many informants explained, in-migrants almost always returned to their villages when the rainy season started in June to do agricultural work.[50] As such, the Burkinabe seasonally driven circular migration pattern very closely mirrored the general understanding of circular migration in the Sudano-Sahelian region. Many of the informants highlighted this, stating that such migration became widely

49 See also Thorsen, "Weaving In and Out of Employment."

50 Nielsen and Reenberg, "Cultural Barriers to Climate Change Adaptation."

practised in Burkina Faso in the early 1970s because of the widespread droughts of the time. Without the levels of agricultural production recorded during the wet 1950s and 1960s, people from all over the country went in search of money for food during the agricultural off season. Since the early 2000s, however, a change has taken place, according to the interviewees, since now "people stay on in Ouagadougou." Almost all informants and interviewees linked this shift to precipitation variability rather than to other factors such as the improved economic situation in Ouagadougou. Many informants highlighted the continued impact of climate variability on the practice of agriculture, which had represented the primary reason for returning to the village in the first place. A young man explained it as such: "Agriculture is not attractive, you work all day and get nothing because of poor rain. It has been like that now for a long time and so people in my village have lost hope for better rain and millet production. So why go back?"

Many informants, like the young man just quoted, referred to the rain when they spoke of climate variability in conversations and interviews. Their complaints about the rain pivoted around the decrease in rainfall since the 1960s and the increased inter- and intra-seasonal variability, but noted in particular the significant length of time for which precipitation variability (including drought, one of its major consequences) has been experienced in the villages (see Section 4.1). Many informants, regardless of their origin, also explained how for approximately the past ten years there had been more rain, but that "more is not necessarily better," as was commonly articulated. They found discouraging the fact that more rain was falling but not in the way that they desired. Rain falling "too heavily," causing crop damage, run-off, and erosion, was a major problem that had been experienced since the early 2000s. Consequently, the hope of improving levels of agricultural production was slim among the informants. "We had always hoped," two older women who had experienced the "better" rains of the 1960s said, "that if only we got more rain it [agriculture] would be better. But the rain is not right. It does not fall right." Return migration was thus almost pointless, because "even with more rain the fields do not produce enough food." Moreover, returning home to retain usufruct rights in the hope that better times were on their way was now regarded by many as "a waste of time," especially considering that the last forty years had been one long "disappointment" due to the fact that the rainfall conditions of the 1960s had not returned, as described by one man in his late sixties. This sentiment was echoed in interviews with younger in-migrants, such as a 39-year-old man from a village just outside Ouagadougou, who reflected upon conditions in the 1960s compared to those of the present, "when it is also raining a lot."

In-migrants therefore made establishing more permanent roots in Ouagadougou their priority. For my interview partners, this was not just a matter of "getting on personally," but was also seen as a better option than returning home to work in the fields to help their village and family members who had stayed behind. This view revolved around two aspects. First, people consistently argued that the growing rural population had made returning home to help with agricultural tasks—such as sowing, hoeing, and harvesting—superfluous. Second, sending money home was now seen as a much better way to help. This has become a cheap and easy option within approximately the last five years, as it can now be done via mobile phone. The ease, speed, and safety of this transfer method were crucial arguments in support of staying in Ouagadougou; as a middle-aged man explained, "in this way I can help my sisters straight away." This money was often used to pay medical bills, transport costs, and school fees or for weddings and other social events,[51] but was mostly used to buy food and agricultural materials such as seeds and fertiliser for family members living in the village. In the words of a young man working in a small roadside garage:

> There is nothing in the village for me, one year it rains, the next it doesn't, and even with more rain it is not right. I also have no land really. Here in Ouaga I can at least make some money, so staying rather than returning means that I, but also my relatives and my village, have a chance to improve life. So I'm staying!

5 Discussion

Work, the chance to live a different life, and continued challenges to agricultural production due to precipitation variability were, as just described, major reasons cited for staying in Ouagadougou. It is obviously difficult to argue from qualitative research alone that such reasons constitute a major demographic trend, since it is impossible to know how many people are represented by each of these reasons. Reliable and good census data is needed to confirm or challenge these findings. Yet, the results do indicate that good reasons for renewed scholarly interest in circular labour migration patterns in Burkina Faso and perhaps elsewhere in SSA can be found. This is particularly the case because

51 See also Nielsen, "Drought and Marriage."

one of the major reasons given by the informants for discontinuing their participation in circular migration was climate variability.

Linking human decision-making to climatic factors and the impacts thereof is not straightforward, even in an area like the Sudano-Sahelian region where massive climate variability has been experienced over the past fifty years.[52] On a very general level, this is due to the fact that the climate is often, if not always, only one of many factors that influence human decision-making.[53] The results presented in this chapter illustrate this. Stories about social problems in the villages and a desire for a different and better life, particularly for women,[54] intermingle with gold prices, the crisis in Côte d'Ivoire, mobile phone technology, a growing rural population, and precipitation variability. Singling out one of these factors as being more important than the others for a change in migration patterns is thus not easy. Nevertheless, it has consistently been argued that rural out-migration hinges to a very large extent upon whether or not there is employment and food available in the village of origin.[55] In the Sudano-Sahelian region, both of these aspects are closely related to precipitation patterns. For rural populations, rain-fed subsistence agriculture constitutes both the main source of food and work;[56] consequently, most, if not all,

---

52 Barbier, B. et al., "Human Vulnerability to Climate Variability in the Sahel: Farmers' Adaptation Strategies in Northern Burkina Faso," *Environmental Management* 43.5 (2009): 790–803; Tschakert, P., "Views from the Vulnerable: Understanding Climatic and Other Stressors in the Sahel," *Global Environmental Change* 17 (2007): 381–396; Mertz et al. "Climate Factors Play a Limited Role"; Nielsen, and Reenberg, "Temporality and the Problem with Singling Out Climate"; D'haen, Nielsen, and Lambin, "Beyond Local Climate."

53 Belliveau, S., B. Smit, and B. Bradshaw, "Multiple Exposures and Dynamic Vulnerability: Evidence from the Grape Industry in the Okanagan Valley, Canada," *Global Environmental Change* 16 (2006): 364–378; Leichenko, R. M., and K. O'Brien, *Environmental Change and Globalization: Double Exposures* (Oxford: Oxford University Press, 2008); Nielsen, J. Ø., and A. Reenberg, "Exploring Causal Relations: The Societal Effects of Climate Change," *Geografisk Tidsskrift-Danish Journal of Geography* 112 (2012): 89–92.

54 See also Bocquier, P., and S. Traore, *Urbanisation et Dynamique Migratoire en Afrique de l'Ouest: La Croissance Urbaine en Panne* (Paris: Editions L'Harmattan, 2000); Le Jeune, G., and V. Piche, "Toward a Reconsideration of Female Migration Patterns in Burkina Faso," *Canadian Studies in Population* 31 (2005): 145–177; Beauchemin, C., "Rural-Urban Migration in West Africa: Towards a Reversal? Migration Trends and Economic Situation in Burkina Faso and Côte d'Ivoire," *Population Space and Place* 17 (2011): 47–72.

55 Harris, J. R., and M. Todaro, "Migration, Unemployment and Development: A Two Sector Analysis," *American Economic Review* 60 (1970): 126–142.

56 Niemeijer, and Mazzucato, "Soil Degradation in the West African Sahel"; Reenberg, A., L.V. Rasmussen, and J. Ø. Nielsen, "Causal Relations and Land Use Transformation in

literature on circular migration in the Sudano-Sahelian region highlights the link between rainfall variation and labour migration for the period since the early 1970s.[57]

This is also the case for Burkina Faso.[58] In one of the most comprehensive studies of this link in Burkina Faso, Henry and colleagues show how rainfall variation and circular migration are closely connected, but also note that "it could be fruitful to examine the effect of environmental conditions of the village of origin on the risk of *returning* to that village."[59]

On the continental scale, research has increasingly argued that climate change, including precipitation variability, makes returning less and less desirable for many rural out-migrants.[60] McGranahan et al. note in a large review on this topic how it is "likely that as a consequence of climate change this movement [rural out-migration] will increase and intensify, and possibly become more permanent."[61] If climate variability makes returning less attractive, as most of the returnees would go back to participate in rain-fed agricultural activities,[62] then the number of returnees could be significantly affected.

The results presented in this chapter have shown that this might be the case. Almost all interviewees emphasised that the continuing problems to do with the rain, the lack of hope for the return of improved climatic conditions similar to those experienced in the 1950s and 1960s, and the fact that agricultural work is no longer a viable option have resulted in far fewer migrants going back to their villages. My informants also linked this state of affairs to what literature on climatic conditions and trends in the Sudano-Sahelian region has called a recovery of rainfall since the early 2000s.[63] According to both the

---

the Sahel: Conceptual Lenses for Processes, Temporal Totality and Inertia," *Geografisk Tidsskrift-Danish Journal of Geography* 112 (2012): 159–173; D'haen, Nielsen, and Lambin, "Beyond Local Climate."

57 Cordell, Gregory, and Piché, *Hoe and Wage*; Rain, *Eaters of the Dry Season*.

58 Bolwig, "Livelihood Practices and Land Use in the Sahel"; Breusers, "Responses to Climate Variability"; Hamshire, "Flexibility in Domestic Organization"; Nielsen, and Reenberg, "Temporality and the Problem with Singling Out Climate."

59 Henry, Schoumaker, and Beauchemin, "The Impact of Rainfall on the First Out-Migration," 244, italics in original text.

60 International Organization for Migration (IOM), *Migration, Environment and Climate Change: Assessing the Evidence* (Geneva: IOM, 2009); Annez, P., R. Buckley, and J. Kalarickal, "African Urbanization as Flight? Some Policy Implications of Geography," *Urban Forum* 21 (2010): 221–234; Parnell, and Walawege, "Sub-Saharan African Urbanisation."

61 McGranahan et al., *Africa's Urban Transition*, 12.

62 Nielsen, and Reenberg, "Temporality and the Problem with Singling Out Climate."

63 Lebel, and Ali, "Recent Trends"; Nicholson, "On the Question of the 'Recovery'."

informants and the scientific data, rainfall has become more abundant since the late 1990s, but because the rain falls too heavily, at the wrong times of the growing season, and with a large intra-seasonal and intra-annual variability, more rain has not resulted in better conditions for agricultural production. There was thus little or no hope among the informants of an improvement or a return to 1960s rainfall patterns, and consequently many saw no point in returning to their villages to farm now or in order to retain the rights to do so in the future. Instead, staying on in Ouagadougou was what mattered, for which a diversity of reasons was given. Improved financial circumstances due largely to high gold prices and the jobs created by migrants from Côte d'Ivoire was one very important explanation.

In recent literature on migration and climate change, it has been argued that research needs to focus less on "*whether* environmental drivers are the *sole causal* factors causing mobility," but should rather look at how migration is used as a strategy to manage the risks associated with changing environmental conditions.[64] Arguably, the literature on circular labour migration in the Sudano-Sahelian region has by and large done precisely this;[65] however, the continued focus in the literature on circular labour migration as a means of securing a viable livelihood in SSA[66] and in the Sudano-Sahelian region[67] means that the ways in which this system is changing are not given much attention. This is problematic because, as the results presented in this chapter indicate, remaining in Ouagadougou rather than returning to the village of origin potentially represents a new migration strategy aimed at managing the risks facing villages in relation to environmental change. For the migrants, their decision

---

64 Warner, K., and T. Afifi, "Where the Rain Falls: Evidence from 8 Countries on How Vulnerable Households Use Migration to Manage the Risk of Rainfall Variability and Food Insecurity," *Climate and Development* 6 (2014): 1–17, 1, italics in original; Morrissey, J. W., "Understanding the Relationship between Environmental Change and Migration: The Development of an Effects Framework Based on the Case of Northern Ethiopia," *Global Environmental Change* 23 (2013): 1501–1510.

65 E.g. Bruijn, M. de et al., *Sahelian Pathways: Climate and Society in Central and South Mali* (Leiden: African Studies Centre, 2005).

66 See also Parnell, and Walawege, "Sub-Saharan African Urbanisation."

67 E.g. Konseiga, A., "Household Migration Decisions as Survival Strategy: The Case of Burkina Faso," *Journal of African Economies* 16 (2007): 198–233; Rasmussen, L.V., and A. Reenberg, "Collapse and Recovery in Sahelian Agro-Pastoral Systems: Rethinking Trajectories of Change," *Ecology and Society* 17 (2012): 14; Van der Land, V., and D. Hummel, "Vulnerability and the Role of Education in Environmentally Induced Migration in Mali and Senegal," *Ecology and Society* 18 (2013): 14; Mortimore, M., "Adapting to Drought in the Sahel: Lessons for Climate Change," *Wiley Interdisciplinary Reviews-Climate Change* 1 (2010): 134–143.

to remain in Ouagadougou was closely related to the lack of agricultural production caused by climate variability as well as to the newfound ease of sending money home. Theories of labour migration note that migration is often enacted as a strategy for managing risk at the level of the household rather than for the individual alone.[68] This was a pronounced rationale among my informants; instead of going back to their villages to assist with agricultural activities, making money and sending it home was seen by many as a better means of helping their village-based family members deal with the impacts of climate variability. From this perspective, permanent migration resembles circular labour migration as a strategy intended to cope with environmental change, but this chapter's results indicate that even well-established human adaptation strategies to climate variability and climate change potentially lose their relevance or are altered fundamentally once such changes become more temporally embedded or increasingly dramatic.

6 Conclusion

The relationship between climate change and variability, on the one hand, and circular labour migration in sub-Saharan Africa, on the other, is well-documented. In the Sudano-Sahelian region, the droughts experienced in the early 1970s and 1980s and the continued challenges arising from rainfall variability in the 1990s and 2000s have consistently been posited as explanations for circular labour migration. Poor subsistence farmers needing money for food that was no longer available from agriculture due to precipitation problems flocked to destinations where employment could be found only to return at the beginning of the next agricultural season. In recent years, the persistence of this practice has been questioned in the literature. Again, the focus is on climate change and variability; now, however, these phenomena are invoked as reasons for the discontinuation of circular labour migration.

Migrants whom I interviewed and with whom I spoke during my fieldwork in Ouagadougou, Burkina Faso, in 2013 persistently underlined that they wanted to stay in the capital rather than return to their villages of origin. A major reason for this was continued problems with precipitation patterns and in particularly the fact that since the early 2000s more rain had fallen but without leading to an increase in levels of agricultural production. Disillusioned

68 E.g. Castles, S., and M. Miller, The Age of Migration: International Population Movements in the Modern World (New York: Palgrave Macmillan, 2009).

by these persistent precipitation problems and the more abundant but wrong type of rain my informants saw no point in maintaining ties with agricultural production. Moreover, taking into account the better economic opportunities in Ouagadougou, they also argued that staying there rather than returning to their villages represented a better strategy for dealing with the impacts of precipitation variability on agriculture.

As such, the data presented in this chapter shows that the predominance of circular migration as a coping strategy in the face of environmental change is being challenged in Burkina Faso. Very little empirical work has been done on this issue in SSA, while no relevant post-2000 statistical census data on this issue exists. Arguing that the results presented here mirror a general trend in SSA, or indeed in Burkina Faso itself, is therefore difficult. Yet, current and predicted climatic trends for this region indicate that rural populations who are dependent upon subsistence agriculture will continue to face challenging times and that new coping strategies may thus be required. More permanent migration to cities and other areas of economic opportunity might be one such strategy.

CHAPTER 6

More of the Same: a Gender Lens on Life in a Changing Climate in Sub-Saharan Africa

Joy Clancy

Abstract

This chapter provides a desk-based review of peer-reviewed and grey literature on gender and adaptation to climate change in the context of rural sub-Saharan Africa using a feminist political ecology approach. The aspects of rural livelihoods reviewed are health, water, and energy. The existing empirical evidence is limited. The literature tends to be deductive, based on past experiences with floods and drought. Adaptation strategies focus on building resilience to the impacts of extreme weather events linked to a changing climate. There are signs that women and men are taking their own steps to adapt based on their knowledge of their natural ecosystems and their past experiences with drought—some of which also lead to shifts in gender roles. The chapter makes the case for taking a gender approach in analysing vulnerability and resilience, since both are gendered states that can offer a more holistic insight into the diverse coping strategies of households and communities. The overall conclusion is that while communities in SSA have recognised the change in weather patterns and the impact it is having on their lives, it is not something new but 'more of the same'. Their priorities for securing their livelihoods can be shaped by the ways in which people attach meaning to the environment and climatic events, such as droughts.

1 Introduction

> I don't think we need to be told that something is wrong with the climate. We can see for ourselves that it is getting hotter and we get very little rain. We can no longer tell when to expect the rain because it does not come at the time we expect it. We are no longer sure when we should plant or whether we should even plant because we waste seeds and our time if the crops fail. We have been disappointed with the rain so many times, and it is painful to see

 | DOI:10.1163/9789004410848_008

> our fields dry when they should be giving us food. So it is obvious that changes are there in the climate.[1]

These words were spoken by a female farmer from the Thelaphi community in South Africa. She is not alone in her perception that our climate is changing. There is a growing global consensus, which includes the scientific community and ordinary citizens like this female farmer, that our climate is undergoing a transformation that is bringing with it heightened extremes of temperature while rain fall patterns remain uncertain and intensity fluctuates. These changes will have impacts on global socio-ecosystems, threatening their sustainability.

These impacts will not fall evenly across the globe but are predicted to be differentiated across regions, generations, classes, income groups, and occupations, as well as between women and men. Indeed, it is considered that it will be people with low or no incomes living in the Global South who will be most affected, with Africa identified as the most vulnerable region.[2] Agriculture is regarded as a highly climate-sensitive sector. The potential impacts in Africa are thus a particular cause for concern due to its reliance on rain-fed agriculture, which supports the livelihoods of about sewenty per cent of Africans, contributes about thirty per cent of the continent's GDP and about fifty per cent of its total export value, and employs some 65 per cent of its labour force,[3] of which it is estimated that women constitute more than seventy per cent.[4]

This chapter uses a gender lens to gain insights into the ways in which rural communities in sub-Saharan Africa are dealing with the effects of a changing climate. There is a strong equity argument for taking a gender perspective on climate change, since this recognises the different contributions that women and men make to their household's sustainability. For instance, resources controlled by women tend to be invested (at the margin) more heavily in children than those controlled by men.[5] Evidence also shows that "the greater the degree of control exercised by women over the family income, the

---

1 Babugura, Agnes, *Gender and Climate Change: South Africa Case Study* (Cape Town: Heinrich Böll Foundation Southern Africa, 2010), 74.

2 Intergovernmental Panel on Climate Change (IPCC), *Climate Change 2007—Impacts, Adaptation and Vulnerability*. Contribution of Working Group II to the Fourth Assessment Report of the IPCC (Cambridge: Cambridge University Press, 2007), 976.

3 IPCC, Climate Change 2007.

4 Babugura, *Gender and Climate Change*.

5 World Bank, *Engendering Development: Through Gender Equality in Rights, Resources, and Voice*. A World Bank Policy Research Report (Washington DC: World Bank, 2001), 70.

greater the proportion of income spent on food."[6] There is some evidence to suggest that when women actively contribute to household income, they have more influence over the ways in which that income is spent. Interventions to support communities with coping strategies linked to climate change should ensure that women also have the opportunity to become involved in income-generating strategies.[7]

Given the role women play in ensuring their household's well-being and hence its sustainability, it is of potential significance to recognise that the impacts of variations in weather patterns are predicted to have a disproportionately greater effect on women than men. Therefore interventions to enable individuals and communities to adapt to a changing climate need to ensure that all affected persons receive the type of support that meets their needs and circumstances. In any given community, women are often poorer and less educated than men as well as being excluded from political and household decision-making processes that affect their lives. Women also tend to have a lower level of assets than men. An individual's assets determine the strategies that can be adopted to respond to changes in weather patterns. Migration may not be a suitable strategy for the elderly, the very young, those with limited resources, or those facing social or religious restrictions on their mobility.[8] This also reminds us that a 'community' is not a binary divide of 'women' and 'men', but that people have multiple identities, which combine gender with age, class, ethnicity, race, religion, etc. These characteristics also shape an individual's assets. If equitable interventions are not made, then climate change is predicted to increase inequalities, including those based on gender.[9] Not addressing the diversity of needs in a community threatens its sustainability.

This chapter is structured as follows: Section 2 describes the analytical approaches used to write the chapter; Section 3 defines the concepts of vulnerability and resilience, which currently shape thinking on how to respond to climate change; Section 4 makes some observations on the nature of the evidence; Section 5 briefly reviews, from a gender perspective, the evidence related to the impacts on and responses by communities to climate change in

6 Rahman Osmani, Siddiqur, *Food Security, Poverty and Women: Lessons from Rural Asia*, IFAD, 2010, accessed October 5, 2017. http://www.ifad.org/gender/thematic/rural/rural_2.htm.

7 Bäthge, Sandra, *Climate Change and Gender: Economic Empowerment of Women through Climate Mitigation and Adaptation?* (Eschborn: Deutsche Gesellschaft für Technische Zusammenarbeit GmbH (GTZ), 2010), 22.

8 Demetriades, Justina, and Emily Esplen, "The Gender Dimensions of Poverty and Climate Change Adaptation," *IDS Bulletin* 39.4 (2008): 24–31.

9 Dankelman, Irene, "Climate Change: Learning from Gender Analysis and Women's Experiences of Organising for Sustainable Development," *Gender & Development* 10.2 (2002), 21–29.

rural SSA; and Section 6 closes the chapter with some general observations on its findings.

2 Methodological Approach

This chapter uses a feminist political ecology (FPE) framework to review the evidence from the peer-reviewed and grey literature on the impacts that the gender–climate change nexus has on sustainable development in the context of rural areas in sub-Saharan Africa. Such an approach combines two powerful analytical tools to provide a more holistic understanding of the dynamics in rural areas that relate to vulnerability due to climate change and related threats to sustainability. First, political ecology recognises the relationships between the natural environment and political, economic, and social processes, as well as their impacts on sustainability. Rural women and men use the natural environment as their primary resource base for their goods and services. However, for rural people their relationship with the natural environment goes beyond its productive function. The natural environment is what Escobar calls the "space for being."[10] The natural world is central to one's identity, and the elements of nature (such as mountains, lakes, rivers, plants, animals, moon, sun, and stars) are endowed with cultural significance and/or spiritual and aesthetic value.[11] The 'space for being' is where the establishment and reinforcement of kinship-based social relationships take place, contributing to the strengthening of identity and culture and allowing for the development of social capital and ties of solidarity.[12] Changes in weather patterns bring changes to the natural environment threatening to undermine the sustainability of rural communities—not only materially, but also culturally.

FPE adds gender analysis to the toolbox. In particular, FPE brings a focus on the social dimensions of sustainability and, in so doing, contributes to a more holistic understanding of vulnerability and coping strategies for responding to

---

10 Escobar, A., "Whose Knowledge, Whose Nature? Biodiversity, Conservation, and the Political Ecology of Social Movements," *Journal of Political Ecology* 5 (1998), 53–82.

11 Ash, Neville, and Martin Jenkins, *Biodiversity and Poverty Reduction: The Importance of Biodiversity for Ecosystem Services* (Cambridge: UNEP-WCMC, 2007), 40; TEEB, *The Economics of Ecosystems and Biodiversity: Mainstreaming the Economics of Nature: A Synthesis of the Approach, Conclusions and Recommendations of TEEB* (London: Earthscan, 2010), 39.

12 Marin-Burgos, Victoria, "Social Environmental Conflicts and Agrofuel Crops: The Case of Oil Palm Expansion in Colombia," in *The Political Ecology of Agrofuels*, ed. Kristina Dietz, Bettina Engels, Oliver Pye, and Achim Brunnengräber (Abingdon: Routledge, 2015), 264.

climate change. The alterations in weather patterns and the increasing occurrence of extreme weather events bring environmental, political, and economic changes to livelihoods, ecosystems, property regimes, and social relations. In this context, FPE provides a framework to analyse the gendered experiences of and responses to these challenges.[13] FPE also draws attention to the point made in the opening section that uneven access to as well as distribution and control of resources depends not only on gender, but also on other social characteristics such as caste, class, and ethnicity.[14] Gender analysis reminds us that the use of gender-neutral terms such as 'community', 'people', or 'households' risks a simplistic analysis with an incomplete picture of the reality of the lives of women and men in rural areas.

FPE also sees women and men not only as passive victims of climate change, but also as agents of change. Attention can be drawn to local agency and creativity, demonstrating the ways in which women (in particular) are (re)defining their situations in response to climate change impacts, often in light of—or in relation to—significant constraints.[15] These constraints can be attributed, firstly, to the difference in asset levels between women and men and, secondly, to gendered power relations. Men often have a better level of and more control over assets such as access to land, credit, agricultural inputs, decision-making bodies, technology and training services, education, natural resources, mobility, equal economic opportunities, as well as information and communication systems. Access to a wide range of assets gives men more options to respond to climate change impacts. Because of their limited control over household resources, women rely on resources held in common property to fulfil their household tasks.[16] The changes that extreme weather events bring to the natural resource base increase women's vulnerability since they have more limited options than men in the same community.

The difference in asset levels has its roots in power relations. The feminist position is to reject the notion of the household as a unified entity pooling resources from which all individuals derive equal benefit. Instead, the household is seen as a place of negotiation in which women and men define their roles

---

13 Hovorka, Alice. J., "The No. 1 Ladies' Poultry Farm: A Feminist Political Ecology of Urban Agriculture in Botswana," *Gender, Place & Culture: A Journal of Feminist Geography* 13.3 (2006), 207–225.

14 Shepherd, Andrew et al., *The Geography of Poverty, Disasters and Climate Extremes in 2030* (London: Overseas Development Institute (ODI), 2013), 88.

15 Rocheleau, Dianne, Barbara Thomas-Slayter, and Esther Wangari, *Feminist Political Ecology* (London: Routledge, 1996), 289.

16 Rocheleau, Thomas-Slayter, and Wangari, *Feminist Political Ecology*.

and relations (mediated by informal and formal institutions). Within a household there can be both cooperation among and conflicts of interest between female and male members as well as competing priorities in relation to labour allocation and the distribution of resources.[17] The capacity to use these resources has important implications for individual outcomes. There are usually men in the household who make decisions about and have control over their own bodies, lives, and resources but also those of other family members. Control is a means to exercise power. The issue of power is a key concept in a political ecology framework drawing on Foucault's definition of 'power' as: "[...] the ability of an actor to control their own interaction with the environment and the interaction of other actors with the environment."[18]

This balance of power between men and women defines the relationship between them and operates at all levels of society: household, community, organisational, national, and international. Shifting the balance of power towards women by transforming gender relations will enable women to exercise agency,[19] namely the ability to make decisions—free of violence, retribution, or fear—about one's own life and to act on them in order to achieve a desired goal.

3 Vulnerability and Resilience

Current approaches to addressing the impacts of the extreme weather events linked to climate change include identifying which individuals and communities are most likely to suffer from these events and classifying them in terms of their vulnerability.[20] Vulnerability represents a threat to the existence of the individual, household, or community—as well as the ecosystems in which they are embedded. Building the resilience of these vulnerable individuals and communities, together with maintaining ecosystems, is seen as an appropriate coping strategy for ensuring their sustainability. Despite the widespread use

17 Social Development Department, Gender-Responsive Social Analysis: A Guidance Note. Incorporating Social Dimensions into Bank-Supported Projects (Washington, DC: The World Bank, 2005), 53.

18 Raymond L. Bryant, and Sinead Bailey, *Third World Political Ecology* (London: Routledge, 1997), 37.

19 Tacoli et al., Building Resilience to Environmental Change.

20 Field, Christopher B. et al., ed., Managing the Risks of Extreme Events and Disasters to Advance Climate Change Adaptation—A Special Report of Working Groups I and II of the Intergovernmental Panel on Climate Change (Cambridge: Cambridge University Press, 2012), 19.

of the concept of vulnerability, there is no universally shared consensus about its definition. In part, this reflects the fact that the study of 'climate change' is the focus of numerous scientific disciplines reflecting different epistemological orientations and methodological practices.[21] Here, however, I draw on the definition by Cardona:

> Vulnerability is an intrinsic predisposition to be affected by or to be susceptible to damage; that means vulnerability represents the system or the community's physical, economic, social or political susceptibility to damage as the result of a hazardous event of natural or anthropogenic origin.[22]

This definition fits in well with a feminist political ecology analysis, since it can be applied to both social and ecological systems. Interestingly, Cardona's definition does not equate vulnerability with poverty. This is somewhat at odds with the prevalent view in the literature that it is the poor in a community who are disproportionately affected by disasters, as their lower level of assets mean that they are the least able to rebuild their lives in the aftermath.[23] Recent research in Ethiopia has found that this may not necessarily be the case depending on the coping strategies a household adopts, such as selling assets and/or reducing consumption.[24]

Designing interventions to support an ecosystem, an individual, a household, or a community to survive shocks and significant disturbances, such as floods, requires an understanding of vulnerability. Again, despite the widespread use of the term, there is considerable variation in definitions thereof.[25]

---

21 Babugura, *Gender and Climate Change.*

22 Cardona, Omar D., "The Need for Rethinking the Concepts of Vulnerability and Risk from a Holistic Perspective: A Necessary Review and Criticism for Effective Risk Management," in *Mapping Vulnerability: Disasters, Development and People*, ed. Greg Bankoff, Georg Frerks, and Dorothea Hilhorst (London: Earthscan, 2004), cited in Jörn Birkmann, "Measuring Vulnerability to Promote Disaster-Resilient Societies: Conceptual Frameworks and Definitions," in *Measuring Vulnerability to Natural Hazards: Towards Disaster Resilient Societies*, ed. Jörn Birkmann (Tokyo: UNU Press, 2006), 12.

23 Béné, Christophe, Stephen Devereux, and Rachel Sabates-Wheeler, *Shocks and Social Protection in the Horn of Africa: Analysis from the Productive Safety Net Programme in Ethiopia*, IDS Working Paper 395 (Brighton: Institute of Development Studies, 2012), 120.

24 Béné, Devereux, and Sabates-Wheeler, *Shocks and Social Protection in the Horn of Africa.*

25 Béné, Christophe et al., *Resilience: New Utopia or New Tyranny? Reflection about the Potentials and Limits of the Concept of Resilience in Relation to Vulnerability-Reduction Programmes*, IDS Working Paper 405 (Brighton, UK: Institute of Development Studies, 2012), 61.

However, in reviewing this range of definitions, Vogel and O'Brien found agreement on three characteristics of vulnerability:[26]

1) multi-dimensional and differential (it varies across physical space as well as among and within social groups)
2) scale dependent (with regard to time, space, and units of analysis such as individuals, households, regions, or systems)
3) dynamic (its characteristics and driving forces change over time).

Weather patterns show annual fluctuations around a mean. In some seasons there can be more rain than in others, which may sometimes lead to floods. In other seasons there is less rain, which can result in shortages and drought. Floods and drought cause shocks to natural and human systems, potentially disrupting their functioning for a period of time before they return to normal. This idea that systems are able to 'bounce back' has led to the concept of 'resilience'. People and their communities, as well as the ecosystems they depend on, can deal with the threats posed by extreme weather events when they are resilient to shocks.[27] As with 'vulnerability', there is a lack of general agreement on a common definition for 'resilience', again a reflection that the concept is used across a range of scientific disciplines.[28] The Intergovernmental Panel on Climate Change (IPCC) uses a definition that is reflected—with slight modifications—in a number of other definitions of the term:

> The ability of a system and its component parts to anticipate, absorb, accommodate, or recover from the effects of a hazardous event in a timely and efficient manner, including through ensuring the preservation, restoration, or improvement of its essential basic structures and functions.[29]

Communities have a capacity for learning from past disasters to reduce their vulnerability to risk and ensure their sustainability. The degree to which a social system is capable of organising itself can be seen as an indicator of its

---

26 Vogel, Coleen, and Karen O'Brien, "Vulnerability and Global Environmental Change: Rhetoric and Reality," *AVISO* 13 (2004). Information Bulletin on Global Environmental Change and Human Security, pp8. http://www.ihdp.unu.edu/docs/Publications/GECHS/Aviso/Aviso13-2004.pdf.

27 Tompkins, Emma L., and W. Neil Adger, "Does Adaptive Management of Natural Resources Enhance Resilience to Climate Change?" *Ecology and Society* 9.2 (2004): 10.

28 Béné, Godfrey-Wood, Newsham, and Davies, *Resilience: New Utopia or New Tyranny?*

29 IPCC, "Glossary of Terms," in *Managing the Risks of Extreme Events and Disasters to Advance Climate Change Adaptation*, ed. C.B. Field et al. A Special Report of Working Groups I and II of the Intergovernmental Panel on Climate Change (IPCC). (Cambridge, New York: Cambridge University Press, 2012), 555–564.

resilience.[30] However, resilience needs to exist across or be built at different, interlinked scales: individual, household, community, region, and ecosystem. Here, the concept of 'scale' is not meant in a Euclidian sense, but in an actor-network theory sense of relational spheres of influence, power, and connectivity.[31] The political ecology approach fits in well with analysing vulnerability and resilience, since both are considered scale dependent.

4 The Nature of the Evidence

It has taken time for the interplay between gender and the impacts of climate change to be of interest to social science researchers. As a consequence, most of the available literature is self-published by international organisations, non-governmental organisations, private foundations, and aid and disaster relief organisations. There is a small but growing body of relevant peer-reviewed literature; nevertheless, the evidence currently available is limited, often anecdotal, patchy, varied, and highly contextual in nature. It also tends to be confined to small-scale qualitative case studies. These do, however, make an important contribution to our understanding of the ways in which women and men experience and understand the phenomenon of 'climate change'. They also help to underline the context-specific nature of these experiences, indicating that there is no 'one size fits all' solution even within the same region of a country.

There is limited sex-disaggregated data available on the actual differential experiences of women and men related to climate change impacts.[32] Indeed, it appears that the literature tends to focus on women's experiences of climate change—although many of the claims are based on a priori assumptions rather than on robust empirical evidence in support of these claims. There is even less data related to children and the elderly, who are considered particularly vulnerable to the health issues exacerbated by climate change (for example, malaria and diarrhoea affect children disproportionally, while elderly people suffer during heat waves).[33] This makes tracking increasing vulnerability and designing

30 Yamin, Farhana, Atiq Rahman, and Saleemul Huq, "Vulnerability, Adaptation and Climate Disasters: A Conceptual Overview," *IDS Bulletin* 36 (2005): 1–14.

31 Henderson et al., "Global Production Networks and the Analysis of Economic Development," *Review of International Political Economy* 9(2002): 4436–4464.

32 Bradshaw, Sarah, and Maureen Fordham, *Women, Girls and Disasters: A Review for UK's Department for International Development (DFID)* (London: DFID, 2013), 54.

33 Otzelberger, Agnes, *Gender-Responsive Strategies on Climate Change: Recent Progress and Ways Forward for Donors* (Brighton: BRIDGE, Institute of Development Studies, 2011), 58.

and targeting appropriate interventions to maintain and build resilience problematic. The 'household' appears to be the standard unit of measurement for development agencies. Taking a gender approach appears to go little beyond presenting data disaggregated by the sex of the household head. Blackden points out that the development literature has a tendency to use the shorthand term 'household' to cover whatever everything that exists between the level of the community and that of the individual.[34] In many settings, households can be seen to share common features (such as co-residence, joint production, shared consumption, and kinship links); however, anthropologists would caution that even within communities there are possibilities for diversity. Households are also dynamic: their composition changes over time, through natural life processes of birth, marriage, and death, but also due to the temporary relocation of their members for a range of reasons such as schooling or employment.

Treating the household as a unified entity fails to capture a more holistic understanding of what is happening between and within households. Female-headed households can be 'de facto' or 'de jure'—if the former are in receipt of remittances from male family members working elsewhere, such households could be in a stronger financial position than the latter. It is tempting to think that a woman at the head of a de jure female-headed household will make the decisions and, as is often presented in the literature, possess greater agency than women in male-headed households. However, this may well be a false assumption, since gender norms can dictate that decision-making responsibilities lie with the eldest son in the household.[35] There is also a temptation to assume that all female-headed households are poor; yet, the empirical evidence shows that this is not necessarily the case.[36]

In Africa, there are also households with no family members older than 18 years of age (referred to as 'child-headed households').[37] Again, it would be an easy assumption to make that these households are resource poor and therefore vulnerable. However, a study in a village in Limpopo province in

34 Blackden, C.M., *Literature Review on Intra-Household Resource Allocation*, ("Framing Paper for the Bill and Melinda Gates Foundation Convening on Intra-Household Resource Allocation," Washington, DC, September 2009).

35 Angula, Margaret, *Gender and Climate Change: Namibia Case Study* (Cape Town: Heinrich Böll Foundation Southern Africa, 2010), 50.

36 Chant, Sylvia, "Re-thinking the 'Feminization of Poverty' in Relation to Aggregate Gender Indices," *Journal of Human Development* 7.2 (2006): 201–220.

37 There is a lack of data on the actual number of 'child-headed households' in Africa. UNICEF estimates that in South Africa there are 150,000 children living in homes with no adults. UNICEF, "Orphans and Vulnerable Children," accessed November 12, 2017. https://www.unicef.org/southafrica/protection_6631.html.

South Africa found that the child-headed households were the least vulnerable in the village. The explanation lay in these households being de facto child-headed households rather than de jure ones. The parents were working elsewhere and sending remittances back to support their children, preferring that they remained in their home village to complete their education.[38]

There are climate change interventions that claim to have taken gender into account, but there have been criticisms of a number of these projects. First, there can be a tendency for 'gender' and 'women' to be used interchangeably in project documents with a focus on roles and assets. Second, women are treated as a homogenous group of passive victims. This 'victimisation' of women also neglects women's capacity for responding to change. It also neglects the underlying cause of why women's capacity in this area is constrained: gender relations.

The next section gives a brief overview of the evidence, from a gender perspective, about how rural communities in SSA are adapting to climate change and ensuring their sustainability. The available evidence on gendered impacts is limited and has focused on sectors that are obviously climate-sensitive and linked to natural resources (food security and agriculture, forestry, and water). There is a wide knowledge gap concerning those sectors where the gendered impacts of climate change are less tangible but play an important role in building resilience (for example, in transport and infrastructure, energy access, housing, and formal and informal employment).[39]

5 Ensuring Community Sustainability: Adaptation to Climate Change

The sustainability of communities in the face of extreme weather events can depend on their capacity to adapt to climate change. The IPCC defines adaptation as the "adjustment in natural or human systems in response to actual or expected climatic stimuli or their effects, which moderates harm or exploits beneficial opportunities." The capacity of people to adapt depends on their possessing of a range of assets such as wealth, technology, education, information, skills, infrastructure, access to resources, and management capabilities.[40]

38 Vincent, Katharine, Tracy Cull, and Emma R. M. Archer, "Gendered Vulnerability to Climate Change in Limpopo Province, South Africa," in *Gender and Climate Change: An Introduction*, ed. Irene Dankelman (London, Sterling: Earthscan, 2010), 284.

39 Otzelberger, Gender-Responsive Strategies on Climate Change.

40 Intergovernmental Panel on Climate Change (IPCC), *Climate Change 2001: Impacts, Adaptation and Vulnerability, Summary for Policymakers* (a report of Working Group II

The importance of having access to such a wide array of assets will disadvantage women compared to men from the same socio-economic group, since women will usually have a lower level of assets. Therefore, we can assume that this difference in access to assets is likely to result in a gender-differentiated capacity to respond to climate change. This means that interventions to build resilience will need to include gender-responsive coping strategies. Experience shows that to ensure that both women's and men's concerns are addressed by community-level adaptation projects, gender needs to be mainstreamed at the beginning of the project cycle.[41]

What follows in this section are, first, some general observations about how rural communities in SSA perceive climate change. Then some evidence is presented from a gender perspective in terms of impacts and coping strategies in SSA in relation to five dimensions of rural life (health, agriculture, water, energy, and non-farm employment) as well as to climate-related disasters. The available empirical evidence, as pointed out above, is limited and betrays a tendency to predict what may happen in the future based on what has happened in natural disasters in the past. These predictions rely on mathematical models, the accuracy of which is dependent on the quality of the available data (that is, how accurately the data describes the event to be modelled). In the context of climate change, model predictions are on shaky ground, since projections of future climate change relate to a state never observed before—so there is nothing to test the model outputs against to validate their accuracy.[42] Scientists working in these fields are well aware of the limitations of their work and recognise that "past experience may not be a good guide for the future."[43] Decision makers may demand greater accuracy, which comes at a cost and may only be possible at some unspecified point in the future. As a result, judgement on dealing with uncertainty is required—in this case, on how to adapt to unpredictable and changing weather patterns. As Dessai et al. point out planning does not necessarily depend alone on scientists being

of the Intergovernmental Panel on Climate Change), 2011, 18, accessed January 6, 2013. http://www.ipcc.ch/pdf/climate-changes-2001/synthesis-syr/english/wg2-summary-policymakers.pdf.

41 UNDP, *Gender, Climate Change and Community-Based Adaptation* (New York: UNDP, 2010), 80.

42 Knutti, Reto, "Should We Believe Model Predictions of Future Climate Change?" *Phil. Trans. R. Soc. A* 366 (2008): 4647–4664. doi:10.1098/rsta.2008.0169.

43 Stainforth, D.A., Allen, M.R., Tredger, E.R., and Smith, L.A., "Confidence, Uncertainty and Decision-Support Relevance in Climate Predictions," *Phil. Trans. R. Soc. A* 365 (2007): 2145–2161, here 2152. doi:10.1098/rsta.2007.2074.

able to accurately predict changes in climate.[44] Outcomes are dependent not only on the climate but also on many other variables (such as globalisation and migration) that cannot be easily adapted to modelling; yet, we do adapt to these processes, albeit not always smoothly or without difficulty. We also live with the threat of other natural events that are hard to predict, such as earthquakes and tsunamis. The response to the former has been to construct buildings with appropriate materials and design features in order to withstand shocks, while in response to the latter early warning systems to facilitate the evacuation of people to safer locations have been installed. My comments here are not intended to denigrate the work of climate scientists, who have done much valuable research to point out firstly, that the climate is changing at an unprecedented rate and, secondly, to prove using robust evidence that the cause thereof is anthropogenic. This has been conducted against a background of hostility from economic and political interests connected to the fossil fuel industry. However, science policy needs to look at where to strike the balance between long-term accuracy and short- to medium-term adaptation measures.

There is more than enough evidence to suggest that rural African communities are aware of changes to the climate, and the quotation which opens this chapter serves as a good illustration thereof. Rural people recognise how these changes influence the weather, which in turn impacts on agriculture,[45] often with recollections showing good correlation with meteorological records. Oral histories enable rural communities to assess 'catastrophic events' such as droughts and to make judgements about changes over time. Rural people's awareness about changes in weather patterns and the corresponding links to the scientific concept of climate change show variations in relation to gender and age. Men (particularly young men) are more likely to be aware of climate change than women. There is evidence to suggest that there is gender differentiation in the ways in which women and men receive information about

---

44 Dessai, Suraje et al., "Climate Prediction: A Limit to Adaptation?" in *Adapting to Climate Change: Thresholds, Values, Governance*, ed. W. Neil Adger, Irene Lorenzoni, and Karen L. O'Brien (Cambridge: Cambridge University Press, 2009), 514.

45 See, for example, Goh, Amelia H.X., *A Literature Review of the Gender-Differentiated Impacts of Climate Change on Women's and Men's Assets and Well-Being in Developing Countries*. CAPRi Working Paper No. 106 (Washington, D.C.: International Food Policy Research Institute, 2012), 43. http://dx.doi.org/10.2499/CAPRiWP106; Swai, Okuli W., Jonathan S. Mbwambo, and Flavianus T. Magayane, "Gender and Adaptation Practices to the Effects of Climate Change in Bahi and Kondoa Districts Dodoma Region, Tanzania," Journal of Sustainable Development 5.12 (2012): 65–77.

changes in weather patterns, which can also be culturally mediated.[46] A study in South Africa found that men had time to listen to radio broadcasts, whereas women preferred to attend meetings with agricultural offices, in part because this enabled them to ask questions. However, in Burkina Faso women were not invited to workshops by male community leaders, since this would reduce the number of places available for men.[47]

5.1 *Health*

The impacts on health arising from climate change are varied. Some are the direct result of high temperatures, floods, or droughts, whereas others are more indirect and act in combination with further factors (that may or may not be induced by climate change). The predicted health effects include an increase in waterborne diseases, potentially higher rates of malnutrition due to food shortages, and increases in heat-related mortality and morbidity.[48] There is very limited epidemiological evidence that relates gender to the effects of climate change and even less does so specifically in relation to SSA. The World Health Organisation (WHO) reports in one of the few studies on gender, health, and natural disasters that the mortality of women and girls due to natural disasters is higher than that for men in countries where there is significant gender disparity in economic, social, and political status which cannot be attributed to biological factors but to socially constructed roles and relations that contribute to gendered vulnerability.[49]

The need for more nuanced analysis of communities is underlined not only by gender differences, but also by age and life cycle status as factors relevant to health. Pregnant women, for example, may be more susceptible to malaria and other complications related to high humidity levels and temperatures, possibly depending on their nutritional status.

Care for the sick within a household is generally a woman's task, which adds to the daily workload. Women use the natural environment as a source of

---

46 Twyman, Jennifer et al., *Adaptation Actions in Africa: Evidence that Gender Matters*. CGIAR Research Program on Climate Change, Agriculture and Food Security (CCAFS). Working Paper No. 83 (Copenhagen: CCAFS, 2014), 36.

47 Goh, *Gender-Differentiated Impacts of Climate Change*.

48 Brody, Alyson, Justina Demetriades, and Emily Esplen, *Gender and Climate Change: Mapping the Linkages—A Scoping Study on Knowledge and Gaps* (Brighton: BRIDGE, Institute of Development Studies, 2008), 27.

49 Neumayer Eric, and Thomas Plümper, "The Gendered Nature of Natural Disasters: The Impact of Catastrophic Events on the Gender Gap in Life Expectancy, 1981–2002," *Annals of the Association of American Geographers*, 97 (2007): 551–566, cited in: WHO, *Gender, Climate Change and Health* (Geneva: World Health Organisation, 2014), 44.

medicinal plants. When biodiversity degrades, the capacity to treat illness for little financial outlay also reduces. Rural communities in Nigeria now report on the need to go to clinics to seek treatment rather than attempt to see to their own medication.[50]

A less acknowledged aspect of the interplay between health and climate change are psychological impacts. In her study of three South African rural communities, Babugura found that the changes to the natural environment brought about by shifting weather patterns had psychological impacts on the women and men she interviewed.[51] The men felt ashamed that they could not provide for their families and guilty about the additional workload this placed on the women. Women felt guilty about the extra time that they spent away from their children, such as when collecting water, and reported being tired when they came home. There has been a clear shift in gender roles, with men assuming responsibility for tasks that were traditionally regarded as 'women's work', for instance growing vegetables in home gardens and collecting fuelwood. How these changes in gender roles will evolve in terms of masculine and feminine identity and behaviour is yet unknown.

5.2 *Food Security*

An individual can be considered to be living in food security when he or she has, at all times, physical and economic access to sufficient, safe, and nutritious food that enables him or her to live a healthy and active life.[52] Food security promoting good health not only removes the overwhelming reason why households move into poverty—namely illness[53]—but also helps to ensure household resilience to the impacts of climate change.

Agriculture in SSA is predominantly the domain of the smallholder (farming three hectares or less of cropland), with approximately 73 per cent of the rural population classified by the International Fund for Agricultural Development (IFAD) as being engaged in farming.[54] Nevertheless, there is a large gender difference in land ownership, although the extent of this imbalance

---

50 Agwu, Julia, and Appollonia A. *Okhimamhe, Gender and Climate Change in Nigeria: A Study of Four Communities in North-Central and South-Eastern Nigeria.* (Ikoyi, Lagos: Heinrich Böll Foundation, Nigeria, 2009), cited in: Goh, *Gender-Differentiated Impacts of Climate Change.*

51 Babugura, *Gender and Climate Change.*

52 World Food Summit, FAO, Rome 1996.

53 CPRC, *The Chronic Poverty Report 2008–09: Escaping Poverty Traps* (Manchester: The Chronic Poverty Research Centre (CPRC), 2008), 164.

54 John F. Morton, "The Impact of Climate Change on Smallholder and Subsistence Agriculture," *PNAS* 104.50 (2007): 19680–19685, here 19680.

varies between countries. In Africa, an average of 15 per cent of landholders are women—ranging from less than five per cent in Mali to over thirty per cent in Botswana, Cape Verde, and Malawi.[55]

Rural women and men play complementary roles in guaranteeing household food security.[56] Women grow vegetables for their family's own consumption and to sell at local markets. Women are also responsible for raising small livestock. Men are generally responsible for growing cash crops and tending larger livestock. Nevertheless, reliance on rain-fed agriculture on small parcels of land is precarious even under good weather conditions. To ensure that a household is food secure requires the input of labour, water, and agricultural resources. Food security is also threatened by a lack of options for keeping food safe for consumption after harvest or slaughter.

Extreme weather events linked to climate change can lead to crop failure and the death of livestock. This can entail reduced food intake, both in terms of quantity and quality, which affects women and men's capacity to work and adopt coping strategies. A study in northern Tanzania found that women would go without food for a day or two.[57] Lwando reports similar findings from western Zambia.[58] There is some evidence to suggest that while adults recuperate from short-term loss of food intake, children's physical development may be affected, with this effect more pronounced in poorer households.[59] During the severe 1994/1995 drought in Zimbabwe, women's body mass was found to be negatively affected whereas men's was unaffected, suggesting that women had a reduced calorie intake while men did not. However, this did not apply to all women. Daughters-in-law were found to be unaffected.[60]

A study in Burkina Faso found that while women had developed management strategies for dealing with food shortages during periods of drought, there was conflict within households over food allocation, not only between women and men but also between generations. In Namibia, meanwhile, gender roles mean that dealing with food shortages is not a matter for negotiation; instead, women have the sole responsibility for finding solutions, which can

55 Senaya Habtezion, *Gender, Agriculture and Food Security*. Gender and Climate Change Capacity Development Series (New York: UNDP, 2012), 34.

56 FAO, *Gender and Development Plan of Action 2002–2007* (Rome: Food and Agriculture Organisation of the United Nations, 2003), 4.

57 Swai et al., "Gender and Adaptation Practices."

58 Lwando, Catherine, "Climate Variability and Gender: Emerging Experiences from Western Zambia," *Environment and Natural Resources Research* 3.4 (2013): 133–146.

59 Goh, *Gender-Differentiated Impacts of Climate Change.*

60 Goh, *Gender-Differentiated Impacts of Climate Change.*

range from selling the small livestock for which they are responsible to selling handicrafts.[61]

Women and men respond to changes in weather patterns by diversifying their crops, a strategy that requires additional investment in time. This adds to the burdens facing women in particular, who have to synchronise working with unfamiliar crops with their other household responsibilities. Gender differentiation in ownership of farming tools can also lead to women being disadvantaged in their climate change coping strategies. In northern Tanzania, both women and men use manure for crop fertiliser and as a soil stabiliser. However, when men own machines, they can use these to transport manure to their fields several kilometres away, whereas women have to carry the manure themselves.[62] In eastern Uganda, female-headed households were found to be less likely to implement climate change adaptation strategies.[63] Reasons cited included a lack of credit, time, and relevant information. However, this is contrary to findings from the Nile basin in Ethiopia, where female-headed households were found to be more likely than male-headed households to invest in climate change adaptation measures.[64] There is also evidence from around Lake Faguibine in Mali that there are changes to land access restricting women, in particular, from agriculture where communities from the northern shore are cultivating land under a shared cropping system around the former lake's southeast or close to the Niger River. This represents a commodification of land with associated power struggles over access by men.[65]

Female smallholders in Malawi are reported to have adopted a climate change adaptation strategy that harnesses shifting rainfall patterns to produce two annual maize harvests.[66] Younger women in Namibia have been experimenting with new varieties of millet that are more drought resistant, although older women reject these in favour of traditional varieties.[67] Men in northern

---

61 Angula, *Gender and Climate Change.*

62 Swai et al., "Gender and Adaptation Practices."

63 Nabikolo, D., B. Bashaasha, M.N. Mangheni, and J.G.M. Majaliwa, "Determinants of Climate Change Adaptation among Male and Female Headed Farm Households in Eastern Uganda," *African Crop Science Journal*, 20.2 (2012): 203–212.

64 Nhemachena, C., and Hassan, R.M., *Microlevel Analysis of Farmers' Adaptation to Climate Change in South Africa*, IFPRI Discussion Paper 714 (Washington, DC: International Food Policy Research Institute, 2008), cited in Nabikolo et al., "Determinants of Climate Change Adaptation."

65 Djoudi, Houria, and Brockhaus, M., "Is Adaptation to Climate Change Gender Neutral? Lessons from Communities Dependent on Livestock and Forests in Northern Mali," *International Forestry Review* 13.2 (2011), 123–135.

66 Rodenberg, Birte, *Climate Change Adaptation from a Gender Perspective* (Bonn: German Development Institute, 2009), 66.

67 Swai et al., "Gender and Adaptation Practices."

Tanzania have adopted methods to reduce water run-off and control erosion, such as building contours and ridge farming.[68]

5.3 *Water*

It is estimated that up to six hundred million people in Africa live in conditions of water scarcity.[69] Providing water for a household's drinking, cooking, washing, and sanitation needs is usually the responsibility of women and girls. When rainfall decreases and water scarcity increases, rural women and girls have to spend more time fetching water. Women in Inhambane province, Mozambique, report that they now walk for six hours per day to collect water, whereas before the current drought they would spend two hours completing the same task. Men also become involved when distance increases the need for wheeled transport such as bicycles and wheelbarrows. Households also have to pay for water, thereby diminishing their financial assets.[70] Boreholes need to be drilled deeper, wherefore women may have to haul water up from depths of up to thirty metres.[71] Women engage in household water conservation techniques, such as reducing bathing frequency. In Ghana, traditional cultural practice can lead women to prioritise their husband's water needs over their own.[72]

Female time poverty due to searching for water can be addressed by improvements in water supply infrastructure, although this also requires a reliable energy supply. It is notable that women in the one community in Babugura's study that had a pumped water supply had far more time available to invest in other coping strategies than women in the other two communities.[73] Women in urban areas fare no better than women in rural areas, with interruptions to water supply resulting in women being forced to queue for long periods.

---

68 Nelson, Valerie, and Tanya Stathers, "Resilience, Power, Culture, and Climate: A Case Study from Semi-Arid Tanzania, and New Research Directions," *Gender & Development* 17.1 (2009): 81–94.

69 UNDP, *Fighting Climate Change: Human Solidarity in a Divided World*, Human Development Report 2007/2008 (New York: UNDP, 2007), 31.

70 Swai et al., "Gender and Adaptation Practices."

71 WEDO (Women's Environment and Development Organization), Abantu for Development, ActionAid Bangladesh, and ENDA (Environmental Development Action in the Third World), *Gender, Climate Change and Human Security: Lessons from Bangladesh, Ghana and Senegal* (report prepared for Hellenic Ministry of Foreign Affairs (MFA) and the Hellenic Foundation for European and Foreign Policy ELIAMEP, 2008), 73.

72 WHO Gender, Climate Change, and Health.

73 Babugura, *Gender and Climate Change.*

A lack of water also impacts on livestock, which are an important part of a household's assets. Men, who are usually responsible for large livestock, will migrate with their cattle in search of water and fodder.[74] Other coping strategies are emerging in the Sahel, where in some places livestock composition is switching from cattle to small ruminants, which have different fodder demands and higher dehydration tolerance levels and are considered most well-adapted to drought conditions.[75] The authors do not state whether this has added to women's workloads, since they are usually responsible for small animals.

5.4 *Energy*

In rural areas of SSA, fuelwood is still the most commonly used energy source for household needs, especially cooking and water boiling, two end uses that are key for maintaining good health. In 21 of 38 countries in sub-Saharan Africa with available data, over ninety per cent of households still cook with solid fuels.[76] Degradation of the ecosystem—for example, through forest clearing to create more land on which to plant a greater area of crops to offset crop losses arising from changed weather patterns[77]—reduces the available biomass. This, in turn, results in similar behavioural changes to water collection: women walk further, often assisted by their daughters (who may be withdrawn from school),[78] while men start to get involved. Swai et al. report in their study on northern Tanzania that male involvement in fuelwood collection began in the 1990s.[79] Men are able to make use of mechanised transport to travel further distances and collect large stores of wood. However, at least in the case of the villages in the northern Tanzanian study, this was leading to the felling of trees, thereby contributing to the undermining of the sustainability of the natural resource base. There was also evidence in these villages of residents switching to poorer quality biomass such as maize cobs to compensate for the reduced availability of fuelwood; nevertheless, this may only be a short-term strategy, since households had also started planting trees within their homesteads to provide fuelwood.

---

74 Angula, *Gender and Climate Change*; Swai et al., "Gender and Adaptation Practices."

75 Djoudi, and Brockhaus, "Is Adaptation to Climate Change Gender Neutral?"

76 UN DESA, The World's Women 2010: Trends and Statistics (New York: UN DESA, 2010), 284.

77 Lwando, "Climate Variability and Gender."

78 Fischer, Amy, *Hope Dries Up? Women and Girls Coping with Drought and Climate Change in Mozambique*. Report prepared for CARE International, accessed October 20, 2017. http://careclimatechange.org/wp-content/uploads/2016/11/El_Nino_Mozambique_Report_final.pdf.

79 Swai et al., "Gender and Adaptation Practices."

Tree planting cannot be taken as an automatic means of ensuring future fuelwood supplies since land and tree ownership are mediated by gender norms. In the Kakamega area of Kenya, men own the village land under traditional law, thereby granting them rights over what is planted on the land; hence, even during a fuelwood shortage, women would not plant trees they would not have control over when and how the wood would be used.[80] However, in a more recent survey conducted in Nyando and Wote in Kenya, where women would traditionally not have had access to or control over trees, both men and women report being involved in tree planting as part of their coping strategies.[81]

Improved cook stoves are seen as a technical solution to lack of fuelwood availability, albeit one with multiple potential outcomes. Increased stove efficiency may reduce the demand for fuelwood, thereby potentially saving time otherwise spent on wood collection; nevertheless, evidence does show that, even in areas of fuelwood stress, collection does not cease altogether.[82] Improved cook stoves are also considered to reduce indoor air pollution related to cooking with biomass. A broad measure of the burden of a disease is the concept of 'disability-adjusted life years' (known as DALYs), which estimates the years of life lost as a result of premature death caused by a disease as well as the years lived with that disease. Of the approximately forty million DALYs globally that are attributable to solid fuel use (including coal), 44 per cent (approximately 18 million) occur in sub-Saharan Africa.[83] Gender disaggregation of this data shows a rather unexpected result. In sub-Saharan Africa, the mortality rate linked to chronic obstructive pulmonary disease (COPD) is 3.2 times higher for men than for women despite the incidence of COPD being 1.8 times higher for women than for men.[84] The explanation offered is the higher prevalence of smoking among men, which results in greater underlying death rates from lung cancer for men. Moreover, it is not only women but also men who are exposed to indoor air pollution (IAP), since families often spend some time together in the kitchen while food is being prepared.[85] Another gender

---

80 Bradley, P.N., *Woodfuel, Women and Woodlots: The Foundations of a Woodfuel Development Strategy for East Africa* (London: Macmillan, 1991), 352.

81 Twyman et al., *Adaptation Actions in Africa.*

82 Masera, Omar R., Barbara D. Saatkamp, and Daniel D. Kammen, "From Linear Fuel Switching to Multiple Cooking Strategies: A Critique and Alternative to the Energy Ladder Model," *World Development* 28.12 (2000): 2083–2103.

83 UNDP, and WHO, *The Energy Access Situation in Developing Countries* (New York: UNDP, and WHO, 2009), 142.

84 Ibid.

85 World Bank, *State of the Clean Cooking Energy Sector in Sub-Saharan Africa* (Washington DC: World Bank, 2012), 52.

issue linked to stoves is that it is men who decide about the acquisition of new household equipment, including for the kitchen.[86]

Energy access, particularly to electricity, can contribute towards building resilience as an input into water purification, increasing food yields, providing better processing and storage, and cooling to reduce heat stress. Grid electricity barely reaches rural people in SSA; in 2014 the average access to the electrical grid in SSA was 18 per cent.[87] While electricity can be expensive for some forms of cooking that require long periods of simmering, it can be a popular means of boiling water for tea and coffee and of powering rice cookers. Electricity can make important contributions to fostering healthy populations by providing refrigeration for food storage and, in rural healthcare centres, for storing vaccines and medicines as well as providing good lighting making the working environments safer for delivering babies and undertaking minor surgeries. Africa has a largely untapped hydropower potential, which is now threatened by increasingly common extended periods of drought. Existing dams are also suffering from silting reducing their power output.[88]

5.5 *Non-Farm Income Generation*

Being able to earn an income forms an important part of coping strategies. Both women and men can be involved in income generation—who within a household varies from context to context. For women, however, income generation may not be an automatic option, since gender relations can restrict female agency. In the Democratic Republic of the Congo (DRC), for example, only 18 per cent of small businesses are run by women, whereas in Rwanda women account for the running of more than 41 per cent. In the former, a husband's permission is required to start a business, while in the latter it is not.[89]

Rural women's livelihoods are drawn primarily from natural resources. Negative impacts on ecosystems reduce women's income from natural produce.

86 Clancy, Joy S., Tanja Winther, Margaret N. Matinga, and Sheila Oparaocha, *Gender Equity in Access to and Benefits from Modern Energy and Improved Energy Technologies.* Background Paper World Development Report 2012, (Leusden: ENERGIA/Norad/World Bank, 2011), 44.

87 OECD/IEA, *World Energy Outlook 2016: Energy and Air Pollution* (Paris: International Energy Agency, 2016), 266.

88 Karekezi, Stephen, John Kimani, and Oscar Onguru, *Climate Change and Energy Security in East Africa.* Policy Paper on Energy, Climate Change and Poverty Alleviation, Global Network on Energy for Sustainable Development (GNESD) (Nairobi: AFREPREN/FWD, 2009); see also McCann in this volume.

89 World Bank, FAO, and IFAD, *Gender in Agriculture Source Book* (Washington: World Bank, 2009), 792.

Women in western Zambia report that the harvest of wild fruits from forests, which provide both a source of income and a supplementary food source, has diminished.[90] Loss of income feeds through into the hindering of other coping strategies, such as purchasing kerosene or liquefied petroleum gas (LPG) as a substitute for collecting fuelwood. Climate change also impacts on fishing, a livelihood often involving both men (as fishermen) and women (as processors and traders). The sandy shoreline along Senegal's Atlantic coast is reported to be being eroded at a rate of 1.25 to 1.30 metres per year, with the loss of infrastructure undermining an important income source.[91] In Namibia, changes to ocean currents are predicted to affect marine ecosystems, thereby reducing fish stocks.[92]

Social class influences women's access to income generation opportunities. Around Lake Faguibine in Mali, for example, Iklan women have adopted charcoal production as a coping strategy, which is rejected by Illelan women as an activity not commensurate with their social position. The Iklan women, despite coming from a group with fewer assets than the Illelan, thus have more opportunity for income generation, since they face fewer socially imposed restrictions on their mobility.

In urban areas, the informal sector is a place of refuge in times of crisis.[93] It is not unreasonable to assume that this is a possible coping strategy for rural migrants in response to the impacts of climate change. Food processing or selling prepared food are popular options for informal sector businesses, particularly for women, since this type of enterprise usually makes use of household equipment and thus requires little or no initial investment.

Social norms of allowing men greater mobility enable them to adopt migration to urban areas as a coping strategy. Men's generally higher level of assets than women means that they are usually able to find new income sources, although they may not accumulate sufficient capital after paying for rent and food and to send remittances home. Men from rural communities can choose to search for casual work while women engage in a range of activities that can be carried out from the home, such as petty trading. Male migration can also leave women to take over men's tasks. For the elderly as well as for widows, however, the potential for earning an income is limited, resulting in the need to beg for food. Where social protection measures are in place, the need for

---

90 Lwando, "Climate Variability and Gender."

91 WEDO et al., Gender, Climate Change and Human Security.

92 Angula, *Gender and Climate Change*.

93 Heltberg, Rasmus, Naomi Hossain, and Anna Reva, ed., *Living through Crises: How the Food, Fuel, and Financial Shocks Affect the Poor* (Washington DC: World Bank, 2012), 306.

begging can be reduced, for example in Namibia, where relief is available for those affected by disasters.[94]

5.6 *Climate-Related Disasters*

Extreme weather events can lead to the destruction of ecosystems, which in turn threatens livelihoods that draw on natural resources. Degradation of the natural vegetation can lead to armed conflicts, which can lead to fatalities among male family members, creating women-headed households with reduced assets. Incidents of this kind are reported among cattle herders in northern Kenya.[95]

Degradation of natural resources deepens female time poverty by extending the distances women walk in search of clean water and fuel. This increased time poverty has consequences for their health and well-being, since women have less time to rest and recuperate. Their daughters may also be withdrawn from school to help. Women may prioritise re-building the family home over participating in initiatives to revitalise communities, which could also mean a lost opportunity to develop skills and knowledge that are crucial for building up resilience.[96]

As in the aftermath of any disaster situation, women have two options, both of which can increase their vulnerability: either to live in displaced persons camps or to stay put. Reasons for remaining appear complex, including possible social constraints, such as requiring a male relative to act as an escort outside the home or a social obligation to look after a less mobile family member.[97]

The provision of energy carriers unlike that of water and shelter has been a neglected commodity for relief agencies. Having to search in unfamiliar surroundings for fuelwood can result in women being subjected to sexual harassment.[98] If their husbands and sons respond by stepping in and taking over fuelwood collection, they in turn can also become victims of physical assault.

---

94 Angula, *Gender and Climate Change*.

95 Omolo, Nancy A., "Gender and Climate Change-Induced Conflict in Pastoral Communities: Case Study of Turkana in North-Western Kenya," African Journal on Conflict Resolution 10.2 (2010): 81–102. Accessed January 1, 2017. http://www.ajol.info/index.php/ajcr/article/viewFile/63312/51195, cited in Goh, *Gender-Differentiated Impacts of Climate Change*.

96 Demetriades, and Esplen, *Gender Dimensions*.

97 Bradshaw, Sarah, and Maureen Fordham, *Women, Girls and Disasters*.

98 Kasirye, Birikadde G., Magi Matinga, and Joy Clancy, "Fuel Security and Supply Dynamics in Internally Displaced Persons' Camps of Northern Uganda," *Journal of Humanitarian Assistance*, accessed October 20, 2017. http://sites.tufts.edu/jha/archives/462.

Social perceptions of gender roles also play a part in determining strategies to respond to hazards. Men are expected to be proactive, but in disasters men's roles and responsibilities can expose them to dangers that can result in their untimely death and their wives being left as head of the household. Widowhood is one of the quickest routes into poverty.[99] On the other hand, more women are victims of flooding than men often due to social barriers that prevent women from learning how to swim and that place restrictions on their movement outside the home.[100] This creates a new category of households about which we have little understanding: male-headed households without adult women.

6 Where Are We Now? Where Are We Going?

As stated above, the nature of the evidence discussed in this chapter is limited and disparate, which makes it challenging to draw any strong conclusions and to effectively compare and contrast different case studies. Nevertheless, the existing evidence does give some useful indicators. In short, these include the fact that there are gender-differentiated climate change impacts, degrees of vulnerability, and levels of resilience that respectively threaten or strengthen the sustainability of households and communities. However, these impacts are not always rigid, straightforward, or predictable, but rather varying with context and may be mediated by a host of other sociocultural, economic, ecological, and/or political and institutional factors. Linear assumptions and conclusions, such as that higher socio-economic classes have more assets and therefore possess greater adaptive capacity, need to be avoided, since they can lead to biased vulnerability assessments of different groups and inappropriate interventions. Vulnerability and resilience are influenced by the intersectionality of gender with characteristics such as class, ethnicity, and age/life course. Likewise, it would be dangerous to assume that women are always more negatively affected than men from the same group in the same situation. Nor should women be considered a homogeneous group of victims of climate change. First, not all women are victims. Second, such a portrayal of women overlooks their agency to be part of the solution of building resilience. However, as gender analysis shows, women's agency can be mediated by gender power relations which exist at all levels of society. While much is made in the literature about women's

99 Narayan, Deepa, *Can Anyone Hear Us? Voices from 47 Countries*, vol. 1 (Washington DC: The World Bank, 1999), 280.

100 UNDP, *Gender, Climate Change and Community-Based Adaptation.*

lack of voice in decision-making at all levels, there are signs that this is beginning to change even within traditional power structures in Africa. The Balete people of Botswana, for instance, chose their first female paramount chief in 2003.[101]

The evidence presented here indicates that there is a need for a more nuanced approach to understanding vulnerability and resilience in rural SSA. Men are also vulnerable in ways that are socially constructed. For example, the role of having to provide food for the family in times of drought or flooding can mean extra stress, possibly leading to negative behavioural changes such as increased gambling or increased consumption of alcohol.[102] Society generally expects men to be proactive in the face of adversity, but during a disaster male roles and responsibilities can expose men to dangers which lead to loss of life and widowhood for their wives—a well-recognised route into poverty. Men and boys can also be victims; moreover, they have to be part of the solution.[103] Based on research in the Morogoro region in Tanzania, Van Aest concluded that disaggregation into male- and female-headed households is too simplistic and leads to overgeneralisations. The status of the woman at the head of a household (single, married, divorced, or widowed) was particularly significant in determining its capacity for livelihood diversification and access to water resources for irrigation.[104]

There is evidence of social change induced by the impacts of changes in weather patterns. For example, households in Mozambique are using bride price to raise cash,[105] whereas evidence from Zimbabwe suggests that rising cattle prices due to livestock deaths from drought is influencing men's decisions about delaying marriage.[106] There are other signs of behavioural change among men as a consequence of responses to changing weather patterns. When this leads to shifts in gender roles and a more equitable share

---

101 Babugura, *Gender and Climate Change*.

102 Blaikie, Piers, Terry Cannon, Ian Davis, and Ben Wisner, *At Risk: Natural Hazards, People's Vulnerability and Disasters* (London: Routledge, 1994), cited in Valerie Nelson et al., "Uncertain Predictions, Invisible Impacts, and the Need to Mainstream Gender in Climate Change Adaptations," *Gender & Development* 10.2 (2002): 51–59.

103 MenEngage Global Alliance, *Men, Masculinities and Climate Change: A Discussion Paper* (undated), 14, accessed December 30, 2016. http://www.menengage.org.

104 van Aelst, Katrien, "Gender, Households and Climate Change: Adaptation Decision-Making in the Morogoro Region of Tanzania" (PhD diss., University of Antwerp, 2016), 319.

105 Fischer, *Hope Dries Up?*

106 Hoogeveen, Johannes, Bas van der Klaauw, and Gijsbert van Lomwel, *On the Timing of Marriage, Cattle and Weather Shocks* (Amsterdam: Tinbergen Institute, 2004), cited in Goh, *Gender-Differentiated Impacts of Climate Change*.

of household tasks, such changes are to be welcomed. However, changes to practical arrangements within a household in relation to 'who does what' also have psychological impacts including challenging an individual's constructed identity of what it means to be a man or a woman in a particular society. There are signs that some men, particularly in older age groups, can find that climate change-induced effects prevent them from being able to demonstrate and fulfil their role as 'provider for the family'. This can appear as a challenge to their masculine identity, causing stress and possibly psychological problems that may lead to alcoholism and gender-based violence. The evidence for this is weak and tends to come from South Africa, where there is considerable social awareness of the problems of alcoholism and gender-based violence.[107] Generating the evidence linking alcoholism to stress arising out of climate change is methodologically difficult, since these are sensitive topics that do not lend themselves to the easy collection of data through interview techniques and may require more time-consuming ethnographic approaches. There may also be other confounding factors involved. While there are signs of changes in gender roles, evidence to show that gender relations are changing with a shift in the balance of power towards women at the grassroots level, in particular, is more difficult to find.

When dealing with rural communities, it is important not to attribute all environmental degradation to climate change (for example, increased urbanisation of small rural towns also leads to land use changes).[108] Furthermore, climate change may not always be negative for smallholders. In Tanzania, for instance, it has been suggested that climate change might be strongly negative for maize, the main food crop, whereas the impacts on coffee and cotton, both significant cash crops, may be positive.[109] As such, climate change might not, on balance, harm a smallholder's food security.

This leads me to the general observation that explains the title of this chapter. There seems to be a temptation to treat the impacts of climate change as 'exotic events'; however, rural people in SSA have been living with 'harsh', fluctuating climatic conditions for millennia, during which they have developed their own coping strategies based on their scientific knowledge of the ecosystems in which they live. For example, pastoralists have traditionally migrated to find water and fodder; however, when faced with the challenges of climate

107 See, for example, Babugura, *Gender and Climate Change*.

108 Morton, "The Impact of Climate Change."

109 Agrawala, Shardul et al., *Development and Climate Change in Tanzania: Focus on Mount Kilimanjaro* (Paris: Organisation for Economic Cooperation and Development, 2003), cited in Morton, "The Impact of Climate Change."

change their coping strategies have not remained rooted in the past and have incorporated 'modern' mechanisms, such as bank accounts. What rural people see is 'more of the same'—periods of drought and floods—but at a more frequent rate than before. Rural people have their own coping strategies based on their indigenous knowledge of the ecosystems in which they live. The factors that impede women and men in their efforts to build up resilience are nothing new, nor are they specific to climate change. Indeed, these difficulties are well-known to the development community: lack of credit (particularly for women), inadequate tools, and insufficient gender-targeted support. Here, it is particularly important to stress the gender dimension, since women and men have different responsibilities within farming systems—with men tending to be more engaged with commercial activities, which receive the attention of agricultural extension services, while women's practices, which are generally related to household food security, receive less attention. Solutions that would help both male and female farmers are readily available but may require investment: training for female farmers (recognising that they often prefer to work in groups[110] and are more restricted in terms of their travel due to family responsibilities) as well as improved communication and transport systems allowing both men and women to sell their products (both farm and non-farm) more easily. The type of information provided as well as the timing thereof is crucial—not only for agronomic reasons but also to synchronise with social norms. For example, in the Kaffrine Region, Senegal, men plant their crops first, with women only being allowed to start when the men have finished, while in polygamous households women plant in order of seniority.[111] Radio broadcasts also need to take into account when women are able to listen to information programmes, such as in the evening.

Targeting women and girls can have benefits when specifically aimed at helping them overcome the inherent disadvantages embedded in gender relations. However, many interventions are not set up with this intention in mind. Without an understanding of gender roles in the context in which they are working, project implementers run the risk of burdening women with additional tasks that they are not always best placed to undertake.

The evidence in this chapter has been presented in a compartmentalised way, which reflects the different academic disciplines responsible for generating the data. However, this might inadvertently be contributing to

---

110 Goudou, D. et al., *Village Baseline Study: Site Analysis Report for Kaffrine—Kaffrine, Senegal (SE0112)*. (Copenhagen: CGIAR Research Program on Climate Change, Agriculture and Food Security (CCAFS), 2012), cited in Twyman et al., *Adaptation Actions in Africa.*

111 Twyman et al., *Adaptation Actions in Africa.*

inappropriate solutions, since linkages can be missed. For example, in periods of prolonged drought, relief agencies tend to focus on providing food and miss the associated water and energy supply problems as well as the trade-offs that are required for households to balance these different needs, a burden that usually falls primarily on women.[112]

When designing responses to climate change, it should be recognised that climate change might not be a priority for poor communities that may be facing more pressing local-level issues. These priorities can be shaped by the ways in which rural communities attach meaning to the environment and climatic events, such as droughts. Rural people will interpret the information they receive about responses to climate change according to their own worldview, concerns, culture, and accumulated experience of climatic events.[113] Understanding these perspectives can lead to more effective interventions—in particular, when recognising the link between identity and place. Disruption—or even more drastically, destruction—of the natural environment undermines identity, weakening or removing the bond that keeps an individual in a place as well as threatening community sustainability.

112 Parker, Helen et al., *Gender, Agriculture and Water Insecurity* (London: Overseas Development Institute (ODI), 2016), 49.

113 Roncoli, Carla et al., "Meteorological Meanings: Farmers' Interpretations of Seasonal Rainfall Forecasts in Burkina Faso," in *Weather, Culture, Climate*, ed. Sarah Strauss, and Benjamin S. Orlove (Oxford: Berg, 2003), cited in Valerie Nelson, and Tanya Stathers, "Resilience, Power, Culture, and Climate: A Case Study from Semi-Arid Tanzania, and New Research Directions," *Gender and Development* 17.1 (2009): 81–94.

CHAPTER 7

Sustainable Mauritius? Environmental Change, Energy Efficiency, and Sustainable Development in a Small Island State in the Indian Ocean

Laura Jeffery

Abstract

The tropical Small Island Developing States (SIDS) typically contribute little to global climate change, yet they are among the countries most vulnerable to sea level rises, variations in temperature, fluctuations in rainfall, and extreme weather events. With high population densities and limited natural resources, they are also susceptible to the challenges of sustainable natural resource management. The Republic of Mauritius is one of four SIDS in the Indian Ocean and one of six SIDS that are also members of the African Union. In 2008 the Mauritian government launched a programme called Maurice Ile Durable (MID, or "Sustainable Mauritius") with the intention of turning Mauritius into a global and SIDS model of sustainable development. This chapter deploys MID as a lens through which to examine debates about the environment and development in Mauritius. MID's remit included environment, employment, education, and equity, but its main priority was energy efficiency. This chapter contrasts the narrow focus of Mauritian government actors, firstly with the critiques of environmentalists—who called for greater emphasis on the preservation of biodiversity and mitigation of climate change—and secondly with the interpretations of marginalised urban citizens, many of whom came to see MID particularly—and consequently the concept of sustainable development in general—in terms of energy efficiency and economic development rather than (also) in relation to environmental sustainability. It suggests that MID could have been more effective if environmental sustainability had been foregrounded, and more inclusive if economic development had been more equitably distributed.

1 Environment and Development: from Ecological Anthropology to Sustainable Development, Political Ecology, and Climate Change

Social scientists in general—and human geographers and social anthropologists in particular—have long theorised the relationship between the

 | DOI:10.1163/9789004410848_009

environment and development. Ecological anthropology in the 1960s posited that natural environments shaped culture by providing both opportunities and constraints; a culture was assumed to adapt to its natural environment in pursuit of social stability and ecological equilibrium.[1] From this perspective, environmental sustainability and socio-economic development seemed to be at odds with each other.[2] Within a few decades, in an era characterised by transnational migration, global interconnectedness, and an awareness of the long arc of anthropogenic environmental change, the ecological anthropology of the 1960s appeared overly culturally bound, ecologically deterministic, and preoccupied with stability rather than change.[3]

In the 1980s, a social science model of sustainable development sought to reconcile environmental sustainability with socio-economic development. In 1987, the World Commission on Environment and Development's (WCED) Brundtland Report correlated poverty with environmental degradation:

> Many parts of the world are caught in a vicious downwards spiral: Poor people are forced to overuse environmental resources to survive from day to day, and their impoverishment of their environment further impoverishes them, making their survival ever more difficult and uncertain.[4]

Scholars have suggested that the authors of the Brundtland Report were seeking simultaneously to address the concerns of environmentalists about environmental degradation in the developing world and the preoccupation of governments in developing countries with seeking socio-economic development.[5] Subsequently, Integrated Conservation and Development Projects (ICDPs) targeted the activities of impoverished people on the grounds that it was these activities that posed the greatest threat to the environment.

Social scientists were quick to criticise the underlying assumption in the Brundtland Report that environmental degradation could be blamed primarily

1 Kottak, C.P., "The New Ecological Anthropology," *American Anthropologist* 101.1 (1999): 23–24; Baer, H., and M. Singer, *The Anthropology of Climate Change* (London: Routledge, 2014), 60.

2 Gray, L.C., and W.G. Moseley, "A Geographical Perspective on Poverty–Environmental Interactions," *The Geographical Journal* 171.1 (2005): 10.

3 Kottak, "The New Ecological Anthropology," 23–25; Baer, and Singer, *The Anthropology of Climate Change,* 61.

4 World Commission on Environment and Development (WECD), *WCED Brundtland Report,* published March 20, 1987, accessed December 16, 2014. http://www.un-documents.net/our-common-future.pdf.

5 Gray, and Moseley, "A Geographical Perspective," 10.

on poor environmental management by impoverished people.[6] First, this assumption ignores the socio-economic structures that cause impoverishment.[7] Second, much anthropogenic environmental change is caused by the activities associated with industrialisation and the consumption patterns of affluent people.[8] Since the 1980s, political ecology—which combines human ecology with political economy—has sought to address both the underlying causes of impoverishment and the contribution of industrialisation and wealth to environmental degradation.[9]

Across the social sciences, political ecology has become associated with arguments about the gulf between those who contribute most to climate change and those who are most vulnerable to its effects. The vast majority of climate scientists now attribute global climate change to rising emissions of the greenhouse gases (carbon dioxide, nitrous oxide, and methane), particularly since 1750, correlating with the industrial revolution and a subsequent dependency on fossil fuels that has increased in intensity and global extent in the second half of the 20th century.[10] Marginalised people in developing countries tend to cause lower greenhouse gas emissions, and yet it is them who suffer disproportionately from climate change due to their concentration in ecologically fragile parts of the world and their dependency on the environment for their livelihoods.[11] Climate change thus magnifies existing inequalities, compounding existing impoverishment, vulnerability, and marginalisation.[12]

In a recent article on the uneven attention devoted to the concept of climate change around the world, Orlove et al. argue that since the 1980s both sustainable development and climate change have been prominent explanatory and

6 Blaikie, P., and H. Brookfield, "Defining and Debating the Problem," in *Land Degradation and Society*, ed. P. Blaikie, and H. Brookfield (London: Methuen, 1987); Bryant, R.L. "Political Ecology: An Emerging Research Agenda in Third-World Studies," *Political Geography* 11.1 (1992).

7 Escobar, A. "Constructing Nature: Elements for a Poststructural Political Ecology," in *Liberation Ecologies: Environment, Development, Social Movements*, ed. R. Peet, and M. Watts (London: Routledge, 1996), 51; Gray, and Moseley, "A Geographical Perspective," 14.

8 Peet, R., and M. Watts, "Liberation Ecology: Development, Sustainability, and Environment in an Age of Market Triumphalism," in Peet, and Watts, *Liberation Ecologies*, 7; Gray, and Moseley, "A Geographical Perspective," 19.

9 Blaikie, and Brookfield, "Defining and Debating," 17; Bryant, "Political Ecology," 13; Peet, and Watts, *Liberation Ecologies*, 4–5; Robbins, P., *Political Ecology: A Critical Introduction* (Oxford: Blackwell, 2004), 14.

10 Baer, and Singer, *The Anthropology of Climate Change*, 11–14.

11 Crate, S.A., and M. Nuttall, *Anthropology and Climate Change: From Encounters to Actions* (Walnut Creek: Left Coast Press, 2009), 10; Baer, and Singer, *The Anthropology*, 72.

12 Crate, and Nuttall, *Anthropology*, 14–17; Baer, and Singer, *The Anthropology*, 21–22, 72–73.

policy frameworks, but in different geographic contexts.[13] First, they argue that sustainable development is more commonly applied in relation to environmental changes attributed to intensive local land use—such as desertification, which reduces land productivity, and mountainside deforestation, which results in soil erosion and landslides—and so the proposed solution is to promote sustainable resource management and alternative livelihoods.[14] Second, they argue that climate change is more commonly applied in relation to environmental changes—like melting Arctic glaciers and associated rises in sea level threatening low-lying islands—that are attributed to global causes such as greenhouse gas emissions. These, they argue, are thus to be addressed on a global scale,[15] although they do recognise that both frameworks are applicable to all these regions.[16] To start with, deserts and mountains also experience environmental changes associated with global climate change, such as changes in temperature and precipitation patterns and the increased frequency of extreme weather events.[17] The Arctic and low-lying islands, meanwhile, also experience environmental changes associated with unsustainable resource use, such as depleted fish stocks as a result of overfishing, loss of biodiversity associated with invasive species, and declining water quality as a result of solid waste disposal.[18]

This chapter interrogates the interplay between environmental change and society in Africa through the lens of the Republic of Mauritius, a relatively small and highly developed island state in the Indian Ocean, where environmental changes are discursively associated both with vulnerability to global climate change and with unsustainable management of local natural resources. My case study treats the contradiction between the Mauritian government's sustainable development programme, *Maurice Ile Durable* (MID, or Sustainable Mauritius), and its actual energy policies designed to meet rising energy demands through further dependency on imported coal and the conversion of waste to energy rather than to promote sustainability and self-sufficiency through reduced energy demand and the development of local renewable sources of energy.

13 Orlove, B. et al., "Recognitions and Responsibilities: On the Origins and Consequences of the Uneven Attention to Climate Change around the World," *Current Anthropology* 55.3 (2014): 252.

14 Ibid., 259.

15 Ibid., 259–260.

16 Ibid., 259.

17 Loc. cit.

18 Loc. cit.

2 Vulnerability to Environmental Changes in Mauritius

The 1992 UN Conference on Environment and Development recognised that "small island developing states are a special case both for environment and development [...] [and] are considered extremely vulnerable to global warming and sea level rise".[19] The Small Island Developing States (SIDS) are scattered around the Caribbean, the Pacific Ocean, the Indian Ocean, the Atlantic Ocean, and the South China Sea. SIDS are diverse in terms of size, resource endowment, isolation, productive capacity, vulnerability to environmental and natural disasters, and economic diversification and development trajectories.[20] Nevertheless, the usefulness of SIDS as a category rests on several general characteristics. SIDS tend to have relatively small populations that are, however, growing relatively quickly. They are also often densely populated, meaning that their limited agricultural land and marine resources are put under considerable strain. SIDS are often highly dependent on exchange with distant markets—in particular for the importation of food supplies and fossil fuels—making them vulnerable to high transportation costs and price fluctuations. SIDS typically contribute little to global climate change; yet, low-lying islands are disproportionately vulnerable to the effects of climate change, such as rising sea levels, variations in temperature, fluctuations in rainfall, and the increasing frequency and severity of extreme weather events such as tropical storms. At the same time, however, as tropical islands they also have considerable natural resources offering opportunities to develop renewable energy supplies from sources such as solar, hydro-, wave, and wind power.

Four of the SIDS—Mauritius, Seychelles, Comoros and Maldives—are in the Indian Ocean. Six of the SIDS—Mauritius, Seychelles, and Comoros, plus Cape Verde, Guinea Bissau as well as São Tomé and Príncipe in the Atlantic Ocean—are also members of the African Union. Malay, Arab, and Swahili traders explored the southwest Indian Ocean over a thousand years ago, but many of the smaller islands in the region—including Mauritius, Seychelles, and Réunion—were unpopulated prior to European colonial expansion from the end of the 15th century onwards.[21] Following Vasco da Gama's trip

19 UN International Year of Small Island Developing States 2014, accessed December 16, 2014. www.un.org/en/events/islands2014/smallislands.shtml.

20 Kothari, U., and R. Wilkinson, "Global Change, Small Island State Response: Restructuring and the Perpetuation of Uncertainty in Mauritius and Seychelles," *Journal of International Development* 25 (2013): 93.

21 Allen, R.B., *Slaves, Freedmen, and Indentured Laborers in Colonial Mauritius* (Cambridge: Cambridge University Press, 1999), 9.

around the Cape of Good Hope Portuguese navigators in 1498 used Mauritius as a stopping point en route between the Cape and India, but did not establish a permanent settlement on the island.[22] From 1598, the Dutch East India Company (Vereenigde Oost-Indische Compagnie, VOC) used Mauritius as a stopping point en route to and from East Asia, but abandoned the island in 1710 due to the challenges of maintaining the small settlement.[23] The French, who had occupied nearby Réunion since 1642, then claimed Mauritius in 1715 and Seychelles in 1742, populating the islands with enslaved labourers, mostly from coastal East Africa and Madagascar.[24] Britain acquired Mauritius and its dependencies—including Seychelles, Rodrigues, Agalega, St. Brandon, Tromelin, and the Chagos Archipelago, but not Réunion—under the Treaty of Paris in 1814; Seychelles became a separate crown colony in 1903.[25] Following the abolition of the British slave trade in 1807 and the emancipation of enslaved labourers in Mauritius in 1835, the British supplemented the population of Mauritius with indentured labourers from India, who comprised two-thirds of the population by 1871.[26] Mauritius became independent in 1968, and the Republic of Mauritius now constitutes the main island of Mauritius plus the outer island of Rodrigues as well as the coral islands Agalega and St. Brandon. Mauritius also claims the Chagos Archipelago, which is currently administered as a United Kingdom Overseas Territory, and Tromelin, which is currently a French Overseas Territory co-managed with Mauritius.

The VOC introduced sugarcane to Mauritius in the mid-17th century, finding the crop resilient and well suited to the rainy and windy climate.[27] The French experimented with other crops—coffee, cotton, indigo, and spices—which turned out to be more susceptible than sugarcane to natural disasters such as cyclones.[28] The British replaced native forests with sugarcane plantations, and Mauritius became a monocrop agricultural economy based on sugarcane.[29]

22 Barnwell, P.J., and A. Toussaint, *A Short History of Mauritius* (London: Longmans, Green & Co., 1949), 3; Toussaint, A., *History of the Indian Ocean* (London: Routledge & Kegan Paul, 1966), 110.

23 Allen, *Slaves*, 9; Barnwell, and Toussaint, *A Short History*, 34–37.

24 Allen, R.B., "The Mascarene Slave-Trade and Labour Migration in the Indian Ocean during the Eighteenth and Nineteenth Centuries," in *The Structure of Slavery in Indian Ocean Africa and Asia*, ed. G. Campbell (London: Frank Cass, 2004), 34, 37; Barnwell, and Toussaint, *A Short History*, 41, 43; Toussaint, *History*, 272.

25 Allen, *Slaves*, 11; Barnwell, and Toussaint, *A Short History*, 123, 125.

26 Allen, *Slaves*, 17; Carter, M., *Servants, Sirdars and Settlers: Indians in Mauritius, 1834–1874* (Delhi: Oxford University Press, 1995), 271.

27 Barnwell, and Toussaint, *A Short History*, 32–33.

28 Allen, *Slaves*, 11–12; Barnwell, and Toussaint, *A Short History*, 55.

29 Allen, *Slaves*, 28; Carter, *Servants*, 13–14.

Mauritius was still almost entirely dependent on sugarcane at independence in 1968, but this made the economy extremely vulnerable to the weather, to fluctuations in global sugar and fuel prices, and to the gradual end of European Union subsidies and preferential trade agreements.[30]

Since the 1970s, Mauritian governments have sought to reduce dependence on sugarcane by diversifying both agricultural production and economic activity. Nevertheless, ninety per cent of the arable land on the main island of Mauritius is still under sugarcane cultivation, while the remaining ten per cent comprises tea, tobacco, and food crops.[31] Mauritius has a high but declining dependence on food imports, which comprised two-thirds of its food supplies in 2010.[32] The main pillars of the Mauritian economy are now sugarcane, manufacturing, tourism, financial services, and ICT.[33] Agriculture and other economic activities on the smaller outer islands differ from mainland Mauritius in that Rodrigues relies on fisheries and livestock export plus tourism, while Agalega relies on its coconut plantations.[34]

As a result of its relatively peaceful transition to independence, its well-functioning democracy, and the diversification of its economy, Mauritius is often referred to as one of Africa's postcolonial success stories in terms of political, human, and economic development. It is the highest-ranking country on the Ibrahim Index of African Governance, which assesses safety and rule of law, participation and human rights, sustainable economic activity, and human development.[35] In 2013, Mauritius ranked 63rd on the UNDP's Human Development Index, which was higher than almost every country in mainland

30 Kothari, and Wilkinson, "Global Change," 94–95; Lim Tung, O., "Development and the Balancing of Interests in Environmental Policy and Law in Mauritius," in *The Balancing of Interests in Environmental Law in Africa*, ed. M. Faure, and W. du Plessis (Pretoria: Pretoria University Law Press, 2011), 261.

31 Ramjeawon, T., "Life Cycle Assessment of Electricity Generation from Bagasse in Mauritius," *Journal of Cleaner Production* 16 (2008): 1727.

32 Luximon, Y., and M.D. Nowbuth, "A Status of Food Security in Mauritius in Face of Climate Change," *European Water* 32 (2010): 5.

33 Kothari, and Wilkinson, "Global Change," 99–102; Lim Tung, "Development," 262; Mohee, R., and A. Mudhoo, "Energy from Biomass in Mauritius: Overview of Research and Applications," in *Waste to Energy: Green Energy and Technology* (London: Springer, 2012), 301.

34 Gemenne, F., and A. Magnan, "The Other Migrants Preparing for Change: Environmental Changes and Migration in the Republic of Mauritius" (Geneva: International Organisation for Migration, 2001), 25, 52. http://publications.iom.int/bookstore/free/the_other_migrants.pdf.

35 Ibrahim Index of African Governance, accessed December 16, 2014. http://www.moibrahimfoundation.org/iiag/.

Africa (except Libya at 55th); the other African SIDS also ranked lower (Seychelles was 71st, Cape Verde was 123rd, São Tomé and Príncipe was 142nd, Comoros was 159th, and Guinea Bissau was 177th).[36] Given that the Republic of Mauritius is indexed as a country of high human development, it is perhaps less a Small Island Developing State (SIDS) than a middle-income Small Island State (SIS), although with a land area of over 2,000 square kilometres and a population of 1.3 million it is also considerably less 'small' than many other SIDS or SIS.[37] Moreover, the main island of Mauritius, an oceanic island with relatively high elevation, is less at risk from sea-level rises than the archipelagic states comprised entirely of low-lying islands: Kiribati, Maldives, Marshall Islands, and Tuvalu;[38] Mauritius was, for instance, relatively unaffected by the 2004 Indian Ocean tsunami.[39]

Nevertheless, like many Small Island (Developing) States, the islands of Mauritius are vulnerable to the effects of climate change and other environmental changes.[40] Rainfall is declining at an average rate of 57 millimetres per decade, resulting in water shortages that cannot meet the increasing demand from the domestic, agricultural, industrial, and tourism sectors.[41] The average temperature is rising at a rate of 0.15°C per decade (and has risen by 0.74–1.2°C since the 1961–1990 long-term mean); coral bleaching events in 1998 and 2009 resulted in loss of biodiversity, degradation of marine and land ecosystems, and a negative impact upon fisheries and tourism.[42] Sea level at the capital Port Louis was rising by an average of 2.1 millimetres per year by the first decade of the 21st century (compared to around 1.5 millimetres per year over the second half of the 20th century); salination of freshwater and soil negatively impacts upon drinking water supplies and agriculture, while erosion

36 United Nations Development Programme, "UNDP Human Development Index," accessed December 16, 2014. http://data.undp.org/dataset/Table-1-Human-Development-Index-and-its-components/wxub-qc5k.

37 Connell, J., *Islands at Risk? Environments, Economies and Contemporary Change* (Cheltenham: Edward Elgar, 2013), 2–3.

38 Ibid., 243; Robertson, M.L.B., and C. Rubow, "Engaged World-Making: Movements of Sand, Sea, and People at Two Pacific Islands," in *Anthropology and Nature*, ed. K. Hastrup (New York: Routledge, 2014), 65; Rudiak-Gould, P., *Climate Change and Tradition in a Small Island State: The Rising Tide* (New York: Routledge, 2013), 2.

39 Gemenne, and Magnan, "The Other Migrants," 21.

40 Mauritius Meteorological Services, accessed December 16, 2014. http://metservice.intnet.mu/climate-services/climate-change.php.

41 Gemenne, and Magnan, "The Other Migrants," 22–23.

42 Ibid., 25; Ramessur, R.T., R.P. Gunputh, and T.S. Ramessur, "Climate Compatible Development: Legal Implications in the Coastal Zone and Inclusive Development for Mauritius," *Journal of Coastal Development* 16.6 (2013): 2.

of beaches contributes to the unmaking of Mauritius as an attractive tourist destination.[43] The occurrence and severity of tropical storms have intensified, and flash floods after storms in 2013 killed eleven people.[44]

At the same time, socio-economic development has also posed threats to the Mauritian environment, including loss of biodiversity, resource depletion, deforestation, erosion, degradation of the ecosystem, contamination of coastal zones and freshwater supplies, air pollution, and solid and hazardous waste disposal.[45] Overfishing results in diminished fish stocks, and undersea tourism damages the coral reefs.[46] Sugarcane plantations can promote soil erosion and cause landslides on hill slopes, while the fertilisers and pesticides used on these plantations as well as the chemicals used in the textile industry run downhill and enter the water system, where they can harm agriculture and marine wildlife.[47] Traditionally, sugarcane fields have been set on fire to burn off the dry leaves before harvesting the juicy canes, but this releases dioxins and wastes the cane trash, which could otherwise be left on the fields to prevent the regrowth of other species and thus reduce the need for herbicides; now, as a result of more efficient manual harvesting techniques and increasing mechanisation, the practice of burning fields is in decline.[48]

Sugarcane is an exceptionally productive species: in addition to commercial sugar products, cane trash, and cane tops (used as an animal feed or exported as a raw material for paper), the extraction of cane sugar also produces by-products including molasses and a fibrous biomass called bagasse.[49] Bagasse was historically incinerated to power the sugarcane factory itself, but more efficient techniques have meant that the Mauritian sugarcane industry has increasingly produced energy beyond its own needs, which is then sold to the national grid.[50] Molasses can be distilled into ethanol, which can be blended

43 Gemenne, and Magnan, "The Other Migrants," 22, 26–27.

44 Ibid., 21–22; Ramessur et al., "Climate Compatible," 2.

45 Foolmaun, R.K., D.S. Chamilall, and G. Munhurrun, "Overview of Non-Hazardous Solid Waste in the Small Island State of Mauritius," *Resources, Conservation and Recycling* 55 (2011): 967; Ramessur et al., "Climate Compatible," 1.

46 Gamenne, and Magnan, "The Other Migrants," 24, 27.

47 Ibid., 23–24; Ismael, F.M. et al., "A Review of Changing Cultural Practices to Improve Productivity of Sugar Cane in Mauritius," *Proceedings of the South African Sugar Technologists Association* 81 (2008).

48 Ismael et al., "A Review."

49 Chummun, J., "The Potential for Using Renewable Sources of Energy in Mauritius," in *Climate-Smart Technologies: Integrating Renewable Energy and Energy Efficiency in Mitigation and Adaptation Responses* (Berlin: Springer, 2013), 210; Mohee, and Mudhoo, "Energy," 311.

50 Chummun, "The Potential," 211; Ramjeawon, "Life Cycle Assessment," 1727.

with petrol or used as a cheaper and cleaner alternative to petrol; this practice is being developed but remains relatively underexploited in Mauritius.[51] The sugar industry is now increasingly referred to as the cane industry in recognition of the potential energy value of its by-products.[52]

3 Maurice Ile Durable (Sustainable Mauritius)

Mauritius has no coal, natural gas, or oil deposits, so its energy needs are met through imported coal and petroleum products supplemented with local renewable resources: bagasse, hydropower, solar thermal energy, photovoltaics, fuelwood, waste-to-energy, and wind.[53] The country's economy and energy requirements both grew at an average rate of five per cent per year over the first decade of the 21st century; meanwhile, the contribution from local renewable energy sources stagnated.[54] Mauritius is thus extremely vulnerable to fluctuations in global energy prices. In response to the global energy crisis and record high oil prices in 2007, then Mauritian Prime Minister, Navin Ramgoolam, launched the concept of *Maurice Ile Durable* (MID, or Sustainable Mauritius) in 2008. MID aimed to "make Mauritius a model of sustainable development"—particularly in a SIDS context—"in which the needs of the present generation are met, without jeopardising the chances of future generations to meet theirs".[55] This clearly draws on the Brundtland Report's oft-quoted (1987) definition: "sustainable development is development that meets the needs of the present without compromising the ability of future generations to meet their own needs".[56]

The MID Fund was originally established under the aegis of what was then called the Ministry of Renewable Energy and Public Utilities (and became the Ministry of Energy and Public Utilities). Initially, the main thrust of MID was to make Mauritius less dependent on imported fossil fuels—with an initial target of 65 per cent autonomy by 2028—by making increasing use of

51 Chummun, "The Potential," 211; Mohee, and Mudhoo, "Energy," 316–317.

52 Ramjeawon, "Life Cycle Assessment," 1729.

53 Chummun, "The Potential," 210.

54 Elahee, K., "Energy Management Challenges in a Small Island Economy: The Case of Mauritius," *Energy and Environment* 21.7 (2010): 803.

55 Ministry of Environment and Sustainable Development, *Maurice Ile Durable Green Paper: Towards a National Policy for a Sustainable Mauritius* (Port Louis: Government of Mauritius 2011), 2, xii.

56 WCED, WCED Brundtland Report.

renewable energy and more efficient use of energy in general.[57] Many of the original projects financed by the MID Fund were entirely focused on energy efficiency: grants for the purchase of 25,000 solar water heaters, subsidies for the purchase of one million compact fluorescent lamps, the replacement of conventional lighting with fluorescent lighting in public buildings, the replacement of incandescent and halogen traffic lights with LED signal lights, and the installation of solar water heaters in public hospitals; MID also contributed to the ongoing renewal of the bus fleet towards reduced emission vehicles.[58] The proportion of Mauritian energy that came from local renewable sources, actually fell annually from 17.5 per cent in 2009 to 15 per cent in 2013, of which 92 per cent came from bagasse, four per cent from hydropower, and the remaining four per cent from wind, landfill gas, photovoltaics, and fuelwood. Correspondingly, the proportion of Mauritian energy derived from imported fossil fuels rose annually from 82.5 per cent in 2009 to 85 per cent in 2013.[59] The optimistic 2008 target of 65 per cent energy autonomy by 2028 was revised downwards; in 2013, the target stood at 35 per cent renewable energy by 2025.[60] In 2016, however, 85 per cent of Mauritian energy was still derived from imported fossil fuels.

In 2010, a national consultation process with members of the public and special interest groups contributed to the expansion of the scope of MID. According to the resultant "Green Paper",[61] responses to the consultations suggested broader understandings of sustainable development as incorporating the environment (particularly in relation to conservation, pollution, and waste management), the economy (particularly energy and transport, but also self-sufficiency and green construction), society (particularly education, employment, health, and social cohesion), and governance. MID was thus expanded to incorporate the so-called "5Es", which were divided into six Working Groups that met in mid-2011: (1) Energy; (2) Environment—preservation of biodiversity

57 Elahee, K., "The Challenges and Potential Options to Meet the Peak Electricity Demand in Mauritius," *Journal of Energy in Southern Africa* 22.3 (2011): 13.

58 Ibid., 10; Jogoo, V.K. et al., "The Political Economy of Transitioning to a Green Economy in Mauritius," in *Transitioning to a Green Economy: Political Economy of Approaches in Small States*, ed. Smith, Nadine, Anna Haiton, and Janet Strachan (London: Commonwealth Secretariat, 2014), 138.

59 Statistics Mauritius, *Energy and Water Statistics*, (Port Louis: Government of Mauritius, 2013).

60 Osman, Mahomed, "Maurice Ile Durable," published by The United Nations, November 1, 2013, accessed December 16, 2014. http://sustainabledevelopment.un.org/content/documents/4074durable.pdf.

61 Ministry of Environment and Sustainable Development, *Maurice Ile Durable*.

and natural resources; (3) Environment—pollution, wastes and environment; (4) Employment; (5) Education; and (6) Equity.[62] As part of this diversification, responsibility for MID was transferred from the Ministry of Energy and Public Utilities to the Prime Minister's Office (PMO) in collaboration with the newly renamed Ministry of Environment and Sustainable Development (MoESD, formerly known as the Ministry of Environment and National Development Unit). The five additional lead ministries were Energy and Public Utilities; Agro-Industry and Food Security; Labour, Industrial Relations and Employment; Education and Human Resources; and Social Integration and Economic Empowerment. All ministries were expected to prioritise the realisation of MID's goals. Drawing on the final reports from the six Working Groups, the UK consultancy Mott MacDonald drafted the MID 'Policy, Strategy and Action Plan', a much-revised version of which was finally implemented in June 2013.

Although MID's stated aim was to reduce dependency on imported fossil fuels through the development of renewable energy sources and more efficient use of energy in general, Mauritius has instead experienced rises in both demand for energy and in the proportion of energy derived from imported fossil fuels. The following subsections ask what is going on and look in more detail at two controversial energy efficiency projects, which are aimed not at reducing demand or developing renewables, but at meeting increasing demand through the importation of fossil fuels and the conversion of waste to energy in ways which apparently contradict the aims of MID.

3.1 *Energy Efficiency: a Proposed New Coal-Fired Power Station*

Especially in the period since industrialisation, human societies have relied increasingly on the extraction of fossil fuels—coal and later also petroleum and natural gas—for the production of cheap and abundant energy to fuel technological advancement and economic growth, but fossil fuels are highly polluting, and demand for them continues to rise even as supplies dwindle.[63] Responses to the prospect of reaching peak oil extraction vary from pessimism about unemployment, the high cost of energy, and energy shortages to optimism about opportunities to increase sustainability and self-sufficiency through the development of local renewable energy sources.[64]

---

62 Ministry of Environment and Sustainable Development, *Maurice Ile Durable*, 12.

63 Baer, and Singer, *The Anthropology*, 11–14; Horta, A. et al., "Socio-Technical and Cultural Approaches to Energy Consumption: An Introduction," *Nature and Culture* 9.2 (2014): 115.

64 Baer, and Singer, *The Anthropology*, 17; Hornborg, A., "The Fossil Interlude: Euro-American Power and the Return of the Physiocrats," in *Cultures of Energy: Power, Practices, Technology*, ed. S. Strauss, S. Rupp, and T. Love (Walnut Creek: Left Coast Press, 2013);

The state-owned Central Electricity Board (CEB) is responsible for sourcing and supplying electricity in Mauritius. Mauritius has long relied on dual-fuel power stations that burn bagasse for six months of the year (during the sugar-cane harvesting season) and imported coal for the other six months of the year when bagasse is not available; without coal, bagasse would be inefficient, but coal is environmentally more damaging than bagasse.[65] In 2011, nearly 54 per cent of the country's electricity supply came from Independent Power Producers (IPPs) operating dual-fuel bagasse/coal cogeneration facilities, nearly 43 per cent came from thermal power stations reliant on imported heavy fuel oil (which costs twice as much as coal to produce), just over twp per cent came from hydroelectric plants, and the remaining one per cent was derived from bagasse and landfill gas.[66] In the context of increasing demand, perceptions of the unreliability and marginality of renewables, and the high cost of heavy fuel oil, the supposed main alternative to coal, the CEB decided to commission a new single-fuel coal-fired power station.[67]

In 2006, the CEB issued a contract to Mauritius CT Power (MCTP, staffed by Indian and Malaysian engineering teams) to build and operate a coal-fired power station with two 50-megawatt turbo-generator units at Pointe aux Caves near Albion, a town on the west coast of Mauritius.[68] In 2011, the Ministry of Environment and Sustainable Development rejected MCTP's application for an Environmental Impact Assessment (EIA) license, but MCTP appealed and was granted its EIA license in 2013. MCTP was initially to own and operate the power station; the CEB would have a 26 per cent shareholding and would receive all electricity generated for the first twenty years, after which the facility would be transferred to the CEB with an expected further lifespan of twenty years.[69] The power station would use bituminous coal sourced from the South African town of Richards Bay, the source of coal currently used by other IPPs

Strauss, S., S. Rupp, and T. Love, "Powerlines: Cultures of Energy in the Twenty-First Century," in *Cultures of Energy: Power, Practices, Technology*, ed. S. Strauss, S. Rupp, and T. Love (Walnut Creek: Left Coast Press, 2013), 13.

65 Mohee, and Mudhoo, "Energy," 312–315.

66 Central Electricity Board, *Annual Report*, (Port Louis: Central Electricity Board, 2011), 27. http://ceb.intnet.mu/CorporateInfo/ar2011.pdf.

67 Kasenally, Swalay, "Coal Power: The Politics of Indecision," *Le Mauricien*, February 28, 2012, accessed December 16, 2014. http://www.lexpress.mu/article/coal-power-politics-indecision.

68 Mauritius CT Power (MCTP), "The Strategy," accessed December 16, 2014. http://ctpowerltd.com/strategy/.

69 Mauritius CT Power (MCTP), "About CT Power," accessed December 16, 2014. http://ctpowerltd.com/welcome/.

in Mauritius.[70] The initial plan was for coal to arrive at the Port Louis docks for transportation by road 15 kilometres south to Pointe aux Caves, but critics questioned the road network's capacity to support the additional traffic load on already congested roads. Ultimately, the government decided that—with Pointe aux Caves being on the coast—MCTP should instead construct a jetty so that coal could arrive by sea.

MCTP proposed to create one thousand local jobs in construction and, after its completion, two hundred local jobs at the power station.[71] Local fishermen based in Albion to the south and Pointe aux Sables to the north raised concerns about effluent. The Mauritius Environment Platform led opposition to the power station on the grounds that it contradicted MID's purported commitments to reducing reliance on imported fossil fuels and to developing local renewable sources. Despite existing capacity for electricity generation from renewables such as bagasse and hydropower, critics were widely concerned that a single-fuel coal-fired power station would divert investment away from alternative sources of electricity, such as installing photovoltaic cells (which are an initially expensive investment but have long lives as well as low running and maintenance costs), retrofitting the existing dual-fuel power stations to use cleaner natural gas (from Mozambique and Tanzania) instead of coal, and the development of wind and wave power.[72] In 2015, the Mauritian government decided not to proceed with the plant after all; MCTP won its Supreme Court appeal in 2016, and the saga continues.

3.2 *Waste Management: a Proposed New Waste-to-Energy Plant*

Increased waste production poses ever-greater challenges for safe waste disposal. Solid waste is increasingly incinerated and converted into electricity through a process known as Waste-to-Energy (WtE), which also addresses the ever-rising demand for energy.[73] Nevertheless, WtE has its critics, including Friends of the Earth and the Global Anti-Incinerator Alliance (GAIA). WtE facilities emit dioxins and other unintentional persistent organic pollutants

70 Mauritius CT Power (MCTP), "Behind the Scenes," accessed December 16, 2014. http://ctpowerltd.com/engineering-team/.

71 Mauritius CT Power (MCTP), accessed December 16, 2014. http://ctpowerltd.com/wp-content/uploads/2013/04/25-04-2013.pdf.

72 Chummun, "The Potential"; St. Antoine, Pamela de, "Engineer Felix Ah-Kee: Mauritius Should Look at Alternatives to Proposed Coal Plant," *Le Mauricien*, April 7, 2013, accessed December 16, 2014. http://www.lemauricien.com/article/engineer-felix-ah-kee-mauritius-should-look-alternatives-proposed-coal-plant.

73 Chummun, "The Potential," 214; Mohee, and Mudhoo, "Energy," 299.

as well as highly hazardous ash as by-products.[74] Furthermore, WtE relies on waste production, and thus appears to promote waste creation rather than waste reduction. It can also divert attention away from waste segregation for composting and recycling as well as from the development of renewable energy sources.[75]

During the 1980s and 1990s, solid waste in Mauritius was deposited in open-air dumping sites, which emitted pollutants into the water table, the waterways, the soil and the air, especially when rubbish was burned.[76] In 1997, the government opened a landfill site at Mare Chicose in the southwest. This landfill posed a multitude of problems for the local community—leachates that polluted the nearby waterways, scavenging pests, and increased traffic, odours, and skin and respiratory problems—who eventually agreed to relocate to the town of Rose Belle.[77] Mare Chicose had a capacity for processing three hundred tonnes of waste per day for 18 years. Economic development, however, has been accompanied by increased waste production, and by 2009 solid waste production was around four times this amount at 1,200 tonnes per day (or about one kilogram per capita per day). As a result, the site could not keep up.[78]

In 2006, the Mauritian government announced that it was considering a WtE facility that would incinerate 300,000 tonnes of mixed waste per year, which would allow it to sell the resultant twenty megawatts of energy to the CEB. The WtE facility would be constructed and managed by a local company, Gamma Civic, in collaboration with an American company, Covanta Energy. La Chaumière lies between the west coast and the conglomeration comprising the densely populated upland towns of Beau-Bassin/Rose-Hill and Quatre-Bornes/Ebène. According to Gamma–Covanta's Environmental Impact Assessment, La Chaumière was selected, among other reasons, because the

74 GAIA, "Keep Mauritius Incinerator-Free," accessed December 16, 2014. http://www.no-burn.org/article.php?id=374; GAIA, "Mauritius Says No to MSW Incinerator," published June 22, 2009, accessed December 16, 2014. http://www.no-burn.org/article.php?id=734.

75 Clark, J.F.M., " 'The Incineration of Refuse Is Beautiful': Torquay and the Introduction of Municipal Refuse Destructors," *Urban History* 34.4 (2007): 276–277; Alexander, C., and J.O. Reno, "From Biopower to Energopolitics in England's Modern Waste Technology," *Anthropological Quarterly* 87.2 (2014): 340; cf. Chummun, "The Potential," 214; Mohee, and Mudhoo, "Energy," 303.

76 Institute for Environmental and Legal Studies, "Solid Waste Management in Mauritius," accessed December 16, 2014. http://iels.intnet.mu/solidwaste_mau.htm.

77 Gemenne, and Magnan, "The Other Migrants," 51.

78 Foolmaun, Chamilall, and Munhurrun, "Overview," 967; Lim Tung, "Development," 268–269, 281; Mohee, and Mudhoo, "Energy," 302.

prevailing easterly winds would minimise the potential impacts on local air quality and because its proximity to the most densely populated part of the island would reduce transportation distances.[79]

Local residents campaigned against the WtE facility on the grounds that it could have similar local impacts as landfill sites, and filed an appeal against the government's approval of Gamma–Covanta's EIA report before the Environment Appeal Tribunal.[80] The political party Lalit questioned whether the WtE facility would be more polluting than the status quo of landfill sites plus the burning of bagasse and coal in cogeneration electricity plants.[81] Furthermore, Lalit also noted that Gamma–Covanta stood to make MUR 250 million from the project, and that the state would lose money if the WtE facility received less than the contracted 300,000 tonnes of waste per year, which could undermine proposals to reduce waste production and sort waste for composting and recycling.[82] Similarly, local environmentalists argued that the government should instead use MID to concentrate its efforts: firstly on waste reduction; secondly on composting organic animal and food waste; thirdly on recycling paper, glass, plastics, and aluminium; and fourthly on the development of renewable energy.[83]

However, another company, Solid Waste Recycling Ltd., does run a composting plant on state land at La Chaumière, which deploys a process and technology developed by an Indian company called Excel Industries. Solid Waste Recycling Ltd. has a contract with the government to receive 180,000 tonnes of municipal solid waste per year for twenty years. The company then sells the treated compost as fertiliser, while Sotravic transports the residual waste (amounting to about half the total received) from the composting plant to Mare Chicose. There, Sotravic reports converting landfill gas into about three megawatts of electricity per month, which has been provided to the national grid since 2011.[84] According to the Institution of Engineers, Mauritius, the

---

79 La Rédaction, "Not in My Backyard," *L'Express*, May 22, 2007, accessed December 16, 2014.

80 Ackbarally, Nasseem, "Environment–Mauritius: Hold Your Fire," published by Inter Press Service, July 15, 2009, accessed December 16, 2014. http://www.ipsnews.net/2009/07/environment-mauritius-hold-your-fire/.

81 Seegobin, Ram, and Lindsey Collen, "The La Chaumière Waste-to-Energy Project Criticized," *Lalit*, April 23, 2009, accessed December 16, 2014.

82 Ibid.

83 Institute for Environmental and Legal Studies, "Solid Waste Management in Mauritius," accessed December 16, 2014. http://iels.intnet.mu/solidwaste_mau.htm.

84 Sotravic, Water—Environment—Energy, "Power Generation," accessed November 23, 2017. https://www.sotravic.net/waste-and-energy/power-generation.html.

amount of waste taken to landfill at Mare Chicose decreased by 16 per cent from 2011 to 2012, resulting in a reduction in landfill gas emissions and water pollution through leachate infiltration into underground water.[85] Additionally, the Institution of Engineers, Mauritius reported that the rerouting of raw municipal solid waste from the capital, Port Louis, and from the densely populated hill towns to the nearby composting plant rather than to the more distant landfill site had also reduced fuel consumption and CO_2 emissions.[86]

4 Perceptions of MID and Sustainable Development

In light of the political and economic challenges facing the Mauritian government in its quest for energy efficiency within a context of sustainable development, this section examines how people understand, engage with, and critique MID. The material I discuss derives from two research projects: one with MID insiders and the other looking at MID from the outside. The resultant material divides into three subsections: perceptions of MID from within, perceptions of MID among engaged professionals, and perceptions of MID and sustainable development among marginalised citizens.

4.1 *Perceptions of MID from Within*

During the process of revising the draft MID 'Policy, Strategy and Action Plan' in 2012, Saeko Kajima conducted research with officials involved in the development and implementation of MID.[87] She conducted interviews with members of various MID committees and working groups, including representatives from four ministries, two international development organisations, two private sector organisations, a conservation NGO, and the University of Mauritius.[88] Kajima found that there was considerable confusion about the division of responsibilities for MID, and all but one of her respondents agreed that inadequate coordination between ministries was a barrier to its successful implementation.[89] Representatives of the other

85 "Solid Waste Recycling Ltd. Project Brief," accessed December 16, 2014. http://www.iemauritius.com/upload/files/iem_swrl_brief.pdf.

86 Ibid.

87 Kajima, S., "Institutional Interactions and the Implementation of Climate Change and Sustainable Development Programmes: A Case Study of Mauritius' 'Maurice Ile Durable'" (MSc diss., University of Edinburgh, 2012).

88 Kajima, "Institutional Interactions," 27, 44, 111.

89 Ibid., 35, 47.

ministries questioned the MoESD's leadership capacity and wondered why it had been selected as the lead ministry for a programme that was supposed to involve all ministries; a proposed solution was for the PMO and the Prime Minister himself to take a more active lead in coordinating MID activities.[90]

Kajima asked her interviewees, "How would you explain *Maurice Ile Durable* to someone who was not aware of it?"[91] She was struck by the wide diversity of responses and by the fact that none of the respondents mentioned all of the above-mentioned '5Es', despite their centrality to the Mauritian government's vision of MID.[92] Of the '5Es', 'energy and the environment' was most frequently mentioned, while 'education and equity' was mentioned only once, and 'employment' was not mentioned at all.[93] There was general consensus among Kajima's interviewees that MID had raised public awareness of and engagement with issues of sustainable development, and that it had publicised the action taken by the Mauritian government in pursuit of social, economic, and environmental sustainability.[94] Despite this optimistic consensus, there were a few voices of dissent: one interviewee suggested a lack of public understanding about the concept of sustainable development, and another questioned the extent of the dissemination of relevant information among those who did not read newspapers or who had no access to the internet.[95] For its part, the Mauritian government admitted that there was inadequate public awareness of and engagement with its commitment to sustainable development.

4.2 *Perceptions of MID among Engaged Professionals*

As part of a wider project on debates about environmental knowledge, I spoke to people about sustainable development during two periods of ethnographic fieldwork in Mauritius in mid-2011 (when I also observed part of one of the MID Working Group meetings) and mid-2013. The people whom I asked specifically about MID included natural scientists, environmental consultants, leading members of three NGOs affiliated with the Mauritius Environment Platform, and political actors, many of whom had engaged professionally with MID as consultants or working group members. Many of the professionals

90 Kajima, "Institutional Interactions," 35, 47–48.
91 Ibid., IV.
92 Ibid., 42.
93 Ibid., 43.
94 Ibid., 49.
95 Ibid., 50.

whom I asked to tell me about MID immediately volunteered some variation on the theme of the then Leader of the Opposition, Paul Bérenger, that MID is an "empty shell" ("coquille vide").

These professionals had three interrelated principal concerns. First, they were critical of the Mauritian government's self-interest in relation to MID's concentration on energy and the economy and its relative neglect of environmental sustainability. Second, they were concerned about the government's vulnerability to foreign interference, complaining that MID was primarily a means for the French government and French companies to increase their influence and economic activities in Africa. The French government provided significant funding and technical support for MID through its development agency, Agence Française de Développement, which had privileged access to the Mauritian government, thus giving French companies a competitive advantage when it came to bidding for business opportunities, such as the management contract for the Port Louis bypass. Third, my respondents told me that corruption was a concern, because the Mauritian government was continuing to award large-scale energy contracts to supporters of the ruling Labour Party even when such contracts contravened the principles of MID. Arguments about the government's narrow focus on energy efficiency rather than environmental sustainability, its vulnerability to foreign interference, and political corruption were most often brought together and illustrated using the two controversial examples of the proposed coal-fired power station at Pointe aux Caves near Albion and the proposed waste-to-energy plant at La Chaumière, both of which are near to where I and most of my research participants lived in west Mauritius.

4.3 *Perceptions of MID and Sustainable Development among Marginalised Citizens*

Most of my time in Mauritius was spent living and working in the disadvantaged urban neighbourhoods on the outskirts of the capital, Port Louis: Pointe aux Sables, Cassis, Roche Bois, and Baie du Tombeau. The people I knew in these neighbourhoods were relatively marginalised, with comparatively low educational backgrounds. Most did not read the daily newspapers or have access to the internet, gaining the majority of their news from television and radio. Given that MID was heavily promoted by the state-owned Mauritian Broadcasting Corporation (MBC), I asked people living in these neighbourhoods to tell me what MID meant to them. Their responses took three forms—first, energy: MID aimed to encourage the use of renewable energy instead of fossil fuels; second, technology: MID promoted the production of durable products and institutions; third, politics and society: MID represented a drive to advance the country by bringing improvements as well as reducing corruption and ethnic disharmony.

None of these respondents repeated Paul Bérenger's comment that MID was an "empty shell", but two individuals levelled similar criticisms to those cited above: one complained about corruption, while the other bemoaned the contradiction of cutting down trees to build roads while claiming to support the planting of trees. This statement was the only response that indirectly associated MID with environmental issues: all other responses focused on energy, technology, politics, or society. Given that 'Environment' is one of MID's so-called '5Es', I was curious about why so few people had drawn a direct association between MID and the environment. Public conceptions of MID as weighted towards energy solutions may be explained, firstly, by the fact that this was indeed MID's initial focus and, secondly, by the continued disproportionate allocation of MID funding and media attention towards renewable and efficient energy projects. Since the concept of sustainable development tends to include ecological sustainability alongside economic development,[96] I decided to probe my respondents' understandings of 'environment' and 'sustainable development' as well as the relationship between the two.

For many of my respondents, the environment—*lanvironnman* in Kreol, from the French *l'environnement*—refers to the totality of one's environs, milieu, surroundings, or physical context, including the social, political, and cultural circumstances thereof. For instance, when I asked Samantha to define the environment, she stated that "the environment is everything that surrounds us; in fact, everything forms part of the environment". Similarly, for Adela, "our environment is our place where we live; my surroundings [*anturaz*, from the French *entourage*] form part of the environment". When I started to probe by asking whether "the word environment is not reserved for …", she interrupted me before I could say the word, replying:

> Nature? No, but nature forms part of it; the authorities use it in relation to nature and cleanliness. The Minister of the Environment is concerned with cleanliness, keeping places clean, green spaces, but for me the environment is not only that: our environment is where we live.

My ethnographic material suggests that these marginalised urban Mauritians are holistic and "contextualist" when it comes to their understandings of

96 Croll, E., and D. Parkin, "Anthropology, the Environment and Development," in *Bush Base: Forest Farm—Culture, Environment and Development*, ed. E. Croll, and D. Parkin (London: Routledge, 1992), 6.

nature and society:[97] they do not see nature as somehow separate from society, but instead regard both as interconnected constituent parts of an anthropocentric lived environment.

I wondered if this holistic understanding of "environment" would correspond to a similarly broad understanding of "sustainable development" as incorporating the environment, the economy, society, and governance (as per the definitions given by members of the public and special interest groups who participated in the government's national consultation process that led to the expansion of the scope of MID). In fact, however, most of these people told me that their first exposure to the concept of 'sustainable development' was through MID. As such, their understanding of 'sustainable development' was shaped by MID's narrower focus on energy efficiency.

'Sustainable development' is usually rendered as *développement durable* in French and *devlopman dirab* in Mauritian Kreol. The word 'development' seems relatively unproblematic in translation between the English, French, and Mauritian Kreol: in all three languages it refers to growth, progress, advancement, and evolution. People in Mauritius whom I asked to define 'development' mentioned job creation, infrastructure, and construction. 'Sustainable', on the other hand, appears more problematic. While the English 'sustainable' is best translated into French as *durable*, the French word *durable* is perhaps better translated back into English as 'durable', the adjective meaning lasting, enduring, and constant. The same applies for Mauritian Kreol, in which *dirab* means hard-wearing and durable. When I asked people what they understood by the concept of 'sustainable development', most gave explanations that reflected these additional connotations of durability—namely, growth and progress through infrastructure and construction designed to last for the long term—but they did not spontaneously mention the environment.

I wondered how—if at all—people conceptualised the relationship between sustainable development and the environment. When I asked what sustainability means in relation to the environment, my interviewees responded by talking about the promotion of reuse, recycling, composting, and

---

97 Hornberg, "Ecology"; see also Croll, and Parkin, "Anthropology"; Descola, P., "Constructing Natures: Symbolic Ecology and Social Practice," in *Nature and Society: Anthropological Perspectives*, ed. P. Descola, and G. Pálsson (London: Routledge, 1996), 99; Hastrup, K., "Nature: Anthropology on the Edge," in *Anthropology and Nature*, ed. K. Hastrup (New York: Routledge, 2014), 1–2; Walley, C., *Rough Waters: Nature and Development in an East African Marine Park* (Princeton: Princeton University Press, 2004); West, P., J. Igoe, and D. Brockington, "Parks and Peoples: The Social Impact of Protected Areas," *Annual Review of Anthropology* 35 (2005): 251–277.

waste management. Thus, even when asked directly about the environment, their responses recall the remit of MID Working Group 3 (Environment—Pollution, Wastes, Environmental Health) rather than that of Working Group 2 (Environment—Preservation of Biodiversity and Natural Resources). In theory, the Mauritian government conceptualised MID broadly as a socio-political project encompassing the '5Es' of energy, environment, employment, education, and equity. In reality, however, it would seem that the narrow focus of MID in practice reinforced the notion that sustainable development is principally concerned with energy efficiency and waste management. From the perspective of my marginalised respondents, the Mauritian government has yet to demonstrate convincingly that its concept of sustainable development (also) includes preservation of biodiversity and mitigation of climate change.

5 Conclusions

This chapter has deployed MID as a lens through which to examine debates about the environment, sustainability, and development in Mauritius. MID was supposed to encompass the environment, employment, education, and equity; yet its primary focus was on energy efficiency. This chapter has contrasted this narrow focus of Mauritian government actors with two other broad categories of citizen: firstly, environmentalists who argue that MID ought to have been able to also incorporate preservation of biodiversity and mitigation of climate change; and, secondly, marginalised urban citizens, many of whom came to see MID in particular—and, consequently, the concept of sustainable development in general—in terms of socio-economic development rather than (also) in terms of environmental sustainability. I would suggest that MID could had been more effective if environmental sustainability had been foregrounded. Furthermore, MID would have been seen as more inclusive if economic development had been more transparently and equitably distributed among the population as a whole rather than being seen as aimed at rewarding big businesses that support the governing parties.

CHAPTER 8

Transformative Learning for Global Change? Reflections on the WASCAL Master Programme in Climate Change and Education in the Gambia

Irit Eguavoen and Erick Tambo

Abstract

African universities could be appropriate settings to start a process of creative re-thinking about alternative economic and climate-friendly development pathways that lead to socially inclusive growth and the alleviation of poverty. Education, capacity building, and a transformative learning agenda have come into academic and political focus. This chapter begins with a brief conceptual overview of the current research and work streams on climate change communication and education, followed by an introduction to climate change activities in The Gambia that sets the stage for an empirical examination of a master programme training West African students in 'Climate Change and Education' at the University of The Gambia. The findings illustrate practical challenges, students' backgrounds and motivations, contestations over the curriculum, as well as the university's restricted bargaining power due to its dependency on a single foreign donor. While embracing inter- and transdisciplinarity, the transformative learning agenda called for by European scholars has not yet been taken up.

1 Introduction

Universities will play an essential role in influencing African economies and societies in the coming decades. During their training, students work on their skills, study up-to-date material, and conduct research. They are, however, also influenced by their zeitgeist, the dominant political paradigms and challenges of their time, as well as by contemporary approaches to facing these challenges. Graduates then carry their knowledge and ideas into their work environments and start translating them into policy and practice. In this way, universities may serve as catalysts for social and economic transformation processes in their countries, many of which face the combined challenge of attaining poverty-alleviating growth that is simultaneously both economically sustainable and socially inclusive.

 | DOI:10.1163/9789004410848_010

Climate change certainly belongs to the dominant political paradigms of the 21st century. Today, humankind lives in a world that faces a number of historically unprecedented, self-manufactured risks, such as nuclear power hazards, the highest global population in our history, rapidly spreading global pandemics, and data abuse via the internet.[1] Anthropogenic climate change, a process that started around the 1790s with the beginning of European industrialisation, belongs to this list of anthropogenic effects that have changed the earth so irrecoverably that sociologist Anthony Giddens, in his lecture on "The Politics of Climate Change" at Durham University in 2012, argued that we now live in "a world that has fallen out of history".[2] Therefore, on the one hand, one may think that human experience is of little help in guiding our future actions and, as climate change is a historically young process, a look at the past may not be very useful; on the other hand, however, there is rich academic research and documentation of environmental change and human responses thereto that may guide our present educational efforts in this regard.

Another way to describe our times—the Anthropocene—in relation to climate change is from the theoretical standpoint of post-normality, which describes a state that is characterised by much uncertainty, high political stakes, values, and vehement demands for urgent decisions.[3] This view may lead to a shift in perspective away from responding to climate change with rational, technical, and economic approaches towards an acknowledgement of the historical contingency of consumers' demands, lifestyles, and practices in industrial societies, which form "the primary problem underpinning climate change [namely] the continuing energy and material intense character of contemporary industrial society".[4] If conditions and habits are historically constituted and not naturally prescribed, they may change again, and science and policy may determine, to some extent, the direction in which changes lead and which pathways are socially suitable. One of the challenges of our time, therefore, consists in undertaking a "fundamental re-thinking of the kind of societies we aspire to and then abstracting from this which technology, which

1 Giddens, Anthony, *The Politics of Climate Change* (Cambridge: Polity, 2009).

2 Durham University, "The Politics of Climate Change" (talk by Anthony Giddens, October 31, 2012), last modified November 7, 2012, accessed August 9, 2015. http://www.youtube.com/watch?v=yCgzxnRKhcU&list=PL1zMD_kTXdjzTG8CnFyh1abezH-FmkP8V.

3 For a critical review of the concept, see Turnpenny, John, Marvis Jones, and Irene Lorenzoni, "Where Now for Post-Normal Science? A Critical Review of its Development, Definitions, and Uses," *Science, Technology, & Human Values* 36.3 (2011): 287–330, accessed June 8, 2015. doi:10.1177/0162243910385789.

4 Healy, Stephen, "Post-Normal Science in Postnormal Times," *Futures* 43.2 (2011): 204, accessed June 8, 2015. doi:10.1016/j.futures.2010.10.009.

science, which economics might best serve this purpose, rather than the other way around".[5]

As universities are places of knowledge production and circulation, they may be relevant actors in social education on climate change in West Africa. First, universities could be appropriate settings to start such a process of creative rethinking in combination with research on the social and technical feasibility of alternative economic and development pathways. This rethinking project is inherently political in character. Second, universities can be places where political visions are initiated and from where the discourse can then be carried into wider society. O'Brien et al. point out that rethinking education and capacity building as a response to climate change, in particular, resembles calling for "nothing less than a revolution" in education, especially by building on interdisciplinary and transdisciplinary approaches as well as through transformative learning.[6] Transformative learning in this context means "synthesiz[ing] and apply[ing] the latest findings from a range of fields, including cognitive science, teaching methods, creativity and collaborative knowledge creation to transform education".[7] The authors are quite optimistic that developing countries are more open-minded towards transdisciplinary approaches than Western universities; however, they acknowledge the "tendency to mimic northern agendas [...] rather than to develop an endogenous narrative and agenda on what is needed in capacity building"[8] because of the unequal distribution of resources and voices in the academic world.

One other keyword in this context is 'leapfrogging'. The statistics on populations and economies in sub-Saharan Africa clearly tell us that the countries of the continent tend to follow economic pathways and dynamics that differ from those taken by the industrialised European countries and the fast-growing Asian economies (see the chapter by Dietz in this volume). This is partly due to their specific colonial and national histories as well as their environmental conditions. However, technological leapfrogging is another variable. Examples thereof include mobile phone technology, which leapfrogged the extension of telephone wires, as well as off-grid wind and solar energy provision. Renewable energy generation in African countries may serve climate change mitigation, on the one hand, and the extension of basic energy services on the other. Low

---

5 Healy, "Post- Normal Science in Postnormal Times," 204.

6 O'Brien, Karen et al., "You Say You Want a Revolution? Transforming Education and Capacity Building in Response to Global Change," *Environmental Science & Policy* 28 (2013): 49, accessed June 8, 2015. doi:10.1016/j.envsci.2012.11.011.

7 Ibid., 50.

8 O'Brien et al., "You Say You Want a Revolution?" 55.

carbon energy generation, efficient transport systems, and low-energy architecture in urban centres, for example, are options that require more political lobbyism and expertise in Africa. As such, knowledge production and circulation are not the only matters of concern here. University education may also deliver visions for transformation and point to values to guide individual behaviour, economic investment, and legal reforms.

The limited number of university programmes on climate change in West Africa do not seem to go very far. Political perspectives are not given high priority in the current curricula; instead, these programmes are dominated by a scientific approach that is geared towards understanding physical climate change and, foremost technical adaptation. Education, however, is perceived by the United Nations Framework Convention on Climate Change (UNFCCC) as one crucial ingredient of adaptation.[9] Padgham et al., for instance, state that investments in "education and training, curricula development, research, and effective practices for communication of research findings [...] are in essence an adaptation response"[10] that needs to be pursued in sub-Saharan countries. They also state that "the potential of African Universities [...] to be active participants and agents for transformative change is not being fully realized".[11] This is an issue of investment into curricula and teaching methods, into research and science communication, as well as a matter of political participation. It is important for the new generation of students to gain more scientific knowledge on climate change as well as knowledge of its ecological, economic, social, and political impact and possible responses thereto. It is also important for future engineers, architects, economists, and other experts to gain an awareness of how their work contributes to environmental degradation and global warming,[12] as well as to develop ideas about alternative work and life models.

Some of the existing post-growth debates popular among actors from industrial societies as well as suggestions for simple and sufficient ways of living in

9 Climate Change Secretariat: UNFCCC, ed., *The Nairobi Work Programme on Impacts, Vulnerability and Adaptation to Climate Change: Action on the Ground. A Synthesis of Activities in the Areas of Education, Training and Awareness-Raising for Adaptation* (Bonn: UNFCCC, 2010).

10 Padgham, Jon, Hassan Virji, and Clark Seipt, "Promoting Climate Change Curricula Development at African Universities," *Environmental Development* 5 (2013): 169, accessed June 8, 2015. doi:10.1016/j.envdev.2012.08.001.

11 Ibid., 169.

12 Leal Filho, Walter, "Climate Change at Universities: Results of a World Survey," in *Universities and Climate Change: Introducing Climate Change to University Programmes*, ed. Walter Leal Filho (Berlin, Heidelberg: Springer, 2010), 2.

order to attain sustainability[13] will probably be contested and resisted by sections of the urban African middle classes, who have successfully left poverty and the subsistence economy behind and have oriented themselves towards the social and economic achievements of the industrialised countries (such as global market integration, consumerism, high mobility, and high energy demand). These classes, however, do also tend to show an increased awareness of environmental issues (including clean environments, healthy food, and sustainability). Meanwhile, for many Africans living in rural areas, the global transition town model[14] is also unlikely to prove compelling, because subsistence lifestyles and low-energy consumption are the norm there anyway and are rather perceived as a development drawback that needs to be overcome. This underlines the important role of African universities as institutions that initiate processes of transformative learning and where locally embedded visions and alternative models of low-carbon yet efficient economies will be envisaged and tested.

Scientific knowledge on climate change will not meet any significant resistance in contemporary academic and political circles. The very high political priority given to climate change by the international community—the urgency paradigm—is, however, contested by several African countries[15] that clearly prioritise other contemporary challenges, such as HIV/AIDS, Ebola outbreaks, national security, or high rates of unemployment and poverty, over adaptation and CO_2 mitigation.

Another challenge for climate change education should not be ignored: uncertainty in climate model projections is an epistemological condition of climatology. The findings that derive from climate projections create visions of the environmental and social realities of the coming decades. Models, however, produce very abstract forms of knowledge that extend beyond the grasp of most people.[16] The science of climate change, therefore, needs to be translated to become meaningful and conceivable for the public. Moreover, since research and polity cannot also but rely on a mutual understanding of numerous

---

13 For the debate in Germany, cf. for example: Schneidewind, Uwe, and Angelika Zahrnt, *The Politics of Sufficiency: Making It Easier to Live the Good Life* (Munich: Oekom, 2014).

14 Transition Netzwerk e.V., "Was ist eine Transition Town Initiative?" accessed October 10, 2018. https://www.transition-initiativen.de/was-ist-eine-transition-town-initiative.

15 Lockwood, Matthew, "What Can Climate-Adaptation Policy in Sub-Saharan Africa Learn from Research on Governance and Politics?" *Development Policy Review* 31.6 (2013): 647–676, accessed June 8, 2015. doi: 10.1111/dpr.12029

16 Climate models and their inherent uncertainties are difficult to understand for non-modelers, since they are technically complex and require expert knowledge to understand the methodologies employed in the process.

diversely interpreted frameworks of climate change, there is even more need for translation work to be done between these different systems of knowledge as well as between different interest groups and across various sectors and scales.

The chapter begins with a brief conceptual overview of the current research and work streams on climate change communication and education, followed by an introduction to climate change activities in The Gambia to set the stage for an examination of a master programme at the University of The Gambia (UTG) that trains students in 'Climate Change and Education'. This programme forms part of the WASCAL Graduate School Programme, which will be described before discussing the Gambian experience in detail.

2 Current Research and Work Streams

2.1 *Science Communication and Public Education*

As in any form of journalism or education, "communication about climate change needs to be informed by the latest research evidence based upon a very clear understanding of the demographics of the target audience, their media preferences and opinion leaders, their political ideologies and cultural worldviews".[17] The skills required to do this kind of translation work are usually taught at the university level in communication science and public health programmes.

Apart from scientific studies on climate change, the international market offers an increasing number of popular science books on the topic as well as books aimed at particular audiences (such as children or youth).[18] The availability of these books in West Africa, however, is still very limited and is therefore not discussed here. Another fruitful area of study is curricula development, but the analysis of farmers' uptake of climate forecasts and media discourse on climate change is probably the most researched area of climate science translation.[19] The number of anthropological studies on

17 Anderson, Alison, "Rethinking Climate Change Communication," in *Culture, Politics and Climate Change: How Information Shapes Our Common Future*, ed. Deserai A. Crow, and Maxwell T. Boykoff (Abingdon, New York: Earthscan from Routledge, 2014), 222.

18 See, for example, the children's book about the importance and restoration of mangrove forests in Eritrea by: Roth, Susan, and Cindy Trumbore, *The Mangrove Tree: Planting Trees to Feed Families* (New York: Lee and Low Books, 2011).

19 E.g. Lewis, Justin, and Tammy Boyce, ed., *Climate Change and the Media* (New York, Bern, Berlin, Brussels, Frankfurt, Oxford, Vienna: Peter Lang, 2009).

climate change communication and its interplay with local cultural models and knowledge systems is also increasing.[20]

2.2 *Cultural Representations and Image Politics*

Only recently have the image politics of climate change—the delivery of politically relevant meaning through visual media—entered the scholarly debate. Schneider and Nocke have shown how the translation of climate science happens and why translation faces the challenge of "visualizing the unimaginable".[21] Translations of climate science have moved beyond the academic sphere and into the global political arena of the UNFCCC (with the IPCC assessment reports representing the most powerful translation exercise in this domain) and, from there, into diverse sectors. The initial engine for this translation and knowledge distribution exercise was Article 6 of the UNFCCC and the resultant Nairobi Work Programme, which was introduced at the Conference of Parties (COP) in Montreal in 2005:

> Under Article 6, Parties are to promote and facilitate at the national and, as appropriate, subregional and regional levels: (a) the development and implementation of educational and public-awareness programmes on climate change and its effects; and (b) the training of scientific, technical and managerial personnel.[22]

These educational sectors included development cooperation, whereby climate change is translated into outreach material, manuals, and toolboxes. The production of this sort of informational material has accelerated over the past five years[23] in West Africa, especially in The Gambia.[24]

Climate change is also appropriated by actors who contribute to other cultural domains, such as arts and popular culture. Illustrations, cartoons, and

---

20 E.g. Rudiak-Gould, Peter, "Promiscuous Corroboration and Climate Change Translation: A Case Study from the Marshall Islands," *Global Environmental Change* 22 (2012): 46–54.

21 Schneider, Birgit, and Thomas Nocke, "Image Politics of Climate Change: Introduction," in *Image Politics of Climate Change: Visualizations, Imaginations, Documentations*, ed. Birgit Schneider, and Thomas Nocke (Bielefeld: transcript, 2014), 10.

22 Climate Change Secretariat: UNFCCC, *The Nairobi Work Programme*.

23 Lockwood, Matthew, "What Can Climate-Adaptation Policy in Sub-Saharan Africa Learn," 647–676.

24 E.g. Shanahan, Mike et al., *Climate Change in Africa: A Guidebook for Journalists* (Paris: UNESCO, 2013); Ampomah, Gifty, and Tahia Devisscher, *Adaptation Toolkit: Guidebook for Researchers and Adaptation Practitioners Working with Local Communities* (Dakar: ENDA, 2013).

photographs visualising the impact of climate change on the lives of humans and other species serve as images that "have started to shape the [Western] imagination of a world under the conditions of climate change".[25] Some of these images have travelled globally[26] and have even developed into iconic representations of climate change (such as the polar bear on its melting ice floe), which in turn tend to create their own realities.[27]

The imagery of climate change in Africa mainly depicts familiar representations of the cultural desertification discourse of deadly dryness, which is iconised in images of cracked soils, starving people, and animal carcasses. As such, there seems to be little visual distinction between drought and climate change. The increase in floods has not yet become part of global public consciousness, nor have coastal erosion, loss of biodiversity, or other physical manifestations of the effects of climate change.[28] Climate change appears to affect mainly poor rural farmers and herders, as well as the poor inhabitants of coastal towns. In contrast, the urban metropolis and the 'modern Western lifestyles' of the African urban middle classes seem to be completely omitted from the current global iconography of climate change in Africa.[29]

On the one hand, communication practitioners require skills and knowledge to perform translation work, including writing ability, technical skills to draw infographics and edit outreach materials, and knowledge of public relations. The curation of exhibitions, photography, and film-making are additional options for raising awareness. All such skills can be taught through training and practical exercises. On the other hand, there is also a demand for scholars to be able to conduct research on existing communication channels, images, and narratives. This means that graduates could also be trained to become experts in the media and communication sciences by undertaking a more academic-focused programme that is sensitive to the power of images for the delivery of political messages when analysing and reflecting on visual discourses.

25 Schneider, and Nocke, "Image Politics of Climate Change," 10.

26 Ibid., 17.

27 Von Storch, Hans, and Werner Krauß, *Die Klimafalle: Die gefährliche Nähe von Politik und Klimaforschung* (Munich: Hanser, 2013), 75.

28 Climate change is of course only one variable contributing to these phenomena.

29 A general indication thereof can be attained from a search for images on Google or other search engines using the keywords "Climate Change Africa", as well as from images on the covers of books and reports about climate change in Africa.

2.3 *Translation and Knowledge Brokerage*

Krauss and von Storch argue that public education on climate science is no longer the issue of the day and that, instead, the main challenge is "to connect scientific climate knowledge to the everyday reality and perceptions of climate and weather of people living in vulnerable areas".[30] According to them, the dialogue between climate science and "cultural climatology approaches"[31] relies on knowledge brokers who mediate between the science and the stakeholders. Furthermore, these knowledge brokers have to translate and connect scientific and local knowledge; as such, they must be conversant with and fluent in both knowledge systems.[32] Were this understanding of translation work to guide curricular development, a different set of knowledge and skills would become a part of it, including courses in social and cultural anthropology/regional cultural studies and science communication.

2.4 *Educational Technologies*

Modern technologies provide new possibilities for supporting public education with regard to climate change in Africa. Educational technology, also termed 'information and communication technology' (ICT) in education or 'e-learning technology', is "the study and ethical practice of facilitating learning and improving performance by creating, using and managing appropriate technological processes and resources".[33] It encompasses hardware, software, and educational theories geared towards enhancing learning. Although the term 'educational technologies' may refer to all kinds of analogue technologies, such as photographs, film, and video, it is usually used specifically to discuss digital computer technology. Depending on the objectives of educational activities and their target groups, numerous methods and tools could be used to strengthen the competencies and capacities of stakeholders who are facing climate change-related challenges in Africa.

Telecommunication-based information systems, audio, and video are mostly used in a non-formalised learning context for public education and to raise public awareness. Interactive voice response systems, as a specific solution offered by telecommunication-based information systems, allow users to call in, select, and receive preconfigured information about different topics.

---

30 Von Storch, and Krauß, *Die Klimafalle,* 216.

31 A term taken from Endfield, Georgina, "Reculturing and Particularizing Climate Discourses: Weather, Identity, and the Work of George Manley," *Osiris* 26.1 (2011): 145.

32 Von Storch, and Krauß, *Die Klimafalle,* 222.

33 Richey, Rita C., "Reflections on the 2008 AECT Definitions of the Field," *TechTrends* 52.1 (2008): 24–25, accessed June 8, 2015. doi:10.1007/s11528-008-0108-2.

Platforms such as Frontline SMS,[34] which operates as an open-source solution for distributing and collecting information via text messages, are already used in several African countries in the context of disaster response and recovery, climate change mitigation, and food security. Community radio stations and participatory videos are also effective tools to reach and empower communities and to raise awareness. The UNESCO project "Empowering Local Radio with ICTs" built up the staff capacity of 32 local radio stations in seven African countries from 2012 until the end of 2014. The Red Cross/Red Crescent Climate Centre utilises videos developed by farmers and local communities for public education.

The evolution of the internet from a static information broadcasting medium to an interactive environment (the so-called Web 2.0) makes it very easy and inexpensive for people to communicate, network, create, share, or exchange information and ideas. Social media improves individual and group communication, synchronous broadcasting, collective authoring, and dissemination of opinions, and documents, and it allows for questions to be raised and discussions to be opened. Social media, which is mainly used by urban youth, is also becoming increasingly important for the delivery of public education. Online communities of practice, such as MyCoop,[35] have emerged in Africa to support online learning. Classical media channels such as community radio, television, CDs, and DVDs are still in predominant use in remote areas for people with low levels of literacy and limited access to internet. However, mobile phones, combined with content in local languages, are increasingly being used to make the transition between these classical channels and the new forms of media.

Self-paced content and learning activities delivered through CD/DVD or online via a digital learning environment are other approaches being used in formal learning contexts (schools and universities) as well as for advanced training (of stakeholders and decision makers). Open educational resources, webinars, and Massive Open Online Courses (MOOCs) are complementary solutions used by African students to improve their knowledge and to complete modules of their curriculum.[36] Many universities exploit the advantages of e-learning or distance learning using modern ICT to improve the quality

---

34 Frontline, "Frontline SMS Africa," accessed August 9, 2015. http://www.frontlinesms.com/tag/africa/.

35 MyCoop, "Managing Your Agricultural Cooperative," accessed August 9, 2015. http://moodle.itcilo.org/mycoop/.

36 For example, the online course offered by the UNFCCC in partnership with the University of Reading entitled "Climate Change: Impact and Responses", developed in cooperation with seven East African universities and launched in 2015, accessed August 28, 2015. http://www.reading.ac.uk/ssc/resource-packs/ccir/.

of their educational programmes. Although technical tasks can be outsourced to technicians, educational technologies do require climate change translators and educators to adhere to suitable formats and content. While each educational technology may be suitable for the teaching of specific content, it may be less suited to other material.

3 Climate Change Adaptation and Mitigation Activities in the Gambia

Despite being a very small African state with only about two million inhabitants, The Gambia has adopted a proactive role and is visible in the international political climate change arena. It has been actively following the UNFCCC road map by preparing and promptly submitting two national communications, a 'National Adaptation Program of Action' and a 'Nationally Appropriate Mitigation Action', to the UNFCCC Secretariat. The national delegation also played an active role in the COP negotiations, especially when the UNFCCC focal point for The Gambia and head of the Gambian COP delegation, Pa Ousman Jarju, was also chair of the Least Developed Country group from 2011 to 2012.

The government has revised its legal framework and established new laws, including the National Disaster Management Act of 2008 and the Renewable Energy Act of 2013 (see Figure 8.1). The most important national policies have also undergone a process of climate mainstreaming.[37] By 2014, the National Climate Change Policy was a work in progress, the realisation of which was being actively pursued. The Gambia has been a member of the West African Science Service Centre for Climate Change and Adapted Land Use (WASCAL) since 2010 (see below). In 2013, the Gambian parliament ratified the WASCAL agreement and constitution, which entails financial responsibilities towards this regional initiative. The government under former president Yahya Jammeh, which was in power until 2017, clearly intends to make use of the funding opportunities offered by the UNFCCC mitigation framework to increase the national power supply and to implement the envisaged Low-Emission Climate-Resilient Development Strategy of 2014.[38] There are indications that

37 Lauer, Hannes, and Irit Eguavoen "Mainstreaming Climate Change Adaptation into Development in the Gambia: A Window of Opportunity for Transformative Processes?" in *Innovation in Climate Change Adaptation,* ed. Walther Leal Filho (Berlin: Springer, 2016), 87–98.

38 Ripplinger, Pascal, "Renewable Energies in Sub-Saharan Africa: A Case Study of the Gambia" (MSc thesis, University of Bonn, 2014).

Draft: D. Badgie, H. Lauer, P. Ripplinger, I. Eguavoen; Design: J. Vajen; WASCAL 2014.

FIGURE 8.1 Historical outline of the climate change adaptation policy process in the Gambia

the democratically elected government under President Adama Barrow continues this policy since 2017.[39]

The Gambia is also one of the first countries to have integrated the term 'climate change' into the name of its national environment ministry. In September 2015, The Gambia submitted the Intended Nationally Determined Contributions (INDC) to the UNFCCC, being among the first countries to have done so.[40]

The country also acts as a front runner in the area of climate change education by hosting the first international university programme of its kind in West Africa. The programme was launched in 2013. In early 2014, the Gambian

39 This chapter, however, describes the Gambian policy environment under the Jammeh government.

40 UN Framework Convention on Climate Change (UNFCCC), "The Gambia Submits its Climate Action Plan Ahead of 2015 Paris Agreement", last modified September 28, 2015, accessed February 05, 2019. https://unfccc.int/news/the-gambia-submits-its-climate-action-plan-ahead-of-2015-paris-agreement.

Ministry of Basic and Secondary Education announced the extension and radical reform of the country's school curricula to respond to climate change and related environmental hazards. The ministry also initiated a cooperation and memorandum of understanding with the NGO Biodiversity Action Journalists Gambia in a bid to become involved in public education campaigns for different audiences, especially media representatives.

4 The West African Science Service Centre for Climate Change and Adapted Land Use (WASCAL)

Climate services can be briefly described as customised information products that are delivered to the public by specialist organisations.[41] Taking the Climate Service Center (CSC) in Hamburg as a role model,[42] the German Federal Ministry of Education and Research (BMBF), which is the main funder of the CSC in Germany, started an initiative to establish similar institutes in the West and Southern African regions.

Reducing knowledge uncertainties about future climate and providing useful recommendations for climate change mitigation and adaptation as well as other climate services in the West African region are central to WASCAL's mandate. By 2018 it had been ratified by eleven West African countries, having received the legal status of an international research institute from the Economic Community of West African States (ECOWAS) in 2013. The aim of the Climate Service Centre is ambitious, as it envisions hosting expert staff and networks as well as a central data management system to serve the region. It aims to facilitate the pooling of climate knowledge in West Africa and strengthen the position of the eleven West African member countries[43] in international climate change negotiations. The organisation also supports politicians and other

41 For a definition as well as a discussion of challenges, see Bowyer, Paul, Guy P. Brasseur, and Daniela Jacob, "The Role of Climate Services in Adapting to Climate Variability and Change," in *Handbook of Climate Change Adaptation*, ed. Walter Leal Filho (Berlin: Springer, 2015), 533–550.

42 The German CSC in Hamburg, also known as the German Climate Agency, was founded in 2009 and works in response to requests by private individuals, organisations, businesses, and political decision makers. It offers news scans as well as many other products and services to facilitate science communication and knowledge transfer. All services are provided by a team of 35 to 40 scientists. One of the four departments within the non-profit organisation is a communication department.

43 The eleven countries are Senegal, Gambia, Mali, Burkina Faso, Niger, Côte d'Ivoire, Ghana, Togo, Benin and Nigeria. By 2015, negotiations over membership were ongoing with four of the remaining five ECOWAS countries (Cape Verde, Liberia, Sierra Leone, and Guinea),

stakeholders in deciding on regional adaptation and mitigation options. The work of WASCAL, which is primarily concerned with producing and delivering climate services, has been accompanied by the WASCAL Governing Board as well as an international Scientific Advisory Committee.

Administration and coordination of WASCAL are located in Ouagadougou and Accra. The organisation's headquarters in Accra hosts a Capacity Building Department, which includes a division of the Graduate Study Programme (see below) and a division of the In-Service (Professional) Training Programme in charge of online learning, short-term training, and other capacity building programmes.[44] In August 2015, the Competence Centre in Ouagadougou had a staff complement of ten researchers as well as eight administrative and technical support members in addition to an executive director. About four years after the establishment of the centre, however, the educational interface was not yet operational. Instead, the dominant idea was to generate research results that would then be translated for and communicated to public audiences. The idea of offering demand-driven climate services, inter alia the provision of customised information products upon request, was not yet prominent.

5 University Education under WASCAL

There are a number of different networks across the countries that form part of the organisation. The WASCAL Graduate Study Programme is run at ten West African universities in nine countries and has a financial volume of €10.2 million (2010–2016).[45] About 160 students from the West African region are enrolled in one of the ten doctoral and master programmes (Figure 8.2), from which the first students graduated in 2014. WASCAL grants all students a monthly stipend as well as research funds during their enrolment period. Doctoral students write up their dissertations in their home countries and also have the opportunity to spend some months at a European university.

The participating universities admit citizens of member countries to the programme according to a quota system (one student per country per academic year). This means that applicants from countries with a lower number

while the other ECOWAS member state, Guinea-Bissau, had already signed the WASCAL constitution. These negotiations also concerned themselves with the question of how to integrate WASCAL into ECOWAS structures.

44 WASCAL, Memorandum CBP/001/2015, Accra.

45 WASCAL, *Report 2010–2014* (Bonn, 2015).

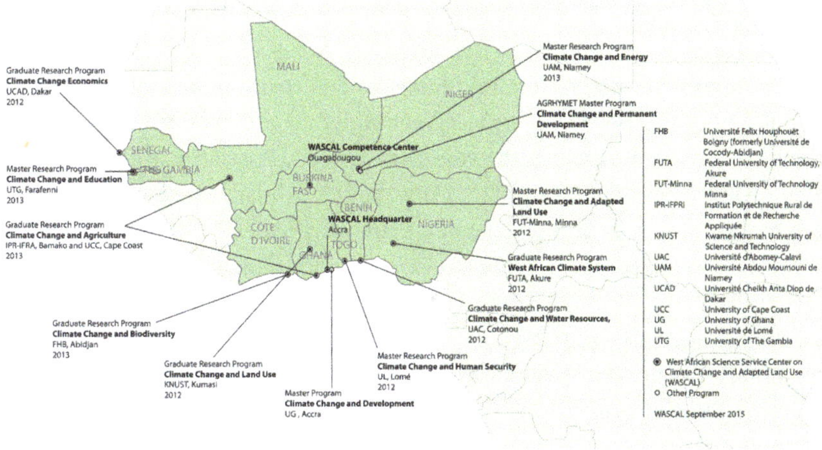

FIGURE 8.2 Capacity building on climate change at West African universities

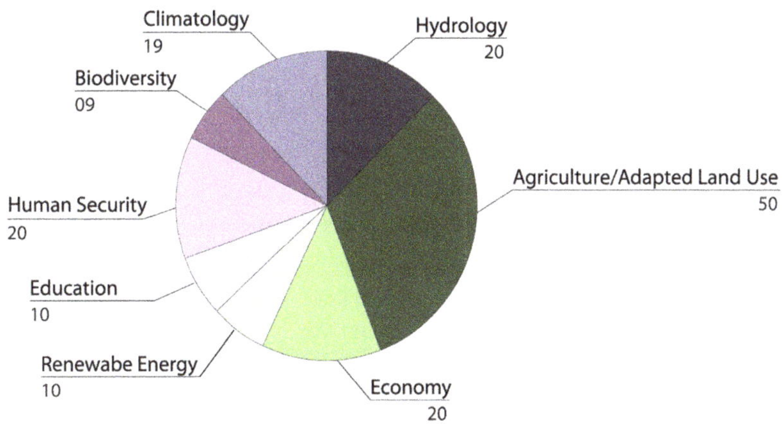

FIGURE 8.3 Areas of capacity building in the WASCAL Graduate Study Programme

of potential candidates (due to smaller national populations and fewer universities), such as The Gambia or Togo, enjoy a comparative advantage over applicants from countries with strong academic landscapes, such as Ghana and Nigeria, where universities are able to develop additional programmes (such as the Climate Change and Sustainable Development master programme at the University of Ghana, which is not part of WASCAL) more easily.

There is a strong natural science bias in the Graduate Study Programme (108 out of a total of 158 students). Three programmes (see Figure 8.3) share a focus on agriculture and adapted land use (together enrolling fifty students), while the economic as well as the human security programmes enrolled a combined total of forty students. The three-year doctoral programme offers students the opportunity to pursue studies in the natural sciences and economics only.

The MSc programme in Climate Change and Education was not included in the initial WASCAL proposal but was later added to the institutional portfolio after the Gambian member of the WASCAL Advisory Board suggested this idea, which received immediate support from the other board members. In a two-year programme, students are expected to become experts who are competent and creative in the translation of climate projections, technological approaches, risk uncertainties, and potential adaptation and mitigation options. However, the programme's purpose and curriculum are not uncontested. Curriculum development and implementation face several challenges, some of which have to do with the programme's current dependency on aid as well as the fact that it is a new course with no existing model to emulate.

6 Climate Change and Education MSc. Programme

6.1 *Observations on Campus*[46]

The MSc in Climate Change and Education is hosted at the University of The Gambia's North Bank Campus in Farafenni, a market town on the border with Senegal. The campus is located in three dozen previously abandoned office buildings, laboratories, and houses; dusty signs indicated their former purpose—that of housing a donor-funded medical research institute that relocated to a more central location some years earlier. The programme director and students from ten West African countries lived on this sleepy campus in very comfortable accommodation.

By the end of the third semester, students still had no access to television, while radios and newspapers had to be sourced individually. The library contained about eighty new and relevant books. The electricity supply was reliable but was limited to a few hours per day and was supported by a diesel generator. Internet access was provided via personal mobile phones and later via wireless LAN. The lack of access to media in everyday life was not optimal, especially for students training in the field of education and communication. Communication was generally a challenge, as international students did not speak the local languages. Although the students tried to interact with the local population

46 These observations were made by the authors during a series of short-term visits to Farafenni in 2014 and 2015, where they were invited to teach. Their observations were complemented by discussions with students and staff, e-mail correspondence, and their own photographs.

through regular visits to the market and by attending Sunday mass, the international programme remained largely invisible to locals.

According to the orientation of the graduate programme, at least 25 per cent of the course content should deal with educational topics. However, most of the master's students' thesis proposals focused on the natural sciences and economics, although the proposals did contain an educational component (namely, the dissemination of research findings among relevant stakeholders using educational models and approaches). Fewer than thirty per cent of students were working on an empirical topic that focused primarily on climate change and education.

6.2 *Divergent Ambitions and a Contested Curriculum*

The MSc in Climate Change and Education commenced in The Gambia in November 2013 with a preliminary first-semester curriculum that included mathematics and statistics modules as well as a module on the 'Challenges of Climate Change Education'. The language of instruction is English. Students joined an intensive three-month English proficiency course in Ghana before moving on to The Gambia. As of March 2014, there was only a vague idea of which modules should be included in the revised curriculum. This was being developed in an ad hoc manner, with much depending on the availability of lecturers. Each module is supposed to offer 39 hours of teaching, in keeping with university regulations.

UTG is a young university with about four thousand students. It was founded in 1998 and operates on several campuses across the country. The Faculty of Education in Brikama is the largest, with eight teaching staff and about four hundred students. UTG mainly offers bachelor's degrees but also runs master's programmes; the WASCAL programme is currently the only international degree on offer. WASCAL supports UTG by funding the programme staff, guest lecturers, and scholarships for international students, as well as the renovation of infrastructure. UTG provides facilities on campus and pays for electricity and maintenance. Currently, however, there is "very limited investment in research and development in The Gambia [...] Research policy development, research investment and funding [...] research collaboration, and partnerships [...] are determined and treated as administrative matters without significant input from scientists."[47]

47 Ozor, Frederick U., "Research Governance and Scientific Knowledge Production in The Gambia," *South African Journal of Science* 110.9/10 (2014): 6, accessed June 8, 2015. doi:10.1590/sajs.2014/20130185.

The small number of UTG teaching staff poses a challenge for the curriculum. Little academic expertise on climate change is yet available in The Gambia, while UTG lecturers are generally occupied with other programmes. As a result, some lecturers teaching in the WASCAL programme are based in other West African countries or in Europe. The director of the master's programme (who holds degrees in chemistry and education) also serves as dean of the Faculty of Education, a job that brings with it numerous additional duties and frequent appointments on other campuses.

During a preparation meeting among lecturers in Germany, important questions arose: What kind of master's programme should be offered? What would the students need to know? These questions had been discussed by the advisory board of the master's programme in 2013, at a time when the board was composed exclusively of West African natural scientists who envisioned a science programme in climate change with an educational component. Moreover, as outlined in the introduction, practice-oriented modules in line with new trends in climate change education—such as action-based learning, problem-based learning, and living laboratories—were not sufficiently included in the curriculum. A module on transformative learning suggested by some of the international lecturers was also absent.

Some external lecturers, however, assumed that the programme would contribute primarily to the arts and humanities (despite being categorised as an MSc programme) by focusing on communication and public education skills that would enable graduates to translate climate science for non-scientific audiences in their future careers. As such, these lecturers envisioned an educational programme with an environmental specialisation.

These divergent visions eventually became apparent in the final curriculum (see Figure 8.4) and in the research proposals by master's students (most of whom proposed science projects with an educational outreach component). The confrontation with the final programme outcome led to disappointment on the part of some of the lecturers, especially among those who had sought to strengthen the role of the social sciences and humanities and had identified the need to develop capacities in public education and science communication within WASCAL. In addition, some students had enrolled in the programme out of a particular interest for public or environmental education, thinking that it was an education and communication programme. Some practical modules on education were integrated into the curriculum (Tables 8.1 and 8.2), but science modules were ultimately dominant. At the end of the teaching modules, neither modules on environmental nor climate change communication had been offered. Instead, the curriculum ended with a module on climate science.

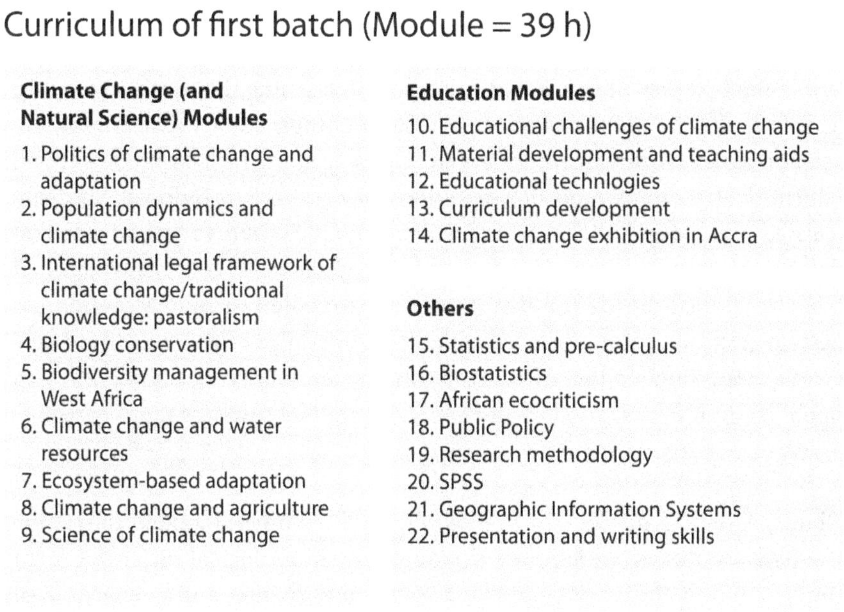

FIGURE 8.4 Curriculum of climate change and education, February 2015

TABLE 8.1 Transdisciplinary learning: Curation of an educational exhibition

The WASCAL students from the programme were invited by the Ghanaian Mmofra Foundation, which runs a community children's park in Accra, to help curate a climate education exhibition. The exhibition was built like a playground parkour for children aged eight to 15 years. The students conceptualised the exhibition, designed its elements and put them in place in Ghana. After opening, they offered guided tours along the parkour during which they explained in chronological order what climate was, as well as the causes and effects of climate change, before showing how people could adapt to climate change and help mitigate further CO2 emissions. The project was supported by the German Embassy in Ghana and the British High Commission.

WASCAL programme students were invited by the Ghanaian Mmofra Foundation, which runs a children's community park in Accra, to assist in curating a climate education exhibition. The exhibition was built to resemble a playground parkour for children aged eight to 15 years. The students conceptualised the exhibition, designed its individual components, and set them up in

TABLE 8.2 Technology-enhanced learning in the context of climate change

While the first part of this module focused on learning theories and models, as well as the design of e-learning courses and their quality assessment, the second part was practically oriented. Open-content license models, open educational resource repositories and e-libraries were introduced. Students learned how to use the Learning Management System to administer and deliver e-courses. Afterwards, the students applied the content which had been developed during the first part to the Moodle Learning Management System, and they peer-assessed the learning modules of the other students. It was important in the African context to familiarise the students with open- content license models and open educational resource repositories.

Ghana. After the opening of the exhibition, they offered guided tours along the parkour, during which they explained in chronological order the definition of climate, the causes and effects of climate change, and how people can adapt to climate change and help to mitigate further CO_2 emissions. The project was supported by the German Embassy and the British High Commission in Ghana.

While the first part of this module focused on learning theories and models as well as the design and quality assessment of e-learning courses, the second part was practically oriented. Open-content license models, open educational resource repositories, and e-libraries were introduced. Students learned how to use the learning management system to administer and deliver e-courses. Thereafter, the students applied the content that had been developed during the first part of the course to the Moodle learning management system and peer-assessed the learning modules of the other students. It was especially important in the African context to familiarise the students with open-content license models and open educational resource repositories.

6.3 *Background and Aspirations of the Students*

In the WASCAL programme in The Gambia, the student gender ratio (six females to every four males) was more balanced than in other WASCAL graduate schools, where male students predominated. Six out of ten students had an educational background in either the natural sciences or economics, although there was a wider diversity in their first-degree disciplines (Figure 8.5).

None of the students had a disciplinary background in education. Only one student had a diploma and postgraduate degree in communication science as well as work experience in public education. Another student had

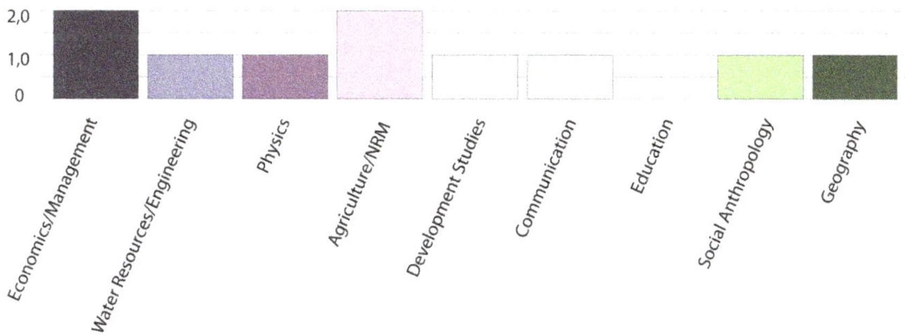

FIGURE 8.5 Educational background of students
NOTE: The selection process for prospective students is organised by the individual WASCAL programmes. In the case of the MSc programme in The Gambia, candidates were ranked by the director and the Advisory Board.

worked for some years at a private company as a communications officer. Two of the ten master's students already had a master's degree before entering the programme.

Although some of the students had an interest in climate change or had even undergone some training in this area, their knowledge at the beginning of the course was quite limited. Nevertheless, all the students were highly motivated to learn about environmental issues and climate change. They were communicative and readily involved themselves in role play, games, and other creative activities. Conversations with the students were a means to investigate their motivations for choosing the programme as well as their professional aspirations. Their comments illustrate that not all the students were eager to take up a profession in the field of climate change communication and public education after graduation. Generally, the male students appeared to have been more strategic in their reasons for selecting the programme. They identified advantages over other master's programmes, such as the provision of scholarships, international exposure and experience, the opportunity to build professional networks and learn English, and the greater chance of being admitted into PhD programmes after graduation:

> I chose this programme first for the topics. The topic was about climate change, and I know climate change is a big deal. It is something very interesting, and I was thinking about doing an international master's after my undergraduate studies.
>
> The programme puts together students from ten countries. So it creates a network among the students. If you finish this programme, you

> already have a strong network. Some students you can collaborate and work with. I chose this programme because, first of all, I love to study the environment … something about environment … and, secondly, because there is a scholarship that can help me to study. […] I have a Master One degree [a BA degree]. After this programme, I am planning to look for an American scholarship for training—one year of training in the USA—because after this programme, I suppose I will have a good background in English. So, I can apply for this programme. And after this programme, I am planning to apply for a PhD from the WASCAL programme.[48]

The female students were more concerned with being able to teach and support people who are affected by climate change. Their potential future roles as climate change translators, however, were described only vaguely. Most students had no professional experience in science communication; hence, there was no obvious role model whom they could use for purposes of orientation:

> I think that if people are more educated, they can face the challenges. So I think education is the basis of everything we can do about climate change.
>
> How to educate people about climate change? The phenomenon … African countries are based on agriculture. They [farmers] don't know the term 'climate change.' But they know that the climate … the events … are changing. So, we have to educate them to adapt. Because the phenomenon … we cannot stop the phenomenon. But we can educate them to adapt to this phenomenon.
>
> I wouldn't want to continue education in the climate change domain. So, I will be an expert in climate change, and maybe I can discuss climate change issues. And I will try […] to put [my BA degree in] natural resources [management] together to raise [foster] Africa's development. Because Africa's development, I think, is the most important. Because you cannot be learning, learning and, finally, we don't use it to develop … to do something good.
>
> According to the IPCC […], my country […] is among those countries which are more affected by climate change. So, what I am looking for is to get the necessary background here and go and help my country easier there.

48 All quotations are by students from the programme's first intake after they had been involved in the programme for a few months.

> When I go back to my country, I know that there are many problems due to climate change. Because there are floods. There is drought. So, how should we work with the environmental people? Look for some adaptation measures in order to help people … those who are most affected.

6.4 *Financial Sustainability of the Programme*

The MSc in Climate Change and Education is highly dependent on German funding, which is also complemented by funding from UTG itself. Current accounting procedures to the BMBF involve four institutions in three countries (UTG, WASCAL headquarters in Ghana, the University of Bonn in Germany, and the so-called *Projektträger* (project-administrating organisation) also located in Bonn). This means that all receipts must be physically passed through all these institutions, a task that represents a logistical challenge given the size of the entire WASCAL organisation and its Graduate Study Programme across West Africa (which multiplies these efforts). These transnational money transfers and the accounting thereof are complex and demand a lot of manpower. Finally, an accountant had to be recruited particularly for this master programme.

In general, for its initial four years, WASCAL's basic dependency on donor funds has meant that the survival of the master's programmes has been completely dependent on decisions taken by the German Ministry. For example, the BMBF requested WASCAL to improve accounting and to reform the Graduate Study Programme, deciding to freeze a substantial amount of money that had already been assigned to the programme until these requests were met in a satisfactory manner. From the perspective of the funder, this is undoubtedly a legitimate request. In practice, however, the African partner universities could not recruit new students during the foreseen intake period or even for a year thereafter. In The Gambia, there was nothing UTG leadership could do to counteract the BMBF decision. The university had made five additional scholarships available for Gambian students in the second intake and had already selected new students for these and the WASCAL scholarships; however, it was now unable to recruit these candidates due to the frozen funding. This had a number of practical implications, thereby decreasing the overall efficiency and quality of the programme.

Extra funding from the German embassy in Ghana and the US Embassy's Science Fellows Program as well as the funders of the practical exercise in Accra supported the MSc in Climate Change and Education. In terms of economic efficiency, however, the programme was not yet efficient enough in its own right given the investment in funding and working hours required to produce ten MSc graduates.[49,50] The only way to balance cost and effect would be to rely on more enduring core funds, to decrease the number of international lecturers by relying more on local teaching staff, and to at least double the number of students per intake by accepting a higher number of Gambian and/or self-financed students. Finally, enrolments should be undertaken every semester or year to reduce staff costs, facilitate exchange across intakes, and allow for more efficient use of the campus facilities.

7 Discussion

The WASCAL Graduate Study Programme attracts students because it offers them an opportunity for an international education as well as serving as a valuable source of income for two to three years. Although there is a need for experts in climate change education who could support the development of relevant communication science, public education, and climate services, not all graduates were interested in pursuing a career in this profession. A lack of role models and the unclear job descriptions in this field seem to be contributing to this reluctance. Climate change education requires a high number of skills, including educational, technical, and cultural skills, as well as a very good foundation in climate change knowledge itself. However, climate change education is a professional field rather than an academic discipline. As such, a definition of what constitutes climate change education still remains vague, especially because the production and delivery of climate services is also a comparatively young professional field.

In general, lectures on how to develop a better model in line with a transformational learning agenda were not part of the curriculum. Such modules could be proposed for the programme in future. It is unclear, however, who at UTG would have an interest in taking up this agenda. Though transformative learning was not emphasised, the programme was nevertheless interdisciplinary and transdisciplinary. For example, the curation of the exhibition in Accra was a very useful transdisciplinary element of the curriculum.

49 From October 2013 to June 2015, the overall expenses for the programme's first intake of students amounted to €315,000 (ca. €35,000 per graduate). The first students graduated in November 2015. Future intakes will prove less costly, since they will benefit from prior infrastructural investment (such as in buildings, the library, and transport).

50 Ultimately, there were only nine graduates from the first intake because one student moved to Saudi Arabia to enrol in another master's programme that offered a more lucrative scholarship.

The transnational education model based on the 'flying faculty approach', which is followed by many excellent graduate programmes in Africa to invite lecturers and practitioners from abroad to teach short-term modules, represents an immense problem for students. The short period in which a module is taught places students under particular pressure. The content is very often condensed, intense, and difficult to digest within the short period of time during which the lecturer is still around to respond to questions. This approach also increases the overall cost of the programme as well as its carbon footprint.

Furthermore, most of the foreign lecturers were not able to meet staff from the respective UTG faculties during their short stays in The Gambia and were unable to learn about UTG's organisational and teaching structures, argue for the relevance of 'Climate Change and Education', or prepare future co-teaching. As Padgham et al. have pointed out, local African faculty members would also need access to resource packs, teaching workshops, and mobility programmes to equip them with the skills to integrate climate change relevant content into their regular courses and degree programmes.[51] In addition, more discussion and exchange of ideas as well as closer cooperation between local and international lecturers, the Faculty of Education, and the university administration would be essential to ensure that the programme can stand on its own two feet in the near future.

UTG's ownership of the programme was limited, however, because decisions over the extension of core funds and student recruitment were not taken within the university's own structures but at higher multilateral levels. A multi-donor funding strategy as well as the opening up of other sources of funding, including moderate tuition fees, are potential strategies to increase UTG bargaining power.

8 Outlook

There are still few master's and doctoral programmes in Africa that specialise in climate change. As such, graduating with a climate change degree suggests that graduates will have a comparative advantage over holders of other more conventional degrees. This is especially the case since climate change, as a cross-sectoral challenge to economic development, ranks high on the agenda of civil society and research organisations working in Africa as well as on that of national and international funders of research. There is thus a job market

51 Padgham, Virji, and Seipt, "Promoting Climate Change Curricula," 170.

awaiting graduates who have learnt to work in interdisciplinary and transdisciplinary environments.

In addition to providing their graduates with international experience as well as language and intercultural skills, programmes such as UTG's MSc in Climate Change and Education can successfully building of expert networks across Africa. These may help to boost the role of African scholars as well as increase the quantity and quality of African research contributions to future assessment reports by the Intergovernmental Panel on Climate Change (IPCC).

As the discussion section has shown, establishing and sustaining an international programme such as UTG's MSc in Climate Change and Education comes with many practical challenges. Besides dependence on a single donor, other pressing challenges include divergent views on the objectives and content of the programme. At UTG (and probably also at the other WASCAL Graduate Schools), transformative learning was not yet a priority when designing the curriculum. However, as the director underlined: "We have learned a lot during the first years. Now we really need to get the opportunity to recruit more batches [of students] and make things work better. We know how to do it now." This is certainly true, irrespective of whether the UTG Faculty of Education will follow the call for a revolution in education and capacity building in response to global change. The observations also illustrate that the expectations expressed in the introduction, which underline the central role that African universities should have in facilitating transformative change and low-carbon development pathways, may be a bit too high, at least in the case of universities that are in a similar situation to UTG.

African ownership of master's and doctoral programmes also means that international partners need to respect the fact that their African counterparts may not subscribe to European educational agendas, as outlined by O'Brien et al.[52] African faculties may set different priorities in curricular development and may define different objectives for the master's programmes than those envisaged by the international funder or by European lecturers. This is fully legitimate, because it underlines the autonomy of the African universities, which despite receiving financial support should not be seen as implementing partners but as agenda-setting institutions in their own right.

Finally, considering the confusing mushrooming of climate change-related initiatives in Africa that aim to increase public education and transnational capacity building, it is also a valuable academic project documenting what pathways these take and how they work in practice.

52 O'Brien et al., "You Say You Want a Revolution?"

PART 4

Prospects

CHAPTER 9

Africa in Transition: What Role for the Environment?

Ton Dietz

Abstract

Africa is not only the fastest growing continent in terms of population (reaching at least two billion inhabitants by 2050), it has also been probably the fastest-growing continental economy of the last decade, with urban centres bursting with energy, (and demand, and expectations), and with growing evidence of agricultural breakthroughs with many crop and animal products. During the last fifty years, Africa's population had already tripled and its expanding agricultural sector—and exploding internal and external demand for firewood and other forestry products as well as for water, minerals, and fossil fuels—had resulted in massive land use change. This brought with it a number of threats to biodiversity and soil quality, all amid the looming backdrop of global climate change and its potential impact on the continent. In this chapter, these recent and predicted processes of environmental change will be unpacked and interpreted in relation to their differentiated impacts on diverse geographical settings (such as humid or arid areas) as well as on manifestations of economic, political, and cultural diversity (such as crop cultivation or pastoralism).

1 Africa Rising

According to a cover story published in *The Economist* in May 2000, Africa could be written off as "the hopeless continent";[1] by 2011, this image had changed dramatically to that of "Africa Rising",[2] a phrase also used in *Time* in 2012.[3] although it should be said that they were optimistic earlier as well, in

1 The Economist, "The Hopeless Continent," *The Economist*, May 13, 2000.
2 The Economist, "Africa Rising. The Hopeful Continent," *The Economist*, December 1, 2011, accessed February 2015. http://www.economist.com/node/21541015.
3 Perry, Alex, "Africa Rising," *Time*, December 3, 2012.

 | DOI:10.1163/9789004410848_011

1998, with a cover saying "Africa Rising".[4] This was only two years before *The Economist* had given up hopes about Africa. In August 2013, *The Economist* opened a debate about contemporary African economic growth with the following introduction:

> Sub-Saharan Africa has made huge leaps in the last decade. Malaria deaths in some of the worst-affected countries have declined by 30% and HIV infections by up to 74%. Life expectancy across Africa has increased by about 10%, and child mortality rates in most African countries have been falling steeply. A booming economy has made a big difference. Real income per person has increased by more than 30%, whereas in the previous 20 years it shrank by nearly 10%. Africa is the world's fastest-growing continent at the moment. Over the next decade its GDP is expected to rise by an average of 6% a year, not least thanks to foreign direct investment. FDI has gone from $15 billion in 2002 to $37 billion in 2006 and $46 billion in 2012.

Nevertheless, it added:

> But Africa has seen booms before, only to crash-land. Violence, corruption and bad governance are still common. Are we seeing a false dawn once again?

In the debate that followed, a small majority of the contributors supported the positive image and argued that indeed 'Africa is rising'. However, a substantial minority was sceptical and identified many barriers on the road to real transformation. The weakness of governance institutions and the lack of employment opportunities, but especially the limitations of industrial development, were cited as the major barriers to sustained growth.[5]

Africa's recent economic growth is very much rooted in the increasing demand for its mineral resources since the start of the 21st century, when Asia became a major importer of African oil, gas, diamonds, gold, platinum,

4 Time, "Africa Rising," *Time*, March 30, 1998.

5 Environmental issues were not mentioned in this debate. In 2011, in contrast, my inaugural lecture as Professor of African Development at Leiden University was entitled "Silverlining Africa" and combined macro- and micro-impressions of recent changes in Africa, including the shifting environmental dynamics. In my lecture, I presented an image of the continent in which "glimmers of hope" were more important than the "images of doom and gloom" that had so dominated writing about Africa in the previous decades.

uranium, copper, bauxite, and many other raw materials for industries elsewhere.[6]

During the past fifteen years, Africa has achieved an annual economic growth rate of around five per cent. The continent's economy is currently much stronger than it was in the 1980s and 1990s. A tipping point for Africa's shifting global image was the fact that six of the ten fastest-growing economies in the world in the period between 2000 and 2010 could be found on the continent. Moreover, the African economy has been growing faster than the Asian economy as a whole since 2010. Economic predictions show that a continuous economic growth rate of five per cent per year would create a continental economy with a total annual income of €14,500 billion or €7,200 per capita in 2050.[7] Although such forecasts were long thought of as being completely unrealistic or even ridiculous, some economists are beginning to take them seriously,[8] although a lot of scepticism still abounds.[9] Furthermore, the economic slowdown in China in the last few years as well as the falling oil price have led to growing anxiety about the African economy.[10]

Africa's rapid recent economic growth should first and foremost be understood in the context of its even more remarkable population growth, or what some authors call the continent's "population explosion".[11] A comparison between Europe and Africa based on the rough population data in Table 9.1 below gives rise to a remarkable picture.

6 KPMG, "Mining in Africa. Towards 2020," last modified 2013, accessed February 2015. https://www.kpmg.com/Africa/en/IssuesAndInsights/Articles-Publications/Documents/Mining%20Indaba%20brochure.pdf.

7 own calculations.

8 Versi, Anver, "Out of Africa: The Rise of the African Economy," *Global Briefing,* January 2013, accessed February 2015. http://www.global-briefing.org/2013/01/out-of-africa-the-rise-of-the-african-economy/.

9 Clarke, Duncan, "Africa 2050 Economic Revolution," *Global Briefing*, April 2013, accessed February 2015. http://www.global-briefing.org/2013/04/africa-2050-economic-revolution/.

10 Bloomberg, "The Impact on Africa from Falling Oil Prices," *Bloomberg Business*, October 6, 2015, accessed February 28, 2016. http://www.bloomberg.com/news/videos/2015-10-06/the-impact-on-africa-from-falling-oil-prices; Financial Times, "African Growth Feels the Strain from China's Slowdown," *Financial Times*, October 27, 2015, accessed February 28, 2016. http://www.ft.com/cms/s/0/ef4d851a-7bc5-11e5-98fb-5a6d4728f74e.html#axzz3ydMKJhCE; BBC, "Africa–China Exports Fall by 40% after China Slowdown," *BBC News,* January 13, 2016, accessed November 20, 2016. www.bbc.com/news/world-africa-35303981.

11 Baer, Drake, " 'The Biggest Change of Our Time' is Happening Right Now in Africa: Africa's Population Explosion Will Change Humanity," *Tech Insider Innovation,* August 18, 2015, accessed February 28, 2016. http://www.techinsider.io/africas-population-explosion-will-change-humanity-2015-8.

TABLE 9.1 Long-term population developments and prospects, Europe and Africa, 1650–2100[a]

	Europe		Africa	
	Population in millions	Percentage of world population	Population in millions	Percentage of world population
1650	80–100	17	80–100	17
1900	650	25	100–150	4–6
2015	740	11	1080–1170	16–17
2050	700–800	8–9	2000–2400	22–25
2100	650–850	6–8	3400–4400	31–36

[a] Own assessments, based on Hayashi, Reiko, "Long Term Population Dynamics in Africa: A Perspective from the Urban Structure" (paper presented at the conference "Fifth African Population Conference," Union for African Population Studies, Arusha, December 10–14, 2007). http://uaps2007.princeton.edu/papers/70296; GGDC, "Growth of World Population, GDP and GDP Per Capita before 1820," last modified September 3, 2008, accessed February 2016. http://www.ggdc.net/MADDISON/other_books/appendix_B.pdf; Heilig, Gerhard, "Demographics," last modified 2015, accessed February 28, 2016. http://www.demographics.at/Growth-Decline/images/PPP2015_Africa_Tot-Pop_1950-2100.gif and http://www.demographics.at/Growth-Decline/fig_ppp2015_africa-europe_tot-pop.htm.

At the height of its economic and geopolitical power around 1900, Europe was home to 25 per cent of the world's population, while Africa had only four to six per cent. In the centuries before 1900, Europe had experienced very fast population growth, while Africa's population had stagnated. Part of this stagnation was caused by the massive forced migration of Africans to the Americas and, to a lesser extent, to the Middle East. In the 20th century, however, it was Africa's turn to grow, which it did at a speed without precedent on a similar scale in world history. Demographic predictions show that at some point during the current century Africa will have a higher percentage of the world's population than Europe has ever had. Africa's total population will start to resemble that of East, South, and Southeast Asia combined. In ever more parts of Africa, population densities will start to resemble those in the densely populated zones of 'intensive anthroposphere' in Asia, as is already visible in parts of Nigeria and the Great Lakes region.

Growth figures for Africa's urban population are even more stunning. When many African countries attained their political independence in the period around 1960, Africa's urban population stood at only 65 million, a fifth of its total population. In 2010, there were 450 million urban residents, or 42 per cent of the total population, with the continent soon forecast to pass the fifty per cent threshold. By 2050, demographers expect an urbanisation rate of seventy per cent, which would mean that 1.4 billion people would be living in Africa's cities. Indeed, it is not unlikely that an urbanisation level of more than eighty per cent will be reached by 2100, which would amount to an urban population of 2.4 billion people. Currently, one out of every seven Africans lives in a metropolis, an urban agglomeration of more than one million inhabitants, of which there were 53 on the continent in May 2014.[12] Of these, six agglomerations already had more than five million inhabitants: Cairo (15.2 million), Lagos (12.5), Johannesburg/Ekurhuleni[13]/Pretoria (10.6), Kinshasa (9.7), Luanda (5.7), and Khartoum (5.1). However, the highest growth rates and the bulk of African urbanites can be found in cities with 500,000 to one million inhabitants, which represent the localities at the forefront of African urbanisation. Moreover, Africa has an extremely youthful population with a median age of below 19 years.[14] African metropolises and cities are thus full of youthful energy, with many of these children and youth coming from rural areas or having parents who originated from rural areas, either in their own countries or further afield. Sociologists are often concerned about the so-called urban youth "explosion", or refer cautiously to its "mixed blessings".[15] In the wake of the so-called Arab Spring, political scientists have tended to be more positive in their assessment of the youth as the drivers and shapers of new, more democratic forms of governance in Africa,[16] regardless of whether they are connected to the "emerging

12 Demographia, *World Urban Areas: Built-Up Urban Areas or World Agglomerations*, 10th edition, 2014-05 revision. (Belleville: Demographia, 2014). http://www.demographia.com/db-worldua.pdf.

13 Formerly referred to as 'East Rand'.

14 Schaffnit-Chatterjee, Claire, "High Population Growth in Africa: A Mixed Blessing," last modified September 9, 2013, accessed February 2015. http://www.howwemadeitinafrica.com/high-population-growth-in-africa-a-mixed-blessing/30043/.

15 E.g. Mago, Stephen, "Urban Youth Unemployment in Africa: Whither Socio-Economic Problems," *Mediterranean Journal of Social Sciences* 5.9 (2014): 33–40.

16 Harsch Ernest, "'Arab Spring' Stirs African Hopes and Anxieties: Will Popular Upsurge in the North Help Spur Reform South of the Sahara?" *Africa Renewal*, August, 2011, accessed November 20, 2016. http://www.un.org/africarenewal/magazine/august-2011/%E2%80%98arab-spring%E2%80%99-stirs-african-hopes-and-anxieties.

middle classes" in Africa's cities.[17] With this in mind, economists talk hopefully of a "demographic dividend".[18]

The rapid increase in Africa's urban population, coupled with rising wealth in its cities, has enabled the swift growth of the continent's food production systems, particularly in the immediate hinterlands of the metropolises.[19] Comparing Africa's population and food production figures from 1960 to 2010, Dietz et al. and other studies conclude that agricultural production has more than caught up with the continent's extreme population growth, and that the available food energy per capita improved, not deteriorated, in those fifty years.[20] While the population grew by 350 per cent during this period, area under cultivation increased by 210 per cent and yields of the major food crops by 180 per cent. In addition, livestock numbers also increased, with the number of chickens in particular burgeoning.[21] However, if we study the dynamics of those fifty years more closely, we can see that the positive developments were mainly restricted to the 1960s and early 1970s and then again to the period from about 2000 onwards. The intervening years were problematic for African agriculture, and most of Africa's population had to cope with hunger, mass malnutrition, and, in some areas,[22] even with famine, starvation, and related deaths. Hunger and malnutrition remain a major problem for many Africans today. Currently, the food situation in the Horn of Africa is rather dramatic again,[23] and even in areas where the food situation is relatively good, like in West Africa, observers note that:

17 Mubila, Maurice, Mohamed Safouane Ben Aissa, and Charles Leyeka Lufumpa, "The Middle of the Pyramid. Dynamics of the Middle Class in Africa," accessed November 20, 2016. http://www.afdb.org/fileadmin/uploads/afdb/Documents/Publications/The%20Middle%20of%20the%20Pyramid_The%20Middle%20of%20the%20Pyramid.pdf; Heeralall, Nirmal, "The World's Fastest Growing Middle Class," accessed February 2015. http://www.uhy.com/the-worlds-fastest-growing-middle-class/.

18 PRB (Population Reference Bureau), "Africa and the Demographic Divided," last modified April 2013, accessed February 2015. http://www.prb.org/Publications/Reports/2013/africa-demographic-dividend.aspx.

19 See Akinyoade, Akinyinka et al., ed. Digging Deeper: Inside Africa's Agricultural, Food and Nutrition Dynamics (Leiden, Boston: Brill, 2014).

20 E.g. Dietz, Ton, "Comparing the Agricultural Performance of Africa and Southeast Asia over the Last Fifty Years," in *Asian Tigers, African Lions: Comparing the Development Performance of Southeast Asia and Africa*, ed. Bernard Berendsen et al., (Leiden, Boston: Brill, 2013), 85–128.

21 By 540 per cent, as indicated in the FAO statistics used in Dietz, "Comparing the Agricultural Performance."

22 Namely, the Sahel and the Horn of Africa in the mid-1970s and early 1980s.

23 World Food Programme, "Horn of Africa Crisis," last modified February 2016, accessed February 28, 2016. https://www.wfp.org/crisis/horn-of-africa.

> malnutrition remains a major challenge for the region: between January and October 2015, almost a million children have been detected and treated for severe acute malnutrition. Whether harvests are good or bad, each year the region must manage between 3 and 5 million people experiencing food insecurity.[24]

Despite these worries, improvements in food provision as well as innovations in agriculture have been impressive since the turn of the century, much of which have happened despite government policies that as of yet still do not prioritise mass poverty alleviation or rural and agricultural transformation.[25]

2 Likely Consequences for the African Environment

Africa's current and expected population growth as well as its increasing rates of urbanisation and economic growth will have many consequences for the continent's environment and for global climate change. For one, the exploration and exploitation of Africa's coal, oil, and gas reserves will probably accelerate, not only to feed the world's hunger for fossil fuels (and China's in particular), but also to support Africa's own industrial and transport expansion. Demand for oil and gas during the years with very high global market prices for fossil fuels has created an explosion of exploration, which has resulted in new oil and gas discoveries and the start of exploitation and related infrastructure development. Overall, however, it can be expected that Africa will also want to diminish its dependence on wood- and fossil fuel-based energy production. Africa's current low rate of energy consumption and relatively low contribution to global greenhouse gas emissions may not remain so modest for long. With its vast potential for hydropower, solar power, wind energy, and geothermal energy, Africa still has the opportunity to choose to base its future energy supplies predominantly on clean energy, while it could also make use of its uranium deposits for nuclear power production. Africa could become the world's leader in solar energy production, with grids connecting the Sahara to Europe and elsewhere. Recurrent phases of low oil and gas prices once again make it more lucrative to shift to

---

24 Bossard, Laurent, "West Africa: Security Crisis and Food Crisis," OECD Insights, accessed February 28, 2016. http://oecdinsights.org/2016/01/06/west-africa-security-crisis-and-food-crisis/.

25 Clarke, "Africa 2050"; Henley, David, *Asia-Africa Development Divergence: A Question of Intent* (London: Zed Books, 2015).

non-fossil fuel exploration and exploitation. On the other hand, finding huge oil and gas fields (such as the gas discoveries along the coast of southeast Africa) will tempt political leaders as well as global and local energy companies to further develop fossil fuel exploitation, instead of deciding that this potential wealth could better be ignored. Much depends on political will, both on a global level and in Africa itself, to shift from fossil to non-fossil fuels. This political will is yet to become apparent, and Africa's political and business leaders will probably hinder rather than foster the mitigation of climate change. An additional problem in Africa is its relatively high dependence on fuelwood and charcoal. Demand for charcoal has created ever wider 'circles of destruction' around areas of demand, which have now crossed the five hundred-kilometre radius around some urban areas. For example, the Kenyan city of Kisumu obtains its charcoal from Turkana, while the Ugandan capital, Kampala, relies on northern Uganda.[26] Furthermore, reducing the use of fuelwood and charcoal (or banning it, as has happened in some countries) will mean that alternatives must be found, and it is quite likely that the energy sources that replace wood, at least in the near future, will be fossil fuels.

The second issue is the impact of African urbanisation on the environment and on climate change. The enormous growth of Africa's cities and metropoles invariably means that land use in and around these centres of human habitation will change rapidly. Urban landscapes will gradually replace nature, and great effort is needed to ensure that Africa's cities become 'green cities'. Areas of rich nature (wildlife and plant biodiversity) near (big) cities will be threatened by urban sprawl, more intensive hunting and agriculture, and the harvesting of wood for timber and fuelwood as well as for charcoal production (see below). Moreover, cities' demand for drinking water and water for industrial and mining use will weigh heavily on the environment of their hinterlands, possibly leading to severe problems due to overuse of existing groundwater reserves and river systems. This may have consequences for the microclimate of areas around cities or even major implications for the global climate. Almost across the continent, Africa's cities are a mixture of

26 IRIN (Integrated Regional Information Networks), "Uganda Charcoal Boom a Bust for Forests," IRIN Humanitarian News and Analysis, February 7, 2010, accessed February 24, 2016. http://www.irinnews.org/report/94810/uganda-charcoal-boom-a-bust-for-forests; MEWNR, "Analysis of the Charcoal Value Chain in Kenya: Final Report" (Nairobi: Ministry of Environment, Water and Natural Resources, Government of Kenya, 2013), accessed February 25, 2016. http://www.kenyaforestservice.org/documents/redd/Charcoal%20Value%20Chain%20Analysis.pdf.

shining architectural modernity—high-rise buildings, fancy hotels, complex motorways—juxtaposed with huge slums, unsanitary housing and streets, and high-volume traffic composed of highly polluting vehicles that cause a significant direct and indirect death toll. If Africa's urban population grows from the current 450 million people to 1.2 billion by 2050 and 2.4 billion by 2100—which may be underestimates—a lot will depend on the ways in which urban housing and transportation are designed. It is possible for Africa to develop healthy and climate-smart cities if the urban political leadership has both the foresight and the means to facilitate this. If current trends continue, however, the opposite will occur. Much will also depend on the future role of industrialisation in Africa's rapid urbanisation.

Currently, Africa's economic expansion is not based on a significant industrial revolution, but this may change during the course of the 21st century. There are signs that international investors, including some from China, see Africa as the new 'factory of the world'. This vision foresees the continent making use of its potentially huge number of young and energetic labourers, of whom more and more have a relatively good level of education and are willing to work for wages and with degrees of discipline that are attractive to both foreign and domestic investors. As such, African cities might also become hotspots of industrial expansion, and a lot will depend on the environmental foresight and care with which these new industrial developments will be designed: dirty and wasteful, or smart, green, and innovative. In parts of Africa, there is growing awareness of the need to make African cities more sustainable, and there is also some willingness to diminish the environmental footprint of urban and industrial development. However, when looking at existing debates about 'sustainable cities' in Africa, it is impossible to escape the conclusion that a lot of the discussion emanates from European, North American, and Japanese donor agencies and companies, and that many authors writing about these issues have non-African backgrounds, although many do feel genuinely attached to Africa. Examples include articles like that by Klopp and Paller[27] or the African Green City Index.[28] However, there are

27 Klopp, Jacqueline M., and Jeffrey W. Paller, "How Africa Can Build Inclusive, Safe and Sustainable Cities," *The Conversation*, October 9, 2015, accessed February 28, 2016. http://theconversation.com/how-africa-can-build-inclusive-safe-and-sustainable-cities-48445.

28 Siemens, "African Green City Index," accessed February 26, 2016. http://www.siemens.co.za/sustainable-development/pdf/African-Green-City-Index.pdf. The names of the Expert Advisory Council for this very interesting overview of fifteen African 'green cities': David Wilk, Marie Jane Ortega, Hiroaki Suzuki, Pablo Vaggione, Sebastian Veit, and Nicholas You.

also sections of the growing urban middle classes in Africa who are beginning to assert themselves by demanding healthier, cleaner, and safer urban conditions. UN-Habitat's overview of sustainable city planning also gives some convincing examples of genuine commitment to this goal by African municipalities, including cities in Egypt, Tanzania, Malawi, Kenya, and even Somalia.[29] Occasionally, African scholars who are connected to African knowledge centres also contribute to this discussion.[30]

My hypothesis is that the most important environmental and climatic effects of Africa's rising population, rapid urbanisation, and growing urban wealth will be felt in the rural areas, particularly because of agricultural expansion and intensification. The anticipated rapid growth in demand for food and a likely increase in average income levels will fuel the further swift expansion of land under cultivation, where food will be grown primarily for consumption in urban areas in Africa rather than for traditional export markets in Europe. This will also mean an incremental growth in crop yields per hectare and per farmer, with harvests for some crops increasing considerably. More and more farmers will sell a rising proportion of their produce to commercial markets ('commodification'), especially to burgeoning urban markets that are located relatively nearby. To sustain their economic outputs, an increasing number of farmers will become more dependent on purchased inputs, tools, and technology, and hence on functioning markets. Many farmers will combine the production of different agricultural products in response to the rapid increase in urban demand for milk, meat, fish, vegetables, fruit, and forest products such as wood and charcoal by intensifying livestock, horticultural, and/or agroforestry production and/or by increasing the number of livestock that they own and/or the area of land under their cultivation. High demand for charcoal will become a major threat to off-farm vegetation. Many farming families will seek to interlink themselves with the emerging non-agricultural economy, not only as farmers, but also as (casual) labourers and as producers of non-agricultural goods and services. Many will profit from remittances sent home by relatives in cities or other countries. Concepts like

29 UN-Habitat, "Planning Sustainable Cities. UN-Habitat Practices and Perspectives," Nairobi, 2010, accessed February 25, 2016. https://portals.iucn.org/library/sites/library/files/documents/Man-Dev-143.pdf.

30 E.g. Nwaka, Geoffrey I., "Planning Sustainable Cities in Africa," *Canadian Journal of Urban Research* 5.1 (1996); Mwangi, Samson Wokabi, "Partnerships in Urban Environmental Management: An Approach to Solving Environmental Problems in Nakuru, Kenya," *Environment and Development*, 12.2 (2000): 77–92; Ogbazi, Joy U., "Alternative Planning Approaches and the Sustainable Cities Programme in Nigeria," *Habitat International* 40 (2013): 109–118.

(relative) "deagrarianisation"[31] and "multi-spatial livelihoods"[32] are probably more applicable to African farming dynamics than "agricultural specialisation in modernising farms".[33] Deagrarianisation signifies the diminishing role of the agricultural sector in the African economy, because the industrial, mining, and service sectors are becoming more important. Furthermore, the relevance of the agricultural sector will abate comparatively, as more elements of food, feed, fibre, and wood production are becoming part of an increasingly complex value chain in which they are no longer counted as agricultural products but as services (such as transport or sale of agricultural produce, but also insurance) and manufactured products (such as tinned food or processed milk). At the micro-level, deagrarianisation also means that farm-based families shift (some of) their sources of income from the agricultural sector to a range of other activities that may take place on-farm or off-farm, locally or further afield. A multi-spatial livelihood is one that is derived from activities at and income from more than one place, which is often then pooled with the rest of the family income. This livelihood strategy frequently arises from a deliberate decision to distribute opportunities and risks. This is also a typical strategy in agriculture: farmers adapt to the vagaries of both the market and nature (including weather extremes) by combining a variety of agricultural activities across a range of settings, if possible also spreading these out over a variety of micro-ecological niches.

It is likely that part of Africa's agricultural expansion will be driven by increasing irrigation, based on both greater use of river water and on more intensive harvesting of groundwater. As much African agricultural land lies in areas of water scarcity, a rising demand for irrigation will compete with a growing demand for drinking water as well as for water for industrial and mining use and hydropower generation. Furthermore, there will be similar competition for energy between urban consumers and the agricultural, industrial, and mining sectors. Africa's rapidly rising energy needs can no longer be met by fuelwood and charcoal nor by fossil fuels, the energy sources that are currently dominant on the continent, although the threat of a 'race to the bottom' for these sources of energy is already evident.

---

31 Bryceson, Deborah, "The Scramble in Africa: Reorienting Rural Livelihoods," *World Development* 30.2 (2002): 725–739.

32 Foeken, Dick, and Sam O. Owuor, "Multi-Spatial Livelihoods in Sub-Saharan Africa: Rural Farming by Urban Households–The Case of Nakuru Town, Kenya," in *Mobile Africa: Changing Patterns of Movement in Africa and Beyond*, ed. Mirjam de Bruijn, Rijk van Dijk, and Dick Foeken (Leiden: Brill, 2001), 125–140.

33 See NEPAD, *Agriculture in Africa: Transformation and Outlook* (Addis Ababa: NEPAD, 2013).

The predicted expansion of land under cultivation from the current seven per cent to probably more than 15 per cent of the continent's total land area by 2050 will occur at the expense of grazing areas, forests, and natural environments such as parks and conservation areas, as already happened between 1960 and 2010, a period during which cropping area more than doubled from 3.4 per cent of total land area in 1960.[34] Metropolises and other urban areas use more space, a trend that occurs at the expense of other forms of land use. In peri-urban areas, a typical mosaic landscape becomes apparent, with a mixture of housing and leisure functions, (semi-)industrial land use, services, and (peri-)urban farming, including forms of intensive and less-intensive livestock production. More intensive forms of agriculture and animal husbandry will mean a higher dependence on fertilisers. Given Africa's often relatively poor-quality and old soils, this will require looking beyond nitrogen, phosphorus, and potassium fertilisers (known by their chemical symbols as NPK fertilisers) and paying far more attention to micro-nutrients than in Asia's 'Green Revolution'. It will also mean a higher demand for water, fuel, and feedstock. In addition, fish farming—and probably also algae farming—may develop rapidly. The expansion of mining and non-fossil energy production will also result in increasing demand for space. Hydropower dams will require the inundation of land (see McCann in this volume), while solar and wind 'farms' will also demand the use of specific land types, all with consequences for ownership and alternative uses of land. Moreover, Africa will not only export but also use its newly discovered gas and oil reserves, thereby contributing much more to global warming than before. The gradual diminishing of Africa's forests is also cause for concern. Between 1990 and 2010, total forested areas on the continent decreased by eleven per cent from 691 million hectares to 616 million hectares. The carbon storage capacity of African forests diminished from 60.9 gigatons to 55.9 gigatons in the same period.[35]

As elsewhere, climate change in Africa will make water-dependent agricultural production more risky and the provision of water for this and other uses less reliable. There will also be the additional risk of more 'extreme weather events', such as storms, floods, and droughts. However, climate change will not only have negative consequences; most plants react positively to higher CO_2 levels in the atmosphere,[36] especially the so-called C3 plants (including

34 Dietz, "Comparing the Agricultural Performance," 86.

35 Ibid., 90.

36 Watkins, Thayer, "The Direct and Indirect Effects of Increased Carbon Dioxide on Plant Growth," San José State University, accessed February 2015. http://www.sjsu.edu/faculty/watkins/CO2plants.htm.

rice, wheat, peanuts, soybeans, cotton, and many others). Unfortunately for Africa, however, many of its crops are C4 plants (such as maize, sorghum, and millets), which respond much less positively to increased CO_2 levels. An additional positive effect of climate change is the potential expansion of cropland into areas that used to be too cold for (intensive) crop production, like some highland and mountain areas in East Africa. Anticipated higher average rainfall in many areas of Africa—although not everywhere and likely marked by large fluctuations—will increase the amount of land that is potentially suitable for crop production, including making a larger area of the semi-arid tropics accessible for crop cultivation. It will also expand the potential area for more intensive forms of livestock production (with a larger number of camels, cattle, goats, and/or sheep per hectare). There are also other reasons not to be too afraid of climate change-induced changes in Africa. In general, Africans can and do make use of their existing repertoires to adapt to environmental and socio-economic change, including skills derived from the very high levels of flexibility, mobility, and adaptability that mark Africa's cultural economy.[37]

In the next section, I reflect on the relevant outcomes of a research project on the registered and potential impacts of climate change in West Africa.[38] This project combined 'hard science' with empirical social research into cultural aspects of livelihoods and agriculture when confronted with climate variability and climate change. The innovations that many African farmers learnt and applied during and after the great droughts of the 1970s and 1980s illustrate the adaptability of the continent's farmers and entrepreneurs. It can be expected that future climate change mitigation and adaptation measures and policies will create opportunities for many African entrepreneurs to experiment and earn extra income, part of which will probably be reinvested in the environment.

3 Reflecting on the Outcomes of an Earlier Research Project on the Impact of Climate Change

From 1997 to 2004, I was responsible for coordinating the 'Impact of Climate Change on Dryland' (ICCD) research project on the effects of climate change

---

37 Read, Hannah, Muyeye Chambwera, and Laurel Murray, *Tried and Tested: Learning from Farmers on Adaptation to Climate Change* (London: IIED, 2013).

38 Dietz, Ton et al., "Climate Change Preparedness in West Africa," in *The Impact of Climate Change on Drylands, with a Focus on West Africa,* ed. Ton Dietz, Ruerd Ruben, and Jan Verhagen (Dordrecht, Boston, London: Kluwer, 2004), 403–408.

on West African agriculture and society, which was funded by the Netherlands Research Programme on Global Air Pollution and Climate Change. The project was a collaboration between the CERES Research School for Resource Studies for Development, Wageningen University-based scholars, the Dutch 'National Institute for Public Health and the Environment' (RIVM), and West African scholars. Among many other achievements, the project resulted in the publication of a book,[39] at the end of which we presented our policy-relevant conclusions. These can be summarised as follows:[40]

- Rainfall variability is evident in the Sahel and elsewhere in Africa, but rainfall trends across the continent are unclear and contradictory.
- There are many different predictions, some of which come to conflicting conclusions.
- People turn to agricultural innovation, livelihood diversification, investment in social networks, and increased mobility in the attempt to adapt to climate-caused stresses.
- Increased vulnerability affects the poor more than the rich, but the poor are more risk-averse and less restricted by taboos in relation to extreme coping behaviour.
- Extreme shocks or disasters can devastate the rich as well as the poor; however, the rich are generally better protected physically, socially, and economically.
- Path dependency is applicable both to people and to geographical areas. Perceptions and collective memories are important drivers of behaviour.

By comparing the Global Climate database provided by the 'International Institute for Applied Systems Analysis' (IIASA) for 1930–1960 with the database provided by CRU/LINK for Global Climatology for 1960–1990, we could identify major shifts in the positioning of regions in Africa and elsewhere as arid, semi-arid, sub-humid, or humid.[41] For Africa, this comparison shows that 1930–1960 was generally more humid than 1960–1990: a considerable number of geographical units shifted from semi-arid to arid, sub-humid to semi-arid, or

39 Ibid. Among many others, I acknowledge the contributions of Ruerd Ruben, Jan Verhagen, Mirjam de Bruijn, Han van Dijk, Fred Zaal, and Marcel Put in the Netherlands, David Millar, Saa Dittoh, Edward Ofori-Sarpong, Francis Obeng, and Richard Yeboah in Ghana, Mohamed Touré and Mama Konaté in Mali, and Asaïta Diallo and Boukary Ouedraogo in Burkina Faso.

40 Ibid.

41 See the map in Dietz, Ton, "Climate Deterioration and Regional Specificity: Regionally Differentiated Portfolios of Options for Drylands," in *The Impact of Climate Change on Drylands, with a Focus on West Africa*, ed. Ton. Dietz, Ruerd Ruben, and Jan Verhagen (Dordrecht, Boston, London: Kluwer, 2004), 337.

humid to sub-humid in the latter period. A few areas in East Africa displayed the reverse trend; however, had we been able to use longer data sets, we would probably have found that 1930–1960 was an exceptionally wet period, and that rainfall conditions during 1960–1990 more closely approximated those during the 1850–1930 period. Since 1990, conditions have become more humid again, but with gradually increasing temperatures it is not clear what effect this will have on the balance between rainfall and evaporation (and evapotranspiration for plants). Using 100-year temperature data for Africa is tricky, as many of the early meteorological stations on the continent were established in areas that had low populations at the time but are now surrounded by urban areas with relatively dense populations. Although the so-called urban heat effect does not appear to influence the interpretation of climate change trends in many parts of the world,[42] there should be more caution in this regard in relation to Africa.

One of the results of the trend that saw most of Africa becoming drier in the period from 1960 to 1990—including some areas that experienced devastating droughts and famines, such as the Sahel droughts in the early 1970s and early 1980s—was mass migration. A map comparing African population densities in 1960 and 1994,[43] for example, illustrates how enormous the shifts in population dynamics have been. Most coastal areas experienced a more than fourfold increase in population density in this period. This occurred not only in cities and even already very densely populated metropolises, but also in rural regions. In contrast, many other regions experienced depopulation, including most of the fringe areas of the drylands in the Sahel-Sahara zone and the drylands in southwestern and East Africa as well as areas hit by large-scale persistent violence, such as Central Africa, parts of East Africa, and scattered areas elsewhere. Population growth in Africa was very high during this period, but mass population redistribution resulted in even higher growth rates in the reception areas, many of which were in more humid regions located near the coast.

Based on predictions by the Max Planck Institute's Global Climate Monitoring Unit, researchers collaborating on the so-called Baseline A Scenario,[44] who also developed the IMAGE model (=Integrated Model to Assess the

---

42 Beck, Coby, "'Warming is due to the Urban Heat Island effect'—No, it isn't," *GRIST*, October 29, 2006, accessed February 2015. http://grist.org/article/warming-is-due-to-the-urban-heat-island-effect/.

43 Dietz, Ton, and Els Veldhuizen, "Population Dynamics: An Important Intervening Variable," in *The Impact of Climate Change on Drylands, with a Focus on West Africa,* ed. Antonius J. Dietz, Ruerd Ruben, and Jan Verhagen (Dordrecht, Boston, London: Kluwer, 2004), 80.

44 See Van den Born, Gertjan, Rik Leemans, and Michael Schaeffer, "Climate Change Scenarios for Dryland West Africa, 1990–2050," in *The Impact of Climate Change on Drylands, with*

Global Environment) for simulating climate scenarios, presented a rather dramatic scenario for West Africa, in which Sahelian West Africa would in all likelihood become even drier. However, records for 1990–2015 seem to suggest the opposite, with floods having been a more frequent threat than drought. In many areas, population movements were once again directed northwards, although recent political and military turmoil in countries such as Mali may have changed this again.

In the book, we included a chapter showing that there were many different scenarios based on a variety of models that predicted completely opposite outcomes.[45] Whereas CSIRO predicts much drier conditions in the Sahel, especially in the central and eastern Sahel, Had CM2 expects much wetter conditions in the eastern Sahel, CGCM1 gives a very varied picture, and CCSR-98 anticipates considerably drier conditions in the western Sahel and much more humid conditions in the eastern Sahel. All of these predictions apply to 2050.

The evidence that we found can be examined by studying farmers' responses in the Sahel and comparing those with farmers' responses elsewhere. All across the parts of the world that are more vulnerable to climate change, including the Sahel, farmers know that they must already cope with 'normal' climate risks. These can be summarised as:

- Risks related to peak rainfall (a high volume of rainfall in short periods, often with high rainfall energy).
- Risks related to peak river discharge, often following peak rainfall in upstream river catchment areas.
- Risks related to severe storms, often near the coast (or the shores of major lakes); often during monsoon periods or hurricane and cyclone seasons.
- Risks related to heat (heat waves).
- Risks related to drought ('normal' dry seasons, dry spells in 'normal' wet seasons, or failing rainy seasons).
- Risks related to frost, particularly if unexpected (e.g. in early autumn or late spring, where these seasonal categories apply).

Farmers have generally developed a whole range of preventive strategies to counter these risks before they arise and to cope with them once they become real and often threatening.

a Focus on West Africa, ed. Antonius J. Dietz, Ruerd Ruben, and Jan Verhagen (Dordrecht, Boston, London: Kluwer, 2004), 43–47.

45 Van Boxel, John, "Uncertainties in Modelling Climate Change," in *The Impact of Climate Change on Drylands, with a Focus on West Africa*, ed. Antonius J. Dietz, Ruerd Ruben, and Jan Verhagen (Dordrecht, Boston, London: Kluwer, 2004), 41.

Furthermore, climate change trends for Africa indicate that there may be additional or more severe risks, which can be summarised as follows:

- There is a gradual shift towards higher temperatures, with higher evaporation and evapotranspiration rates as well as higher heat risks, but also lower frost risks.
- There are changing rainfall patterns, often with more rain and flood risks, but above all with more variability and unpredictability, also in relation to the start of the rainy seasons in monsoon climates.
- In coastal areas, residents must cope with gradually rising sea levels and increased coastal erosion, but above all with more severe coastal storms and more frequent extreme weather events. As so many people have moved to cities, and poor immigrants often live in the riskiest areas in and around these cities, they are the ones who are most confronted with these threats. However, near-coastal zones incorporating transport and urban hubs as well as areas of intensive agriculture and mass tourism are often a country's economic powerhouses, areas where the cost of climate change-triggered disasters can be devastatingly high.
- There is a gradual shift in ecozones, agro-ecozones, and biodiversity/crop niches, a trend that has an impact on the livelihood options of local residents as well as on wildlife and natural vegetation.

The risks for societies coping with the increased chance of climate-induced disasters can be high. Alongside the threat of species extinction and forest and vegetation destruction, there are more health risks for people and animals, with correspondingly higher human and animal death rates. Crop losses and storage losses as well as damage to property and physical infrastructure are a major threat. Livelihoods can be endangered, while livelihood options may become more limited. Because of more economic hardship, economic growth can be put at risk and government budgets strained. Moreover, major risks can also extend to psychological anxiety as well as social and political instability, which may cause secondary effects such as social unrest, mass migration, and violence that can be far more devastating than the climate change-induced threats themselves. As such, the very fabric of society can be undermined, which can result in lower levels of resilience and innovative capability as well as fewer (insurance) buffers. This may also result in a collapse of social trust, which in itself can lead to violence and threatened livelihoods (including harmful effects such as the emergence of 'no go areas', 'climate wars', or 'climate refugees'). Agricultural production may be threatened by drought, floods, heat, and/or trade and transport problems. This may create more food insecurity, hunger, and malnutrition, more environmental refugees, higher food prices and other forms of inflation, and mass social unrest. However, the question of whether this doom scenario will come to pass is still up for debate. There are many alarmists, among them some scholars, who seem certain that this negative scenario will become the norm, particularly in vulnerable regions like the Sahel. Here, Harald Welzer's recent book on 'climate wars'[46] comes to mind. The relative abundance of funding for research that focuses on (adaptation to) climate change has served to stimulate this sort of hype and the alarmist tendencies within it.

My position is that climate change is indeed happening, and that it is the source of a number of serious risks. However, it does also create opportunities. In addition, there are many other processes in human societies that could reinforce or counter the risks induced by climate change. It is important to link climate change to other pertinent issues with which it may be interconnected but that often have their own causes and momentum, such as population growth, urbanisation, economic difficulties related to global trade, biodiversity loss, rising or diminishing poverty, inequality, and unemployment, and the various forms of intolerance related to religious, social, and/or ethnic differences. For example, it would be foolish to think that the recent and current problems in the Sahel (like the Tuareg revolt in Mali, Islamist terrorism, like Boko Haram) are directly linked to climate change, as each of these phenomena are caused by unique drivers with their own very deep historical roots. However, it is useful to undertake solid empirical research into long-term dynamics, in which land use changes, livelihood changes, population growth, migration, and climate change are viewed as longue durée undercurrents of more short-term political turmoil and social anxieties that occasionally manifest themselves in outbreaks of severe violence. Explanations of violence that link such episodes with periods of extreme weather must be argued very carefully. Examples of such an approach include the range of studies presented by Nils Petter Gleditsch in a recent edition of the *Journal of Peace Research*[47] as well as a study by Adano and colleagues.[48] Using detailed long-term data connecting rainfall variability and deadly violence in a very violent pastoral area in northern Kenya, the authors try to prove that violence tends to erupt more often in periods of abundance than in times of scarcity. Thus, whereas increasing rainfall may mean better prospects for animal husbandry, agriculture, and population

46 Welzer, Harald, *Climate Wars: What People Will be Killed for in the 21st Century*, trans. Patrick Camiller (Cambridge: Polity Press, 2012).

47 Gleditsch, Nils Petter, "Special Issue: Climate Change and Conflict," *Journal of Peace Research*, 49.1 (2012).

48 Adano, Wario Roba et al., "Climate Change, Violent Conflict and Local Institutions in Kenya's Drylands," *Journal of Peace Research* 49.1 (2012): 65–80.

growth, it might also result in greater insecurity and more serious violence, and thus the opposite of what many people—including scholars—anticipate.

In 2004, we presented policy recommendations based on our findings in the ICCD project. These seem to be as relevant now, more than ten years later, as they were when they were formulated by a panel of African experts in 2002.

- Develop better early warning systems and more effective channels to communicate its findings.
- Better integrate knowledge about changing natural environments and changing behaviour.
- Develop and share more adaptive agricultural, pastoral, silvicultural, and horticultural practices.
- Give more attention and support to social security networks (including formal and informal social insurance systems), especially in relation to diversified livelihood profiles, migration, and remittances.
- Pay more attention to entitlement changes (e.g. land, water, and forest use rights) and to conflict prevention between groups with different identities (e.g. cultivators versus herders).

4 Following up and Reflecting on Some Recent Trends

The uptake of the ICCD book was slow in the beginning, but gradually picked up, as can be seen in Graph 1, which is based on an analysis using Google Scholar. In mid-2014, 170 academic colleagues had cited the book and/or one of its chapters. Many scholarly books struggle to be noticed at first but then experience an increase in attention that, however, soon dies down. The ICCD book experienced a very slow reception that picked up only after about five years, but since then interest in the book has increased rather than diminished (Figure 9.1). Nevertheless, only recently have the ideas discussed in the book received the attention they deserve, which is also a reason to use some space in this chapter to highlight its most important findings and reflect on their current significance.

Unlike during the period from 1997 to 2004 (when the ICCD book was conceived), since about 2010 many African countries have established agencies responsible for climate change mitigation measures and for participating in global climate change negotiation meetings. There is growing awareness in Africa in general and among these agencies in particular of the need for systematic thinking about climate change adaptation, although this is mainly oriented towards countering extreme events (floods, storms, or droughts) and is subsumed into poverty reduction policies and practices. Parts of Africa have also adopted the 'sustainable cities' concept and now include mitigation and

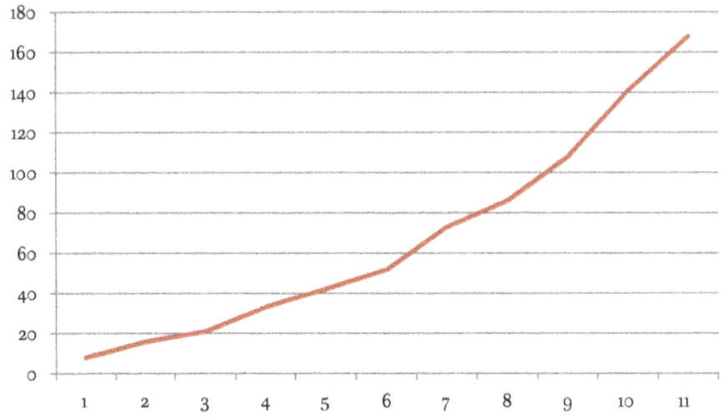

FIGURE 9.1 Citations of the ICCD book between 2004 and 2014

adaptation measures in town and regional planning.[49] Much work in this regard, however, is driven by a major external push by Northern aid agencies. For example, the Climate and Development Knowledge Network (CDKN), which is funded by Great Britain and the Netherlands, presents "climate compatible development" on its website as:

> Development that minimises the harm caused by climate impacts, while maximising development opportunities presented by transitions to a low emission, more resilient future [...] CDKN wants to combine research in partnerships with NGOs, private sector and knowledge centres on climate compatible development.[50]

Another major think tank is the United Nations University Institute for Environment and Human Security (UNU-EHS) located in Bonn. As of 2016, only four out of its 37 experts had an African background, and then often only in junior positions.[51] Except for South Africa, there are not many African examples of REDD, the United Nations Collaborative Programme on Reducing Emissions from Deforestation and Forest Degradation in Developing

49 E.g. Cape Town, see CoCT, "Showcasing Excellence: Mayor's Portfolio of Urban Sustainability," City of Cape Town/Isixeko Sasekapa/Stad Kaapstad, accessed February 24, 2016. https://www.capetown.gov.za/en/EnvironmentalResourceManagement/publications/Documents/CCT_Mayors_Portfolio_of_Urban_Sustainability_2014-06.pdf.

50 CDKN (Climate and Development Knowledge Network), "Helping Developing Countries to Design and Deliver Climate Compatible Development," last modified January 22, 2016, accessed February 28, 2016. http://r4d.dfid.gov.uk/PDF/Outputs/CDKN/CDKN-english-brochure-lores-sept2011.pdf.

51 UNU-IEH, "Experts," accessed February 28, 2016. http://ehs.unu.edu/experts/researchexperts.

Countries,[52] and new compensation regimes or Greenhouse Gas trading. However, there have been other areas in which progress has been made on the continent, such as in water provisioning, although many of those who are involved in enhancing urban water services realise that this remains an uphill task. For example, the city of Ouagadougou, which is located in a semi-arid section of the Sahel, saw its population increase from 59,000 inhabitants at independence in 1960, when the city became capital of the then Upper Volta (now Burkina Faso), to its current total of 1.9 million inhabitants.[53] Despite significant investment in constructing additional dams, many at some distance from the city, and much emphasis on urban water harvesting and improving water institutions, Ouagadougou's water availability per inhabitant decreased from 57 litres per day in 1978 to 39 litres per day in 1986 to only 26 litres per day in 1993, all significantly below the recommended quantity of 100 litres per person per day.[54] Water supply appears to have since stabilised at 1993 levels as a result of water development programmes that have been regarded as successful examples of urban water provisioning.[55]

The sudden increase in global food prices and food price speculation in 2007–2008 resulted in a renewed emphasis on agriculture and on the need for an 'African Green Revolution';[56] after which aid agencies also began to adapt to the public sector's assumption of a greater role in creating better conditions for supporting farmers to produce more and better food. Food and other basic commodity prices started to go down in 2009 and 2010 because of the financial crisis in the West, but from 2011 there has been a steep rise in prices once again. While providing increasing opportunities for market-oriented farmers and food traders, this has come with potentially dire consequences for net food consumers and those without the means to purchase sufficient food. After 2008, many investors from abroad as well as some local investors began

52 MEWNR, "Analysis of the Charcoal Value Chain in Kenya. Final Report," Nairobi: Ministry of Environment, Water and Natural Resources, Government of Kenya, 2013, accessed February 25, 2016. http://www.kenyaforestservice.org/documents/redd/Charcoal%20Value%20Chain%20Analysis.pdf.

53 Demographia, World Urban Areas.

54 Dietz, Ton, and Fred Zaal, "The Provisioning of African Cities, with Ouagadougou as a Case," in *Re-Aligning Actors in an Urbanizing World: Institutions and Governance in an International Perspective*, ed. Isa Baud, and Johan Post (Ashgate: Aldershot, 2002), 263–284.

55 In this regard, I recommend reading the interesting article on water provisioning in Burkina Faso on the English-language Wikipedia. Wikipedia, "Water Supply and Sanitation in Burkina Faso," accessed in February 2015. http://en.wikipedia.org/wiki/Water_supply_and_sanitation_in_Burkina_Faso.

56 A campaign led by former UN Secretary General Kofi Annan; see www.agra.org.

acquiring land with the intention of producing food on a large scale, partly for the Asian market. Land acquisitions, however, often threaten existing land use by poor farmers and livestock owners and can also cause environmental harm. Yet, partly due to the global economic crisis, these 'land grab' threats have diminished in recent years, and many investors have yet to establish large-scale farm operations.

At about the same time, a lot of attention was suddenly placed on biofuels as a way of overcoming global energy scarcity and combining climate change mitigation and adaptation,[57] an option that has nevertheless been filled with controversy. Partly related to this hype surrounding biofuels, there has been enormous critical attention on land acquisitions by foreigners for purposes of biofuel production.[58] That has also had a geopolitical and psychological element, as Asian investors were suddenly cast as "the new imperialists, who came to Africa to rob the country of its natural resources" or described in words to similar effect. If we look at Africa today, many land acquisition plans for biofuel production are on hold, or, where land has been acquired, little action has occurred. There are also more optimistic views on what the demand for African resources could mean in terms of better opportunities to negotiate the emerging global multi-polarity.[59]

5 Conclusion

Africa needs to link different policy agendas that are crucial for its future development and that can only become sustainable development agendas if integrated. Six areas of concern should be combined:

- High rates of population, urban, and economic growth.
- 'Inclusive development', mass job creation, social security, inequality reduction, and mass poverty alleviation.
- Food, energy, and water security.
- Climate change mitigation and adaptation (as well as protecting biodiversity and avoiding deforestation).

57 E.g. the jatropha genus of plants; see Muys, Bart et al., "Integrating Mitigation and Adaptation into Development: The Case of Jatropha Curcas in Sub-Saharan Africa," *Global Change Biology, Bioenergy* 6 (2014): 169–171.

58 Richards, Michael, *Social and Environmental Impacts of Agricultural Large-Scale Land Acquisitions in Africa—With a Focus on West and Central Africa* (Washington, DC: Rights and Resources Initiative, 2013).

59 Dietz, Ton et al., ed., African Engagements: Africa Negotiating an Emerging Multipolar World (Leiden, Boston: Brill, 2011).

- Land use planning and conflict mitigation (access to and use of natural resources as well as fairer distribution of benefits, including the benefits of oil, gas, mineral, and biofuel exploitation).
- Governance of diversity.

Climate change mitigation and adaptation agendas should form part of this broader strategy rather than being treated as the most urgent priority or that in most need of attention and (research) funding. Africa's current and potential environmental problems are a result of many additional developments that will probably have a more important impact on the future of the continent and its people than climate change alone.

Interlinking these different agendas is an urgent task for scholars, who often work in isolated research bubbles with few points of connection to related research. However, it is even more urgent for policymakers and politicians to understand that disconnected and often lukewarm policy intentions will not solve Africa's problems, but will probably only serve to make them even worse.

CHAPTER 10

Africa's High Modernism: Historical Ecologies of Climate Change and Hydrologies of Watersheds (Blue Nile and Zambezi)

James C. McCann

Abstract

This chapter pleads for making watersheds a focal point of research on Africa's ecological past and future, helping us to understand the regional effects of climate change and political ecology. It discusses how Africa's hydrological systems are changing—and will continue to do so—under two major influences. First, dam building has massively altered riverine environments for many decades and is currently experiencing another boom. Second, regional and global climate change have severe effects on hydrologies. This chapter explores the relative effects of climate change and historical trends in water management on water resources in two African contexts: the Blue Nile and Zambezi watersheds.

1 Introduction

African watersheds are, and should be, focal points for our understanding of the ecological and political stages of Africa's longue durée and its long-term future. Sub-Saharan Africa is now in the midst of two massive sets of changes in its hydrological systems. The first consists of the building of a new wave of high modernist hydroelectric/irrigation dam projects on its main river systems (the Congo, the Niger, the Zambezi, the Blue Nile, the Wele, the Volta, and the White Nile). The second are the expected regional and global changes in climate that will have a quite distinctive set of effects on different African hydrologies.

There are any number of settings in which the future dynamics of hydrology schemes and climate change will play out in Africa and across the globe. For the Zambezi watershed alone, the Intergovernmental Panel on Climate Change (IPCC) anticipates a 10 to 15 per cent decline in rainfall in its catchment area.[1]

1 Intergovernmental Panel on Climate Change, "Working Group II: Impacts, Adaptation and Vulnerability," last modified June 6, 2017, accessed July 11, 2017. http://www.ipcc.ch/ipccreports/tar/wg2/index.php?idp=384; see also Sullivan, Leann M., "Five Challenges to the Future

 | DOI:10.1163/9789004410848_012

The annual distribution (seasonality) of that decline will be the decisive factor in determining its human impact. Eastern zones in the rainfall shadow of the Ethiopian highlands' escarpment will likely be hydrological losers. Ethiopia's Blue Nile watershed may well record higher total rainfall than other areas, but it will nevertheless experience an overall decline from its historical patterns. How should historians approach these changes in the near-term future and the historical record?

Watersheds can serve as a foundation for understanding the regional effects of climate change and political ecology. Formally defined, watersheds are both geographical and historical markers that frame a landscape:

> A watershed is the area of land where all of the water that is under it or drains off of it goes into the same place. John Wesley Powell, scientist geographer, put it best when he said that a watershed is: 'that area of land, a bounded hydrologic system, within which all living things are inextricably linked by their common water course and where, as humans settled, simple logic demanded that they become part of a community.'[2]

Africa's distinctive annual wet–dry seasonal oscillation suggests that the effects of overall aridity on the shape of seasonal water patterns will likely be even greater than in other parts of the world, even as water storage interventions (dams) attempt to regulate and control seasonal flows—with all of the inevitable engineering snafus and miscalculations that this entails.

African farmers, fisherfolk, and residents of mushrooming urban ecologies have evolved historically along with their longer-range seasonalities and metabolisms. Moreover, the transformation of Africa's historical patterns of epidemic waterborne tropical and sub-tropical diseases (dysentery, malaria, onchocerciasis, bilharzia, cholera) are fairly certain effects, although these are not part of engineers' plans.[3] Other infectious diseases of human-natural ecological contact—like influenza and Ebola—derive from genetic mutations as viruses leap from one species to another. More complex engineering plans, like disease interactions, teach us about ecological complexity.

---

of Transboundary Water Governance," *Issues in Brief* 29 (2014), The Frederick S. Pardee Center for the Study of the Longer-Range Future.

2 United States Environment Protection Agency, "Water Topics," last modified April 12, 2017, accessed August 10, 2014. http://water.epa.gov/type/watersheds/whatis.cfm.

3 Pearce, Fred, "Will Huge New Hydro Projects Bring Power to African People?" *Yale Environment 360*, May 30, 2013, accessed July 11, 2017. http://e360.yale.edu/features.

Can the past foretell the longer-range future? In the context of climate change and high modernism, future hydrologies are a moving, dodging target. Watersheds are hydrological features of moving water, but they are also markers of cultural and economic identity (like bayous, coastlines), and sites of political conflicts over resources.

As they currently stand, planning studies for Africa's new generation of watershed management infrastructure (aka hydroelectric dams) make neither a meaningful assessment of an inherent historical context, nor do they fully incorporate the likely futures of Africa's hydrological systems in terms of climate change. Climate science, as often as not, falls prey to political agendas, geographies, and engineering oversights. The 'run-of-the-river' (ROR) systems and reservoirs built into the design of Africa's dams have global analogues. These would include watersheds like the Mekong, Colorado, Rhône, Tigris/Euphrates, and Yangtse, although Africa's seasonal climate patterns mark the continent's rivers as distinctive in their past and in their future.

This chapter examines the relative effects of climate change and historical trends in water management on water resources in Africa by exploring two cases: the Blue Nile and Zambezi watersheds. The discussion begins with an argument for making watersheds a primary focus for understanding historical changes in African political ecologies. In the longue durée, Africa's variations of global climate change will take place alongside a new generation of water management schemes—dam building—and changes in the anthroposphere (the domain of human ecology) that will shape the effects of climate change.

2 Climate as Africa's Historical Stage

Let us begin with the deeper background for Africa and its water. Beyond the fundamental topographic shapes offered by its geology, African landscapes reflect the continent's changing patterns of climate—a circulation of wind, moisture, and temperature—that mark Africa as distinct from other global land masses. Unlike in temperate zones—Europe, North America, and Central Asia—in which growing seasons and cycles of life respond directly to fluctuations in temperature, Africa's rhythms of life primarily reflect the distribution of moisture in the form of rainfall affecting, in different ways, river deltas and floodplains. Africa's annual patterns of rainy and dry seasons, the length of each year's growing (or fishing) season, as well as humidity and soil moisture levels all take their cue from the annual rhythms of cyclonic winds, ocean temperatures, and the earth's rotation and movement around the sun. Being bisected by the equator, the tilting of half of Africa's land mass towards the sun in summer and away from the sun in winter prompts the anticyclonic and trade winds that set the yearly cycle between rainy and dry seasons.

In fact, Africa's position on the globe means that the continent's land mass and human landscapes experience the clearest separation between wet and dry seasons of any region on earth. The shifting cloudy mass of rain-bearing turbulence that gathers around the equator—known as the Inter-Tropical Convergence Zone (ITCZ)—establishes a two-part pattern of seasons which in turn sets the pattern of rainfall found across almost the entire continent—summer wet and winter dry. The basic summer wet/winter dry pattern for Africa also exhibits also subtle variations from year to year and within particular geographical settings, which can result in short or delayed rainy seasons with notable effects on the landscape. Several years of short or delayed rains along the ITCZ's edges result in drought, historically a common occurrence in many parts of Africa away from the equator, such as the Sahel region at the southern edge of the Sahara or equivalent grassland areas in Southern Africa. In a new world of human-induced climate change, Africa's wet and dry seasons will continue to be sharply divided, albeit perhaps with even greater year-to-year swings.[4]

In more recent times, scholars of African climate history have used sources such as lake levels, geological stratigraphy, and the still limited archaeology of human settlement to identify climate patterns in the 800–1600 AD period, during which the Sahelian states of Ghana, Mali, and Songhay flourished and then declined. Historian George Brooks and historical meteorologist Sharon Nicholson offer somewhat contrasting conclusions about the nature of these climate periods and their effects on political change. Nicholson uses sources such as lake level stratigraphy and scarce local records from lake bed soil sediments to reconstruct climate epochs. Brooks attempts a similar task of relating climate to changes in political hegemony and patterns of trade.[5]

As it transpired, the years 800–1300 were relatively wet, before being followed by a drier interlude that lasted until 1450. Thereafter came another wetter period, which included the so-called Little Ice Age (1500–1850), during which observers reported that the peaks of both Mt. Kilimanjaro and Ras Dashan in

4 McCann, James, *Green Land, Brown Land, Black Land: An Environmental History of Africa, 1800–1990* (Portsmouth/Oxford: Heinemann/James Currey, 1999), 1–22.

5 Brooks, George, "A Provisional Historical Schema for Western Africa Based on Seven Climatic Periods," *Cahiers d'Études Africaines* 26.101 (1986): 43–62; Nicholson, Sharon E., "The Methodology of Historical Climate Reconstruction and its Application to Africa," *Journal of African History* 20.1 (1979): 31–49.

Ethiopia were snow-capped.[6] In the late 19th century and throughout the 20th century, Kilimanjaro's ice cap receded while Ras Dashan's disappeared completely. Conclusions about the interplay between climate and politics are, of course, subject to debate as further data emerges from archaeological studies of sediment and pollen variations that may have affected human activities. However, at the very least, it appears that the overall dynamics of climate over time seem to be the context for human history in a way that frames Africa's distant past as well as in what climate science anticipates to be the coming age of climate change.

This chapter places Africa's distinctive regional patterns of wet–dry rainfall oscillation in the context of human-induced watershed metabolism. New historical scholarship that explores the confluence of nation-building with ambitious, high modernist visions for the control of nature will frame economic policy as well as our understanding of current political debate that revolves around watershed management. The chapter will thus place the new era of dam construction in Africa within the deeper context of climate change that may lie ahead.

3 A New High Modernism: Africa and the New Global Dam Era

James Scott, high priest of the critique of high modernism, describes several of its characteristics:

1) a strong confidence in the potential for scientific and technological progress, including a reliance on the expertise of scientists, engineers, bureaucrats, and other intellectuals;
2) attempts to meet human needs by mastering nature, also including attempts to control and change human nature;
3) an emphasis on rendering complex environments (such as old cities) or concepts (such as various social dynamics) legible, most often through spatial ordering (e.g. city planning on a grid);
4) a disregard for historical, geographical, and social context in development.[7]

Africa is at the cutting edge of global water management's new era of dam building. The Democratic Republic of the Congo has just initiated the building

6 Cf. Nicholson, Sharon, "A Climatic Chronology for Africa," 75–81, 251–54, cited in James Webb, *Desert Frontier: Ecological and Economic Change along the Western Sahel*, 1600–1850 (Madison: University of Wisconsin Press, 1995), 4–5.

7 Scott, James C., *Seeing Like a State: How Certain Schemes to Improve the Human Condition Have Failed* (New Haven: Yale University Press, 1998), 4–5.

of a 4,800-megawatt (MW) dam called Inga III that will generate more power than Egypt's Aswan High Dam. For some, this represents an environmentally friendly venture, perhaps because it is a run-of-the-river dam that will not require a large reservoir but will rely instead on sustained river flow.[8] Smaller hydroelectric dams—such as the newly opened 250 MW dam on the White Nile at Bujagali in Uganda, Ethiopia's 300 MW dam on the Takazze River, a tributary of the Nile, and a 185 metre-high structure (Africa's highest) in northern Ethiopia to capture river water that flows through a deep canyon before its westward descent into the greater Nile system, are all examples of this type. In Equatorial Guinea, the 120 MW Djibloho Dam, which was completed in 2012, provides ninety per cent of this small country's power needs.[9]

Yet, these projects are miniscule compared to the gigantic world-leading schemes planned for elsewhere in Africa in the next decade. Ethiopia is now in the latter stages of diverting water from the Blue Nile into a reservoir near the Sudan that will allegedly make the monster 6,000 MW Grand Renaissance (Hedasse) Dam Africa's largest and the world's seventh largest—at least for now. The World Bank's support for an Ethiopia-Kenya high-tension power link to carry hydropower to distant cities is planned for a 2019 opening. At a capacity of 1,800 MW, the Gilgil Gibe III project on the Gibe-Omo basin will provide irrigation to Ethiopia's southwestern margins and hydroelectric power to the growing towns of East Africa, but it will also likely halve the size of Lake Turkana in northern Kenya and drastically diminish the lake's fish protein stocks and aquatic biodiversity.

In energy-thirsty Southern Africa, there are already two major dams that capture the lifeblood of the mighty Zambezi River. These include the Kariba Dam, which was opened in 1960, and its mammoth lake reservoir as well as the Portuguese-built Cahora Bassa Dam, completed in 1975, that inundated Mozambique's rural landscapes in the flooded Tsonga homelands, but fed Southern Africa's regional power grids. A smaller 1,600 MW dam downstream from Victoria Falls on the Batoka Gorge on the Zambezi and the planned Mphanda Nkuwa Dam in Mozambique (a river-run scheme) will require a new formulation of the upstream river flow calendar.

8 The Congo River, near the equator, has a lower seasonal change in water flow than other African rivers.

9 Information on ongoing dam projects in Africa can be obtained from the webpage of the organisation International Rivers, which is critical of dam construction; accessed August 2, 2017. https://www.internationalrivers.org/programs/africa.

4 Climate Change and the Transnational Setting: Impacts and Drivers

Africa is in the midst of a period of rapid change in the global movement of people and commodities when compared to the pace of historical patterns. Africa's climate has changed in the past, such as during the Little Ice Age, but its seasonal wet–dry patterns have continued to mark the continent's climate as distinctive (see above). Given the severe climate events envisioned to occur across the world in the latest IPCC report (drought, floods, shifts in seasonality, among others) as well as future anthropogenic interventions in African hydrology (dams, diversions, reservoir storage, urban growth), we can anticipate that there will be winners and losers within the population. The following sub-sections discuss some key issues to watch in this regard. What will be the markers of these futures and pasts?

4.1 *Crop Changes and New Crop 'Value Chains'*

Global capital flows and markets in Asia and Europe have already encouraged investment in land leasing ('land grabs') for production of food crops and oil seeds, including chickpeas, lentils, sesame, and maize, which will all require chemical fertiliser and a reliable supply of water. These crops will need infrastructure so that they can be delivered to domestic urban and international markets. Maize, for example, will be key for export demand for livestock/poultry consumption habits of the expanding urban middle classes.[10]

4.2 *Urban Footprints (Metabolism)*

Africa is the most rapidly urbanising part of the globe. Demand from these new urban populations are creating 'urbansheds', where demand for new infrastructure for the supply of water and energy will outstrip current capacity and transform watersheds' physical landscapes for the purposes of water storage, delivery, and seasonal distribution.

4.3 *Hydroelectric Transmission Infrastructure*

Plans for hydroelectric power production in key watersheds will be regional (Ethiopia/Sudan/Kenya/Uganda) and therefore also transnational (Angola, Zambia, Mozambique/South Africa). As water capture networks become

10 Nalepa, Rachel, "The Global Land Rush: Implications for Food Fuel, and the Future of Development," *Pardee Papers* 13 (2011): 1–36.

concentrated in dams and as part of irrigation schemes, competition over the legal and de facto control of watersheds will increase tensions over access to scarce resources between nations, cities, social classes, and international actors seeking to extract value from crops, labour, distribution, and/or services.

4.4 *Headwater and Delta Ecologies*

Climate change and new hydrologies, in particular watersheds, will raise tensions over the shifting seasonality of water flows and storage, the transformation of former floodplains into reservoirs, and the ecology of river deltas. In most global cases for watersheds, headlands versus delta ecologies have fallen under different political authority, influencing and promoting colonial rule, conflicts as industrial agriculture and urban growth have strained political relations. Each of these conflicts has had its foundations in local hydrological and human ecologies as well as in regional politics.

4.5 *Watershed Health*

Research on the correlation between waterborne diseases, irrigation dams, and their reservoirs is not new, but systematic study thereof has been sparse.[11] Moreover, what has been even less visible is the effect of hydroelectric dam ecologies on health concerning diseases like schistosomiasis, lymphatic filariasis, onchocerciasis (river blindness), and malaria. Vectors and hosts for these infectious diseases differ, but all depend for their transmission on watery settings, vector life cycles, and human–disease vector contacts.[12] Disease ecologies, including those for chronic and epidemic infectious diseases, are therefore important aspects of climate change, the new waterscape transformations that lie ahead as part of development planning, and the ways in which climate change affects particular plans for water storage. Again, winners and losers will emerge in Africa's changing climatic and hydrological patterns.

---

11 See e.g. Scudder, Thayer, *The Ecology of the Gwembe Tonga* (Manchester: Manchester University Press, 239–41). Scudder makes precisely this point about limited research, but lists a number of infectious and vector-borne diseases as endemic to the Zambezi River valley; Kaufman, Frederick, "The Man Who Stole the Nile: An Ethiopian Billionaire's Outrageous Land Grab," *Harper's Magazine*, July 2014, 36–42.

12 Schistosomiasis, whose host is a snail, has become an indicator disease for other water impounding diseases. See Gergel, Diana, "Water Resources Development: Engineering the Future of Global Health," *Issues in Brief* 27 (2013): 2–3, The Frederick S. Pardee Center for the Study of the Longer-Range Future.

5 North and South of the Equator: Watersheds of the Blue Nile and the Zambezi

In general, African watersheds are poorly studied compared to those in other parts of the world. Their hydrological and social histories often relate directly to their colonial pasts. These colonial footprints have now taken a newer transnational form in the regional schemes for access to irrigation and hydroelectric grids that will serve areas like the Nile Valley, the Zambezi, the Congo, the Lesotho highlands, the Niger, and the Volta. Each has a different hydrological profile and history of management in the past century and will be subject to distinctive potential effects of climate change in the coming era. The following subsections discuss two examples.

5.1 *The Blue Nile*

The Blue Nile provides the greater portion (60 per cent) of the Nile waters that descend the long slope past its confluence with the White Nile at Khartoum and downstream towards Egypt. Moreover, the Ethiopian highlands in general provide over 83 per cent of the total Nile waters that reach the Aswan Dam. There is a remarkable difference between the body of data available on the Nile's outflow in Sudan and Egypt and what we know about the Blue Nile, where the flows appear only sporadically in the historical record.[13] Under its 20th-century imperial governments, Ethiopia's long-time policy was to use the 'potential' for controlling the flow of the Blue Nile as a political cat's paw (rather than as a hydrological one) in negotiations with the colonial powers in London, Paris, and Rome.[14] The first dam on Ethiopia's Nile was not built until 1960, and even then this was only a small (11.5 MW) dam at the Blue Nile Falls (Tissisat) that was intended to provide hydroelectric support for a new industry planned at Bahir Dar, the river's egress from the lake.

Given high seasonal and annual variability in rainfall and run-off data from Sutcliffe and Parks' 1997 report, shows rather clearly the patchy data base for engineering both hydropower and agricultural calendars.[15] High modernist empiricism was in many ways an illusion until the current era. Data gaps noted in this time series mask interannual and season fluctuations felt by farmers, travellers, and planners. Topography and recent land use patterns in

---

13 Sutcliffe, John. V., and Parks Y. P., *The Hydrology of the Nile* (Wallingford, Oxfordshire: International Association of Hydrological Sciences (IAHS), 1999), 130–134.

14 McCann, James, "Ethiopia, Britain, and Negotiations for the Lake Tana Dam Project, 1922–1935," *International Journal of African Historical Studies* 14.4 (1981): 667–899.

15 Sutcliffe, and Parks, *The Hydrology of the Nile*, 135.

the Lake Tana Basin and watershed give us a snapshot and a sense of trends, but these need to be correlated with deeper patterns of change. The Blue Nile watershed and the river's legendary source in the Ethiopian highlands has been a point of contention in the region and in Europe ever since the Nile became the subject of European imperial visions.[16] The watershed's importance has been both geopolitical and hydrological, although the river's deep, sweeping gorge has limited its potential for irrigation. This restriction has lifted most recently with the opening of the World Bank-funded Tana Beles project, which captures the Beles River tributary and reroutes it through a tunnel away from its natural course towards the Sudanese border into floodplain irrigation and then to the west and away from its natural drainage into Lake Tana. The Blue Nile basin has now taken on its high modern form as a hydrological, financial, and fiduciary reality. How does that high modern vision match the set of unintended consequences that lie beneath the engineers' optimism and vanity? The Finnish consultancy firm NIRAS describes the project's goals as:

> The Program Tana Beles Integrated Water Resources Development is part of a cluster of programmes aimed at decreasing poverty through sustainable land use practices in the Tana Beles Growth Corridor in Ethiopia. The Programme is linked to the Nile Basin Initiative, and financed through a loan from the World Bank.[17]

The project statement of goals means as much for what it does not say as for what it offers as a description of its intent. Construction ended in 2012 with the opening of the 460 MW power plant. The secondary goal here is agricultural development, which is expected to attract migrant labour (into an endemic malarial zone). The overall aim, however, was fundamentally a human-induced redirection of part of the watershed. This plan consisted in draining water from Lake Tana along its southwestern edge, channelling this water from the hydroelectric plant across the watershed and draining it into the Beles River, which flows back westwards into the Nile basin. Is a run-of-the-river design vulnerable to the vagaries of climate change? That remains a question mark.

---

16 Cf. Tved, Terje, "Hydrology and Empire: The Nile, Water Imperialism and the Partition of Africa," *The Journal of Imperial and Commonwealth History* 39.2 (2011): 173–194.

17 NIRAS, "Ethiopia Tana Beles Integrated Water Resource Development Programme," accessed August 10, 2014. http://www.niras.com/business-areas/development-consulting/references/water/tbwme.aspx.

This Tana Beles part of Blue Nile development is a new initiative that draws on a recent agreement between the upstream Nile countries (Uganda, Kenya, and Ethiopia) and is opposed by the downstream countries (Sudan and Egypt). Important questions remain to be answered: How will this diversion of water from the shallow Lake Tana affect the lakeside ecology since the agricultural plans would introduce new additions of nitrogen into the Beles River and draw down lake levels? Which other water resource projects are in store for the Ethiopian highlands? Do we have adequate data on *changing* patterns of land use in this region? An IFAD survey cited in zur Heide offers an undated snapshot of a system in the process of change;[18] however, it excludes assessments of a damaging lake flood in 2006 and of changes in cropping as the 'urbanshed' and burgeoning city of Bahir Dar rapidly increases its consumption of market-garden crops from lake wetland mini-deltas and fish from the lake. This land cover survey from 2007 is now substantially outdated as the new decade has unfolded under momentum from international capital and local responses, both demographic and ecological. Historical elements of change that occurred over the course of the 20th century would include, inter alia, the growth of new urban market centres, the introduction of new crops (maize, market garden crops), and road construction.

5.2 *The Zambezi*

Its history of colonial-era water management schemes as well as forecasts of rainfall decline make the Zambezi watershed a valuable model for understanding the history and future of African hydroecologies. Two water management schemes on the Zambezi—the Kariba Dam (and Lake Kariba) and Mozambique's Cahora Bassa Dam (and Cahora Bassa reservoir)—are both markers of the high modernism of the late colonial period. The 2007 IPCC report forecast a decline in rainfall of 10 to 15 per cent (see above). What implications will this have for existing watersheds and the ones to come?

Overall, planners anticipate that in the coming decade a total of 13,000 MW of dam projects will be completed along the course of the Zambezi, from northeast Angola, through Zambia, Zimbabwe, and to its delta in Mozambique. None of the studies on the Zambezi watershed have analysed the risks presented by a changing climate and its effects on river flow.[19] Ironically, the Zambezi River has drawn significant scholarship on the meaning and impact

18 Zur Heide, Friedrich, *Feasibility Study.*

19 Pearce, "Hydro Projects," 4. Pearce quotes University of Wisconsin-Madison hydrologist Richard Beilfuss.

of dams along its course. In addition to technical engineering reports on flow rates and storage capacity, there are historical assessments.[20] These follow a long tradition of social impact studies from the Rhodes Livingstone Institute by Thayer Scudder and Elizabeth Colson.[21] In particular, the recent Isaacman study details the "challenge of sources," namely the lopsided nature of data on the social and local economic implications of Zambezi water management compared to older feasibility studies.[22] These are warning signs on the horizon.

Record flooding of the river delta in Mozambique in 2000 and again in 2014 offer a dystopian view of what may come. Recent reports from Zimbabwe and Zambia of cracks in the wall of the Kariba Dam that threaten downstream dams point to the potential knock-on threats to the power grid and to human livelihoods along the course of the river and in the watershed's hinterlands. Dams face recurrent costs for repair and redesign arising from new conditions of climate change and upstream engineering.

A chart of the seasonal flows of the Zambezi before and after the construction of the Cahora Bassa Dam and Kariba Dam suggests an impending radical shift in the watershed in terms of the river's seasonality of flow with the construction of those water management structures. These engineering designs for reservoirs (or run-of-the-river) as well as climate surprises and crises raise questions about the deeper effects on local economies of the arrival of kilowatt hours and flooding.[23] Both the climate change warning and the recent downstream flood emergencies may be the proverbial canaries in the coalmine for the Zambezi watershed.

6 The MIMES Model: Livelihoods and African Watersheds, Past and Future

Is there a model of watershed ecologies that would help us to systematise watershed changes over time that relate to climate change and more localised

20 Tischler, Julia, *Light and Power for a Multiracial Nation: The Kariba Dam Scheme in the Central African Federation* (Cambridge: Palgrave McMillian, 2013); Isaacman, Allan F., and Isaacman, Barbara S., *Dams, Displacement, and the Delusion of Development: Cahora Bassa and Its Legacies in Mozambique, 1965–2007* (Athens: Ohio University Press, 2013).

21 Scudder, Thayer, *The Ecology of the Gwembe Tonga* (Manchester: Manchester University Press, 1962); Colson, Elizabeth, *The Social Consequences of Resettlement: The Impact of the Kariba Resettlement upon the Gwembe Tonga* (Manchester: Manchester University Press, 1971).

22 Isaacman, and Isaacman, *Dams, Displacement,* 19.

23 Quoted in Isaacman, and Isaacman, *Dams, Displacement,* 41.

human-induced dynamics? One model we might propose for data-rich ecologies that may help us to compare future African watershed dynamics of political ecology is the MIMES (Multi-Scale Integrated Model of Ecosystem Services) approach, which is already used to understand water/landscape interactions in the relatively data-rich Gulf of Maine fisheries and the potential effects of wind farming in New England as used by the CHANS project of Boston University's Pardee Center for the Study of the Longer-Range Future.[24]

MIMES is a form of ecosystem 'accounting' that organises data sets to form a multi-layered portrait of Coupled Human and Natural Systems (CHANS), their geographic setting, policy changes over time, and inexorable movements of hydrology. By placing this data into a common geographic information system (GIS) framework and using overlapping mapping, it may be possible to show how the elements of an ecosystem (such as an extended watershed) intersect, including physical regimes, biological habitats, and human activities. We might otherwise call these overlapping phenomena historical conjunctures where demography, economy, nature, and culture result in a new landscape (or waterscape) that drives human livelihood choices, biodiversity, and regional politics. This method thus shows dynamic intersections between natural and human systems.

Understanding these dynamics is critical regarding the formation of policy or making sense of unintended consequences of development interventions, like a dam, or climate change. According to a recent method for analysing CHANS in the Gulf of Maine, a team of scholars from Boston University and Conservation International has argued that "exploring long-time horizons is essential for understanding sustainability since system states can shift into phases that are not easily moved back to a more desirable state."[25] In other words, by assembling data from overlapping human livelihoods and the physical world of a watershed it may be possible to anticipate and respond to knock-on effects of watershed management such as dams, hydroelectric schemes, monocropping on watershed soils, and agricultural landscape change. Climate change also needs to be a part of this planning, though it is rarely so.

The Gulf of Maine case may provide us with an example that may lead us forward in African cases we are considering as engineers and economists

---

24 Partners in this project include Conservation International and the MacArthur Foundation.

25 Altman, Irit et al., "An Ecosystem Framework for Marine Ecosystem-Based Management," in *The Sea: Marine Ecosystem-Based Management*, vol. 16, ed. Michael Fogarty (Cambridge: Harvard University Press, 2014), 245–276, accessed July 14, 2017. https://www.researchgate.net/publication/262674557.

implement their plans. In the Gulf of Maine example, data collectors assembled information on the physical setting (a water column in the Gulf of Maine) from which human actors derive a livelihood. These included fisherfolk dependent on various fish species, boat owners, lobstermen, tour guides, and operators of wind turbines. The model uses past data to trace their income offtake over time. Might this model also apply to watershed/hydraulic landscapes?

With this data-rich model in mind, we can imagine a way of measuring and forecasting the effects of a watershed intervention on the various stakeholders involved in specific types of agriculture, on infrastructure construction, fish protein, human diets, livestock rearing, and, perhaps, on disease outcomes—all by season or in particular types of climate changes, including drought, flood, temperature rise effects on disease vectors, or crop choices. It may also be possible to project outcomes from particular interventions like types of dams, calendars of downstream water release, or hydroelectric strategies. Such analyses could be retrospective in the historical reconstructing of effects of dams, crop change, urbanisation, or new markets. Such an approach will require a major investment by environmental historians to assemble time series and unpublished field reports in colonial archives as well as early UN agency surveys. In many cases, retrospective modelling may be the only alternative.

The challenge at hand for historians is to explore past conditions of human settlement and climate settings and to adapt a historical model for understanding past events such as dam construction or irrigation. This approach could include plotting historical patterns on income streams from a range of livelihoods particular to a watershed and the downstream effects on social institutions, livelihoods, gendered income balance, diet, or other measures of development.

7 Future African Watersheds: an Era of Climate Change and Water Management

The MIMES model of profits/income from levels of the Gulf of Maine water column shows the relationship between incomes (profits) from various activities in a marine ecology. Could such a model also serve as a didactic tool for understanding African watershed stakeholders at the local and regional scale? Here we see economic activities along a marine water column. We might plot the economic activities around a dam or reservoir or a run-of-the-river system that supports fisheries, floodplain agriculture, pastoralism, and trade. What would be the effect of various engineering strategies on local or regional economies or social systems? In the Gulf of Maine, the key intervention was the placement of wind turbines and the restrictions that this placed on various income streams and economic activities. A dam reservoir or run-of-the-river scheme might be analogous.

For any particular watershed or dam ecology, the activities to be measured would differ, but the MIMES model offers a focus for data collection and valuating local livelihoods. Significant areas for assessment might include the following:

(1) relative effect on income streams for local livelihoods or state agencies;
(2) health (e.g. malaria, schistosomiasis, waterborne diseases, or chronic diseases);
(3) nitrogen flows from upland agriculture into reservoirs, seasonal streams, and rivers;
(4) protein (fish/livestock) versus carbohydrate-based (grain/legumes) incomes/diets;
(5) crop changes in irrigated versus rain-fed areas of the watershed;
(6) rates of geological change, erosion, or silting.

The central task here is to understand the interaction, over time, between climate change and anthropogenic changes in water flows, storage, and offtake by sector. The underlying question is whether watersheds as a unit for measuring complexity can serve as a basic tool for policy analysis and/or planning in ways that enhance human development. As the analysis of watershed complexities moves forward it must also include the longue durée of these ecologies.

CHAPTER 11

Increasing Urbanisation and the Role of Green Spaces in Urban Climate Resilience in Africa

Bertrand F. Nero, Daniel Callo-Concha, and Manfred Denich

Abstract

Africa's urban population presently totals 439 million people or forty per cent of its total population, a share that is expected to rise to 1.1 billion or 56 per cent by 2050. Urbanisation is constituted not only in terms of demographic and land expansion, but also in the transformation of the environment and social conditions in the growing urban centres. As the world's shelter and economic hubs, urban areas account for more than seventy per cent of global greenhouse gas emissions. Alongside numerous climate change challenges, urban areas are also characterised by warmer temperatures due to the urban heat island effect as well as rising inequality and poverty. Achieving resilient urbanisation is therefore a multifaceted issue. In this chapter, we discuss African urbanisation and urban green spaces in the context of urban climate resilience.

Combining case studies across cities in Africa, we found that urban areas are growing at an annual rate of five hundred hectares, an expansion occurring at the expense of agricultural and natural land cover. Together with increasing consumption of fossil fuels as well as growing infrastructural deficiencies, waste disposal sites, and industrial facilities within ecologically sensitive zones, African cities are becoming increasingly vulnerable to the perils of climate variability and change. The effects of urban heat islands, poor air quality, flash floods, heatwaves, increasing public health hazards, rising regional mean temperatures, and droughts will escalate further if African urbanisation proceeds at its current pace without any intervention. On the other hand, the culture of creating and preserving green spaces such as street trees, sacred groves, cemetaries, public parks, gardens in residential lots, vegetation in institutional compounds, and urban farm- and grasslands by urban authorities and private landowners is a promising indicator that cities can be resilient to the local impacts of changing climatic conditions through the adoption of adequate adaptation and mitigation measures. For instance, urban green spaces in the Ghanaian city of Kumasi store as much as 168–1,737 tons CO_2 sequestered from the atmosphere per hectare. The management of Africa's urban green spaces, however, poses challenges distinct from those facing cities in the Global North. The weakness of enforcing institutions, despite the occasional existence of appropriate legislation, as well as the

 | DOI:10.1163/9789004410848_013

potential for society to endorse either favourable or detrimental changes concerning green spaces are key in urban green space management. It is important to either amend or, where necessary, enact urban environmental policies to help make an increasingly urban Africa more climate-resilient.

1 Introduction

Besides rapid urban population growth and urban land expansion, cities in Africa are characterised by high levels of informality and poverty, poor infrastructure, and consequently a high susceptibility to environmental hazards. Climate change and its impacts are exacerbating the environmental pressures on cities. Hence, climate-resilient conditions must emerge as principles for the sustainable development of cities.[1]

In an array of options to improve the overall resilience of cities, the importance of urban green spaces (UGS) is widely recognised. Urban green spaces strengthen a city's ability to cope with and recover from stress. However, both urbanisation and changing climatic conditions threaten the existence of green spaces in cities. Despite this, their continued existence can be ensured through specific management and planning efforts that lead to urban vegetation being adequately integrated into the city matrix, thereby also sustainably supporting urban climate resilience. These strategies include ecosystem services that aim to reduce the impact of floods and heatwaves, improve air quality, and contribute to food security.

Although increasing global urbanisation and its environmental, economic, and social problems have been intensively discussed, approaches to enhancing urban climate resilience in developing countries demand further scientific investigation. As such, this chapter (1) elaborates on urbanisation in Africa, (2) discusses urban land expansion and environmental challenges in the context of climate change, and (3) examines the potential of UGS to contribute to urban climate resilience and addresses managerial challenges hindering UGS conservation in Kumasi (Ghana), a mid-sized sub-Saharan African city.

Resilience in socio-ecological systems is defined as the ability of a system to absorb shocks and avoid crossing thresholds that would irreversibly lead to

1 Taylor, Anna, and Camaren Peter, "Strengthening Climate Resilience in African Cities: A Framework for Working with Informality," *Climate & Development Knowledge Network (CDKN) Working Paper* (2014), 18.

new states, instead maintaining the system's capacity to reorganise itself while change takes place.[2] According to Meerow et al.:

> Urban resilience refers to the ability of an urban system and all its constituent socio-ecological and socio-technical networks across temporal and spatial scales to maintain or rapidly return to desired functions in the face of a disturbance, to adapt to change, and to quickly transform systems that limit current or future adaptive capacity.[3]

Climate resilience in natural and human systems depends strongly on their ability to adapt to climate change (equilibrist resilience) as well as to mitigate the effects of climate change (evolutionary resilience).[4] In this context, adaptation refers to reactive adjustments in response to actual or expected climatic stimuli and their effects or impacts, whereas mitigation comprises proactive anthropogenic interventions to reduce the sources or enhance the sinks of greenhouse gases (GHG s).[5] Ultimately, both approaches aim to minimise the undesirable effects of climate change[6] and, accordingly, play a crucial role in urban climate resilience.

2 Urbanisation in Africa

About 54 per cent of the world's population now lives in urban areas, and it is projected that many more people will move to cities in the coming decades. At the turn of the 20th century, Africa contributed less than two per cent of the world's urban population, a figure that has now reached twelve per cent. The continent's urban population has soared from ten per cent of its total population in the 1950s to forty per cent (439 million) today and is projected to reach

2 Walker, Brian et al., "Resilience Management in Social-Ecological Systems: A Working Hypothesis for a Participatory Approach," *Conservation Ecology* 6 (2002): 14.

3 Meerow, Sara, Joshua P. Newell, and Melissa Stults, "Defining Urban Resilience: A Review," *Landscape and Urban Planning* 147 (2016): 38–49.

4 White, Iain, and Paul O'Hare, "From Rhetoric to Reality: Which Resilience, Why Resilience, and Whose Resilience in Spatial Planning?" *Environment and Planning C: Government and Policy* 32.5 (2014): 934–950.

5 Banuri, Tariq et al., "Climate Change 2001: Mitigation, A Report of Working Group III of the Intergovernmental Panel on Climate Change (IPCC)" (Technical Summary, Third Assessment Report, 2001).

6 Parker-Flynn, James E., "The Intersection of Mitigation and Adaptation in Climate Law and Policy," *Environs, Environmental Law & Policy Journal* 38.1 (2014): 1–48, here 47.

56 per cent (1.1 billion) by 2050. The number of African megacities is expected to double by 2030, while that of medium-sized cities will quadruple.[7] Africa will become the most rapidly urbanising continent between 2020 and 2050, while by 2050[8] its cities are expected to account for 21 per cent of the world's urban population.

2.1 *Historical Background*

As early as 3200 BC, large settlements already existed in Africa, initially confined to North Africa before later emerging in sub-Saharan Africa (SSA) too.[9] Ancient urban centres and peri-urban areas were characterised by their ability to produce agricultural surpluses, specialised craftsmen, and monumental architecture.[10] Compared to contemporary urban centres, however, cities in ancient Africa were sparse, low in population, and small in size, although vibrant politically, economically, and socially.

Contact with Arab and later European merchants and missionaries fostered the rapid spread and development of cities.[11] Colonisation, economic expansion, and international politics changed the face of African cities for both colonialists and indigenes. Cities, especially those along the coasts of the Indian and Atlantic oceans, were renowned for their trade in slaves and natural resources from the African continent. To facilitate trade, contemporary European and Arabian architecture, technology, and culture were gradually infused into African cities and culture. Indeed, the history, shape, and character of most African cities today can be traced to colonial city plans, designs, policies, and functions.[12] For instance, urban plans for Lusaka (Zambia) and Kumasi were based on Howard's 'Garden City' concept—the archetypical car-oriented, elitist European city. Colonial master plans, however, did not

7 United Nations, "World Urbanization Prospects: 2014 Revisions, Highlights," (ST/ESA/SER.A/352) (2014), 32.

8 Satterthwaite, David, "The Transition to a Predominantly Urban World and its Underpinnings," *Human Settlements Discussion Paper Series: Urban Change* 4 (2007): 99.

9 Özden, Kemal, and Enwere Chigoyie, "Urbanisation and its Political Challenges in Developing Countries," *Eurasian Journal of Business and Economics* 5.10 (2012): 99–120.

10 Coquery-Vidrovitch, Catherine, *The History of African Cities South of the Sahara: From the Origins to Colonization*, trans. Mary Baker (Princeton: Markus Wiener Publication, 2005).

11 Kitto, Elizabeth, "Before European Colonialism was Africa Essentially Rural?" *Retrospective* online (2012).

12 Watson, Vanessa, and Agbola Babatunde, "Who will Plan Africa's Cities?" *Africa Research Institute, Counterpoints* (2013): 16.

anticipate the sprawl that has redefined the shape and configuration of modern African cities.[13]

2.2 *Current Situation and Trends*

Cairo (Egypt), Kinshasa (DR Congo), and Lagos (Nigeria) are the only megacities (ten million inhabitants or more) on the African continent, together contributing 15 per cent of its total urban population. About 66 per cent of Africa's urban population lives in medium-sized cities with one to five million residents. Of these, 16 cities are in West Africa (e.g. Abidjan, Accra, Dakar, Kano), nine in Southern Africa (e.g. Harare, Johannesburg, Lusaka, Maputo), seven each in North Africa (e.g. Algiers, Fez, Rabat, Tripoli) and East Africa (e.g. Addis Ababa, Mogadishu, Dar es Salaam, Nairobi), and four in Central Africa (Brazzaville, Douala, Mbuji-Mayi, and Yaoundé). Small cities (fewer than one million inhabitants) are generally regarded as the fastest-growing cities in the world (2.4–6 per cent per year).[14]

Although Africa remains the least urbanised continent, subregional variations are significant. Excluding Mozambique, Zambia, and Zimbabwe, Southern Africa is the most urbanised subregion with 61 per cent of its population living in urban areas, followed by North Africa with 51 per cent. In Central and West Africa, the urban share is about 44 per cent, higher than in East Africa, where it is 25 per cent. Most African urban populations are concentrated in coastal areas (e.g. Gulf of Guinea), alluvial plains (e.g. Nile River), or lacustrine plains (e.g. Lake Victoria).

Urbanisation is most rapid in East and West Africa, where by 2050 the urban population will rise to about 328 million and 390 million respectively (Figure 11.1).[15]

2.3 *Causes: Migration and Natural Increase*

Recent urbanisation in Africa has been attributed to demand for labour during the continent's early period of industrialisation from the late 19th century until the middle of the 20th century, which triggered the exodus of rural people

13 Quagraine, Victor K., "Urban Landscape Depletion in the Kumasi Metropolis," in *Future of the Tree: Towards Growth and Development of Kumasi*, ed. Kwasi K. Adarkwa (Kumasi: KNUST University Press, 2011), 212–233.

14 United Nations, "World Urbanization," 20–21.

15 UN-Habitat, "The UN-Habitat Urban Population Database: UN-Habitat for a Better Future," accessed August 27, 2015. http://urbandata.unhabitat.org/download-data/#/countries/1.

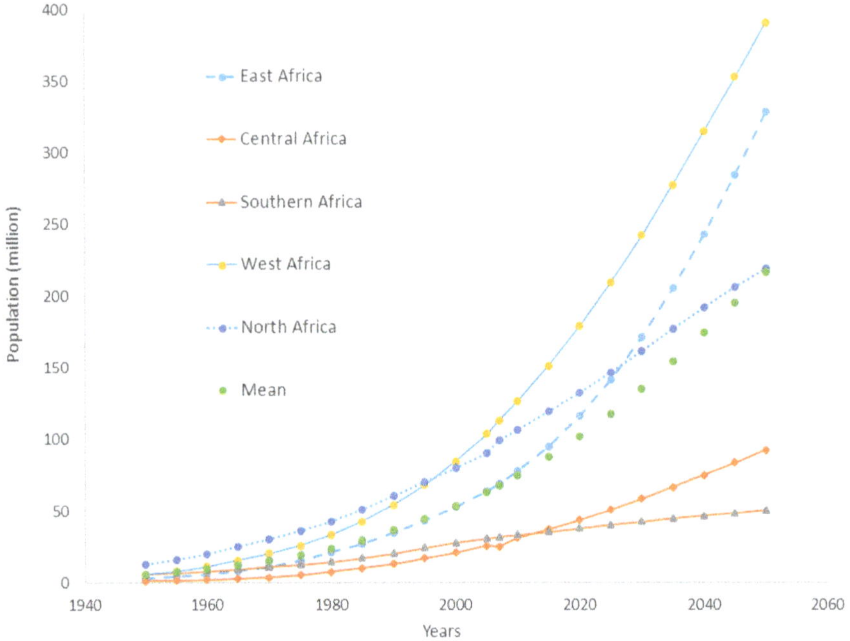

FIGURE 11.1 Urban population growth in the African subregions

into towns and cities.[16] Additionally, civil unrest and violence due to political instability, natural disasters provoked by climatic events (such as drought- or flood-induced famine), alienation of individuals from tribes and indigenous communities, and simply the desire to experience urban life have all been causal factors.[17]

Natural population increases through high birth rates and/or the reclassification of former rural areas as urban now provide the most striking explanations for current urbanisation patterns in Africa.[18] The availability of better

16 Caldwell, C. John, *African Rural–Urban Migration: The Movement to Ghana's Towns* (Canberra: Australian National University Press, 1969), 257.

17 Henderson, J. Vernon, Adam Storeygard, and Uwe Deichmann, "Is Climate Change Driving Urbanization in Africa?" Policy Research Working paper 6925, The World Bank (2014): 50; Østergaard Nielsen in this volume; Satterthwaite, "The Transition to a Predominantly Urban World," 28.

18 Kessides, Christine, *The Urban Transition in Sub-Saharan Africa*, (Washington: The Cities Alliance, 2006), 202; Potts, Deborah, "Whatever Happened to Africa's Rapid Urbanization?" (London: Africa Research Institute, Counterpoints, 2012), 17, accessed November 26, 2017. http://africaresearchinstitute.org/newsite/wp-content/uploads/2013/03/Whatever-happened-to-Africas-rapid-urbanisation.pdf.

healthcare systems in cities has increased natality while reducing mortality rates. In resource-rich countries like Ghana, Côte d'Ivoire, or Nigeria, proceeds from the export of natural resources are disproportionately invested in developing urban goods and services, in turn further accelerating urbanisation.[19]

However, due to a lack of opportunities in cities, high living costs, and cultural discomfort experienced by migrants, rural–urban migration is decreasing.[20] In Burkina Faso and Côte d'Ivoire, for example, counter-urbanisation processes have been observed. Nevertheless, inter- and intra-city mobility among urbanites remain widespread.[21] The pressing question now becomes how urbanisation in Africa will influence urban life economically, socially, culturally, and environmentally.

2.4 *Side Effects of Urbanisation: Poverty and Cultural Erosion*

Between 1981 and 2004, the number of urban poor (income below USD 2.15/day) was rising even faster than the global urban population.[22] By 2050, forty per cent of the 1.1 billion African urbanites will live in income poverty, representing the largest and fastest-growing group of impoverished people in the world.[23] As a consequence, Africa's urbanisation is characterised by an inadequate asset base (such as lack of infrastructure), unavailability of and uneven access to services, amenities, education, and human capital development, as well as worsening environmental conditions.[24] Thus, African cities are not always the engines of economic growth that cities on other continents typically are; instead, they are often both a cause and symptom of various crises.[25]

---

19 Jedwab, Rémi, "Why is African Urbanization Different? Evidence from Resource Exports in Ghana and Ivory Coast," (London, Paris: London School of Economics, Paris School of Economics, 2012), accessed December 17, 2018, https://www.researchgate.net/publication/267386204_Why_Is_African_Urbanization_Different_Evidence_from_Resource_Exports_in_Ghana_and_Ivory_Coast.

20 Beauchemin, Cris, "Rural–Urban Migration in West Africa: An Economic Situation in Burkina Faso," *Population, Space and Place* 17 (2011): 47–72.

21 Simone, Abdou Maliq, "The Urbanity of Movement: Dynamic Frontiers in Contemporary Africa," *Journal of Planning Education and Research* 31.4 (2011): 379–391.

22 Chen, Shaohua, and Ravallion, Martin, "Absolute Poverty Measures for the Developing World, 1981–2004," *Proceedings of the National Academy of Sciences of the United States of America* 104.43 (2007): 16757–16762.

23 Baker, Judy L., *Urban Poverty: A Global Overview (No. UP 5)* (Washington: The World Bank, 2008), 27.

24 Satterthwaite, David, "Reducing Urban Poverty: Constraints on the Effectiveness of Aid Agencies and Development Banks and Some Suggestions for Change," *Environment and Urbanization* 13.1 (2001): 137–157.

25 The World Bank, *Entering the 21st Century: World Development Report 1999/2000* (New York: Oxford University Press, 2000).

Although these trends are changing—with many African cities now acting as hubs of innovation—the pace of change is slow, and most of the continent remains marked by a scarcity of development opportunities and a high proportion of unskilled labour. Most cities, meanwhile, are primarily centres of administrative and political power that remain short of skilled manpower.

Although research on a continental scale is lacking, country-level studies indicate that the majority of Africa's urban poor live in East, Central, and West Africa.[26] For example, thirty per cent of the population in Mombasa (Kenya) live below the absolute poverty line (USD 2.15/day),[27] fifty per cent of the population of Maputo (Mozambique) is poor and thirty per cent characterised as destitute,[28] while 95 per cent of informal economy participants in Kinshasa (DR Congo) have low incomes while ninety per cent have no formal jobs.[29]

Cities also perform the role of hubs of foreign culture and technology that influence local ones. For instance, the large extended family system and communal way of life are giving way to a nuclear family system, individualism, and a struggle for survival as conflict increases in poor urban neighbourhoods. One such example is Mombasa, where frequent clashes occur between coastal residents and up-country immigrants due to economic inequality fueled by a political system that is ethnically inclined.[30] Technology and mechanisation have replaced manual and animal traction. Organic waste, formerly an important resource as fertiliser, has become adulterated with non-biodegradable materials and is thus a menace to public health in cities. In many African cities, waste is improperly disposed of. Waste and landfills pollute soil, air, and water, produce foul smells, provide breeding grounds for disease vectors, and attract dangerous animals such as rodents and snakes.

26 Mabogunje, Akin L., "Global Urban Poverty Research Agenda: The African Case," (paper presented at the seminar "Global Urban Poverty: Setting the Research Agenda," Woodrow Wilson International Center for Scholars, Washington D.C., December 15, 2005), accessed December 17, 2018, 7. https://www.wilsoncenter.org/event/global-urban-poverty-research-agenda-the-african-case.

27 Rakodi, Carole, Rose Gatabaki-Kamau, and Nick Devas, "Poverty and Political Conflict in Mombasa," *Environment and Urbanization* 12.1 (2000): 153–170.

28 Jenkins, Paul, "Urban Management, Urban Poverty and Urban Governance: Planning and Land Management in Maputo," *Environment and Urbanization* 12.1 (2000): 137–152.

29 Misilu, Mia Nsokimieno Eric, Earl Bailey, Tushiminine Martin Mpinda, Abraham Carleen, Jiangfeng Li, and Liqin Zhang, "Ongoing Informal Settlements in Democratic Republic of Congo: Implementing New Urban Policy for Creating Sustainable Neighborhoods," *Journal of Sustainable Development* 7.5 (2014): 254–265.

30 Rakodi et al., "Poverty and Political Conflict in Mombasa," 158.

3 Environmental Impacts: Land Use and Climate Change

3.1 *Urban Land Expansion and Land Use Change*

Globally, urban areas occupy about three to four per cent of the earth's surface and are growing at a rate twice that of the global population.[31] With 256 cities (more than 100,000 inhabitants each) and a total population of 131.6 million, urban areas in SSA occupy an area of approximately 13,000 square kilometres, while 115 cities in North Africa with a population of 53 million occupy an area of 5,342 square kilometres. Together, all these urban areas are predicted to increase in space by 590 per cent by 2030.

Urban expansion occurs unevenly across the continent, mostly concentrated in the Gulf of Guinea region of West Africa, along the Nile River in Egypt, on the northern shore of Lake Victoria in Kenya and Uganda and stretching into Burundi and Rwanda, in the Kano region in northern Nigeria, and in greater Addis Ababa in Ethiopia.[32] Losses in natural land cover to urbanisation are higher in East, North, and West Africa than in Southern and Central Africa.[33]

Generally, megacities and medium-sized cities are undergoing the most rapid land expansion with significantly higher growth rates of 743 and 620 hectares per year respectively (Table 11.1).[34] However, small cities with fewer than

31 Angel, Shlomo et al., "The Dimensions of Global Urban Expansion: Estimates and Projections for All Countries, 2000–2050," *Progress in Planning* (2011): 53–107; Seto, Karen C., Michail Fragkias, Burak Güneralp, and Michael K Reilly, "A Meta-Analysis of Global Urban Land Expansion," *PLOS One* 6.8 (2011): 1–9.

32 Seto, Karen C., Burak Güneralp, and Lucy R. Hutyra, "Global Forecasts of Urban Expansion to 2030 and Direct Impacts on Biodiversity and Carbon Pools," *Proceedings of the American Academy of Sciences* 109.40 (2012): 16083–16088.

33 Otunga, Charles, John Odindi, and Onisimo Mutanga, "Land Use and Land Cover Change in the Fringe of Ethekwini Municipality: Implications for Urban Green Spaces Using Remote Sensing," *South African Journal of Geomatics* 3.2 (2014): 145–162; Nguh, Balgah S., "Land Tenure and Land Use Dynamics in Limbe City, South West Region of Cameroon," *Agriculture Science Development* 2 (2013): 14–24; Sebego, Reuben J., and Thando D. Gwebu, "Patterns, Determinants, Impacts and Policy Implications of the Spatial Expansion of an African Capital City: The Greater Gaborone Example," *International Journal of Sustainable Built Environment* 2.2 (2013): 193–208.

34 Adepoju, M. O., A.C. Mellington, and K.T. Tansey, "Land Use/Land Cover Change Detection in Metropolitan Lagos (Nigeria): 1984–2002," (paper presented at the "ARPS Annual Conference," Reno, May 1–5, 2006); Ayila, Adzandeh E., Fabiyi O. Oluseyi, and Bello Yakasai Anas, "Statistical Analysis of Urban Growth in Kano Metropolis, Nigeria," *International Journal of Environmental Monitoring and Analysis* 2.1 (2014): 50–56; Hassan, Ahmed Abdelhalim M, "Changes in the Urban Spatial Structure of the Greater Cairo Metropolitan Area," *ISPRS—International Archives of the Photogrammetry, Remote Sensing and Spatial Information Sciences* XXXVIII-4/(C21–133) (2011): 133–136; Kamusoko, Courage, Jonah Gamba, and Hitomi Murakami, "Monitoring Urban

TABLE 11.1 Changes in urban land use characteristics in mega, medium-sized, and small cities in Africa (average values)

Citytype	City area (km^2)	Period of change (years)	Increase in built-up area (km^2) Percentage change (%)	Decrease in green space (km^2) Percentage change (%)	Urban expansion rate (ha/year)	Green space coverage (%)	Green space per 1,000 inhabitants (ha)
Mega	972	23	170 (93%)	165 (46%)	743	42	0.47
Medium-sized	844	21	135 (199%)	120 (22%)	620	39	12.1
Small	214	24	22 (148%)	22 (21%)	129	57	23

one million inhabitants are numerous and therefore growing rapidly in terms of their share of the total population, with an average land expansion rate of 129 hectares per year.

Expanding cities encroach upon environmentally sensitive zones such as wetlands, protected nature areas, agricultural land, and open parkland, causing a decline in vegetation cover and primary production.[35] Cities in SSA have historically experienced a growth in built-up areas at the expense of agricultural and forest land, causing 12 to 77 per cent loss in tree cover and 18 to 50 per cent loss in farmland within an average period of 22 years.[36] However, this

Spatial Growth in Harare Metropolitan Province, Zimbabwe," *Advances in Remote Sensing* 2.0 (2013): 322–331; Mohammed, Eiman A., Hanan M. Alawad, Khalid A. Elsayed Zeinelabdein, and Abdelhafi G. Ali, "Urban Expansion and Population Growth in Omdurman City, Sudan Using Geospatial Technologies and Statistical Approaches," *American Journal of Earth Sciences* 2.1 (2015): 1–7; Ujoh, Fanan, Olarewaju O. Ifatimehin, and Adams N. Baba, "Detecting Changes in Land Use/Cover of Umuahia, South-Eastern Nigeria, Using Remote Sensing and GIS Techniques," *Confluence Journal of Environmental Studies* 6 (2011): 72–80; Vermeiren, Karolien et al., "Urban Growth of Kampala, Uganda: Pattern Analysis and Scenario Development," *Landscape and Urban Planning* 106.2 (2012): 199–206; Wafula, Munialo E., and Chege Moses Gichuho, "The Pattern of Expansion of Urban Residential in Eldoret Municipality, Kenya," *International Journal of Science and Research* 2.4 (2013): 586–589; Weber, Christiane, and Anne Puissant, "Urbanization Pressure and Modeling of Urban Growth: Example of the Tunis Metropolitan Area," *Remote Sensing of the Environment* 86.3 (2003): 341–352; Sahalu, Atalel G., *Analysis of Urban Land Use and Land Cover Changes: A Case Study in Bahir Dar, Ethiopia*, (MSc diss., Universidade Nova de Lisboa, Lisbon, Portugal, 2014), 79; Al-Sharif, Abubakr A. A., Biswajeet Pradhan, Helmi Zulhaidi Modh Shafri, and Shattri Mansor, "Spatio-Temporal Analysis of Urban and Population Growths in Tripoli Using Remotely Sensed Data and GIS," *Indian Journal of Science and Technology* 6.8 (2013): 5134–5142.

35 Ramankutty, Navin et al., "Global Land-cover Change: Recent Progress, Remaining Challenges," in *Land Use and Land-Cover Change—Local Processes and Global Impacts*, ed. Eric F. Lambin, and H. Geist (Berlin: Springer, 2010), 1–8.

36 Aduah, M., and P. Baffoe, "Remote Sensing for Mapping Land-Use/Cover Changes and Urban Sprawl in Sekondi-Takoradi, Western Region of Ghana," *The International Journal of Engineering and Science (IJES)* (2013): 66–73; Fanan, Ujoh, Dlama Kwabe Isa, and Oluseyi Ifatimehin Olarewaju, "Urban Expansion and Vegetal Cover Loss in and around Nigeria's Federal Capital City," *Journal of Ecology and Natural Environment* 3.1 (2010): 1–10; Forkuor, Gerald, and Cofie Olufunke, "Dynamics of Land-Use and Land-cover Change in Freetown, Sierra Leone and Its Effects on Urban and Peri-Urban Agriculture—A Remote Sensing Approach," *International Journal of Remote Sensing* 32.4 (2011): 1017–1037; Mundia, C. N., and M Aniya, "Analysis of Land Use/Cover Changes and Urban Expansion of Nairobi City Using Remote Sensing and GIS," *International Journal of Remote Sensing* 26.13 (2005): 2831–2849.

trend is not universal; in Bamako (Mali), non-forest green spaces increased between 1986 and 2006 due to bare land conversion.[37]

While many underlying factors influence urban land expansion, in Africa this is primarily driven by population growth, evidence for which is demonstrated by the correlation between population growth and increase in urban land area across the continent.[38]

Vegetation loss due to urban land expansion exacerbates climate change impacts. By replacing trees with grey infrastructure, the carbon stored in vegetation and soil is released into the atmosphere via several processes. At an average urban built-up area expansion rate of five hundred hectares per annum (Table 11.1), SSA cities emit about 100,000 tons of carbon per year due to urbanisation-driven forest loss. It is estimated that urban expansion-related deforestation in Africa will emit approximately 490 million tons of carbon by 2030. However, urbanisation-driven vegetation loss and its effects on greenhouse gas (GHG) emissions and climate change require further investigation at higher resolution given the wide ecological and socio-economic disparities among African cities. Furthermore, land use change may have much more drastic thermal effects in urban areas than climate change. A simulation study on land surface temperature in Addis Ababa (Ethiopia) and Dar es Salaam (Tanzania) revealed converting vegetation to a built-up or bare area may result in a land surface temperature difference of up to 25°C compared to typical climate change-related increases of about 1.5°C requiring several decades to occur.[39] Also, the sealing of soil surfaces in cities redefines water flow paths, impedes infiltration, and facilitates overland flow, thereby causing more frequent flash floods and major flood events.

However, urbanisation does not necessarily exacerbate loss of vegetation. In arid areas, the environmental conditions in cities can actually provide

37 Diallo, Boubacar A., and Bao Zhengyu, "Land Cover Change Assessment Using Remote Sensing: Case Study of Bamako, Mali," *Researcher* 2.4 (2010): 7–17.

38 Kayembe wa kayembe, Matthieu, Mathieu De Maeyer, and Eleonore Wolff, "The Mapping of the Urban Growth of Kinshasa (DRC) through High Resolution Remote Sensing between 1995 and 2005," in *Remote Sensing—Applications*, ed. Escalante Boris (Rijeka: InTech, 2008), 463–479; Sylla, L., D. Xiong, H. Y. Zhang, and S. T. Bangoura, "A GIS Technology and Method to Assess Environmental Problems from Land Use/Cover Changes: Conakry, Coyah and Dubreka Region Case Study," *Egyptian Journal of Remote Sensing and Space Science* 15.1 (2012): 31–38; Wu, Weicheng, Marie-Francoise Courel, and Jeannine Le Rhun, "Application of Remote Sensing to the Urban Expansion Analysis for Nouakchott, Mauritania," *Geocarto International* 18.1 (2003): 17–24.

39 Lindley, Sarah J. et al., "Green Infrastructure for Climate Adaptation in African Cities," in *Urban Vulnerability and Climate Change in Africa*, ed. Stephan Pauleitet et al. (Heidelberg, New York: Springer, 2015), 107–152.

favourable conditions for plant growth and hence induce city greening due to elevated CO_2 emissions, soil nutrient improvements from wastewater irrigation and organic waste disposal, and higher temperatures.[40]

Sixty per cent of households in African cities use fuelwood for cooking.[41] Fuelwood, together with charcoal, is also the predominant source of energy in African cities, particularly for the poor, and thus a notable contributor of CO_2 emissions.[42] Fuelwood usually comes from forests in peri-urban and neighbouring hinterlands. In the Dar es Salaam area, the radius of exploitation areas for fuelwood and charcoal in the surrounding forests expanded at rates of nine and two kilometres per year respectively, with a reduction in carbon storage and species richness experienced within a radius of up to 220 kilometres from the city centre.[43] Urbanisation in Africa is therefore expected to further exacerbate deforestation in hinterland areas and weaken the potential for terrestrial carbon sequestration, consequently affecting temperature, relative humidity, and precipitation patterns in the respective urban areas.

3.2 *Impacts on Quality of Life and Health*

Human activities lead to an increase in urban temperature, a phenomenon known as the urban heat island (UHI) effect. This refers to the difference in temperature between urban and suburban or rural areas[44] and is attributable to several factors. These include the physical properties of urban land cover, such as dark and compact surfaces (with their absorption of solar radiation, low albedo, and high heat capacity), the increase in high-rise buildings with multiple sunlight-reflecting and -absorbing surfaces as well as wind-blocking effects (caused by their canyon-like heat-trapping structures), and the decrease in area covered by vegetation (resulting in a decline in evaporative cooling).[45]

40 Pouyat, Richard V., Ian D. Yesilonis, and David J. Nowak, "Carbon Storage by Urban Soils in the United States," *Journal of Environmental Quality* (2006): 1566–1575.

41 The World Bank, *Wood-Based Biomass Energy Development for Sub-Saharan Africa* (Washington: The World Bank, 2011), 46.

42 Abd'Razack, Nelson T. A., and Ahmad Nazri bin Muhamad Ludin, "Wood Fuel Consumption and Ecological Footprint of African Cities," *International Journal of Education and Research* 1.2 (2013): 1–18.

43 Ahrends, Antje et al., "Predictable Waves of Sequential Forest Degradation and Biodiversity Loss Spreading from an African City," *Proceedings of the National Academy of Sciences of the United States of America* 107.33 (2010): 14556–14561.

44 Peng, Shushi et al., "Surface Urban Heat Island across 419 Global Big Cities," *Environmental Science & Technology* 46.2 (2012): 696–703.

45 Grimm, Nancy B. et al., "Global Change and the Ecology of Cities," *Science* 319 (2008): 756–760.

A recent evaluation of UHI effects in cities worldwide, including 47 in Africa, revealed differences in day- and night-time temperatures between urban and suburban areas of 1.5°C and 1.1°C respectively. Moreover, the effects of rising air pollutant concentrations in cities interact with UHI, exacerbating health burdens on cities.[46]

Other climate-related vulnerabilities include the prevalence of waterborne diseases such as cholera, malaria, dengue, and yellow fever,[47] aggravated by floods and drought.[48] In relation to the latter, over ninety per cent of the victims of precipitation-related disasters are comprised by the poor in informal settlements and slums.[49]

3.3 *Urban Climate Resilience: Adaptation to and Mitigation of Climate Change*

By taking into account their impressive concentrations of human capital, cities can, on the one hand, contribute to climate change or, alternatively, adapt to or also mitigate the effects of changing climatic conditions.[50] Together, adaptation and mitigation strategies strengthen urban resilience to climate change and its impacts. Adaptive aspects of climate change resilience in cities include modification of the built environment and lifelines, promotion of urban green infrastructure, preventative action to reduce social vulnerability, fostering of climate change governance, and strengthening of strategic urban planning as well as the conducting of regular risk assessments while making such information available to decision makers

46 Harlan, Sharon L., and Darren M. Ruddell, "Climate Change and Health in Cities: Impacts of Heat and Air Pollution and Potential Co-Benefits from Mitigation and Adaptation," *Current Opinion in Environmental Sustainability* 3.3 (2011): 126–134.

47 Unger, Alon, and Lee W. Riley, "Slum Health: From Understanding to Action," *PLoS Medicine* 4.10 (2007): 1561–1566; WHO, "Climate and Health," accessed May 5, 2015. http://www.who.int/globalchange/news/fsclimandhealth/en/.

48 IPCC, "Summary for Policymakers," in Climate Change 2007: The Physical Science Basis. Contribution of Working Group 1 to the Fourth Assessment Report of the Intergovernmental Panel on Climate Change, ed. Dahe Qin et al. (Cambridge, New York: Cambridge University Press, 2007), 1–18.

49 ActionAid, "Climate Change, Urban Flooding and the Rights of the Urban Poor in Africa," (ActionAid, 2006), 1–8; Amoako, Clifford, and David Korboe, "Historical Development, Population Growth and Present Structure of Kumasi," in *Future of the Tree: Towards Growth and Development of Kumasi*, ed. Kwasi K. Adarkwa, (Kumasi: KNUST University Press, 2011), 35–54.

50 Fong, Wee Kean et al., *Global Report on Human Settlements 2011: Cities and Climate Change* (London, Washington: UN Human Settlement Program, 2011), 1–16; Grimm, "Global Change and the Ecology," 756–760.

and urbanites.[51] Major efforts to promote urban climate change resilience in Africa have so far focused on strengthening the infrastructure base of local income communities, enacting policy reforms, engaging in reactive solutions like drainage construction after flood events, initiating poverty reduction programmes, and learning from the experiences of other cities. For instance, in Dakar (Senegal), which is prone to the coastal inundation and erosion that affects many other coastal cities in Africa, 3,000 reinforced housing units were constructed in 2005 for the relocation of flood victims, retention ponds were constructed in the floodplains, and floodplain settlement was prohibited.[52] Experience from Kampala (Uganda) indicates that effective climate change adaptation requires community-based, infrastructural, and institutional interventions.[53] However, in many African cities, the lack of capacity, know-how, and resource availability within national and local government stifles the effective implementation of climate resilience measures.[54]

Although climate change adaptation measures must be given highest priority, urban areas also have the potential to contribute to climate change mitigation. Mitigative strategies to reduce GHG emissions include, most importantly, promoting public transport systems, using renewable sources of energy, implementing improved waste management concepts, and maintaining or expanding UGS.[55]

---

51 Fohlmeister, Sandra et al., "The Way Forward: Climate Resilient Cities for Africa's Future," in *Urban Vulnerability and Climate Change in Africa: A Multidisciplinary Approach* (Heidelberg, New York, Dordrecht, London: Springer, 2015), 369–399.

52 Wilson, Robert H., and Todd G. Smith, "Climate Change Resilience in Urban Africa Part II: The Resilience Policy Agenda" (Policy Research Report #184, Austin: University of Texas, 2015), 244; Wilson, Robert H., and Todd G. Smith, "Urban Resilience to Climate Change Challenges in Africa," (Policy Brief #30, Austin: University of Texas, 2015), 6.

53 Lwasa, Shuaib, "Adapting Urban Areas in Africa to Climate Change: The Case of Kampala," *Current Opinion in Environmental Sustainability* 2.3 (2010): 166–171.

54 Shemdoe, Riziki, Gabriel Kassenga, and Stephen Mbuligwe, "Implementing Climate Change Adaptation and Mitigation Interventions at the Local Government Levels in Tanzania: Where Do We Start?" *Current Opinion in Environmental Sustainability* 13 (2015): 32–41.

55 Nowak, David J., and Daniel E. Crane, "Carbon Storage and Sequestration by Urban Trees in the USA," *Environmental Pollution* 116.3 (2002): 381–389; Nowak et al., "Carbon Storage and Sequestration by Trees in Urban and Community Areas of the United States," *Environmental Pollution* (2013): 229–236; Davies, Zoe G. et al., "Mapping an Urban Ecosystem Service: Quantifying Above-Ground Carbon Storage at a City-Wide Scale," *Journal of Applied Ecology* 48.5 (2011): 1125–1134; Edmondson, Jill L. et al., "Land-Cover Effects on Soil Organic Carbon Stocks in a European City," *Science of the Total Environment* 472 (2014): 444–453.

Clearly, these adaptation and mitigation measures reflect the engineering, planning, and social policy ramifications of urban living, but their effectiveness is yet to be empirically substantiated. Conversely, there is now ample evidence that cities that were once considered mere sources of atmospheric carbon dioxide (CO_2) can also take measures to act as carbon sinks. Estimates from cities in Europe and North America highlight the storage of considerable carbon stocks in UGS. In Leicester (United Kingdom), a city with an area of 73 square kilometres, aboveground vegetation stores about 231,000 tons of carbon dioxide. In the USA (where urban areas represent three per cent of the total land area), carbon storage in urban trees is estimated at 630–700 million tons. In Africa, UGS provide a feasible—albeit often neglected—alternative means of combatting the local consequences of climate change.

4 Case Study: Urban Green Spaces and Climate Resilience in Kumasi

4.1 *Background*

Kumasi is the second largest and fastest-growing city in Ghana, with a land area of 254 square kilometres and approximately 2.5 million inhabitants. The population density stands at about 8,000 inhabitants per square kilometre, while the annual population growth rate is 4.8 per cent.[56] Not only is Kumasi a central point for transiting travellers from within and beyond the borders of the country, the city also plays host to the largest open market in West Africa, making it an important economic hub.[57]

In the early 19th century, urban Kumasi had a population of about 1,500 inhabitants concentrated on a land area of approximately two square kilometres.[58] Its growth and development took place after the Asante Kingdom was defeated in the late 19th century by the British, who subsequently established Western infrastructure in the town. The construction of a railway opened up Kumasi to merchants from the coastal belt, while brisk trade, a cocoa boom, and the building of offices fostered new and refined infrastructural

56 Ghana Statistical Service (GSS), *2010 Population and Housing Census Final Results Ghana Statistical Service* (Accra: Sakoa Press Ltd, 2012), 103; KMA, *The Composite Budget of the Kumasi Metropolitan Assembly for the 2013 Fiscal Year*, (Kumasi Metropolitan Assembly, 2013), 151.

57 Adarkwa, Kwafo Kwasi, "The Role of Kumasi in National Development: Kumasi as a Central Place," in *Future of the Tree: Towards Growth and Development of Kumasi*, ed. Kwasi K. Adarkwa, (Kumasi: KNUST University Press, 2011), 14–34.

58 Amoako, and Korboe, "Historical Development," 35–54.

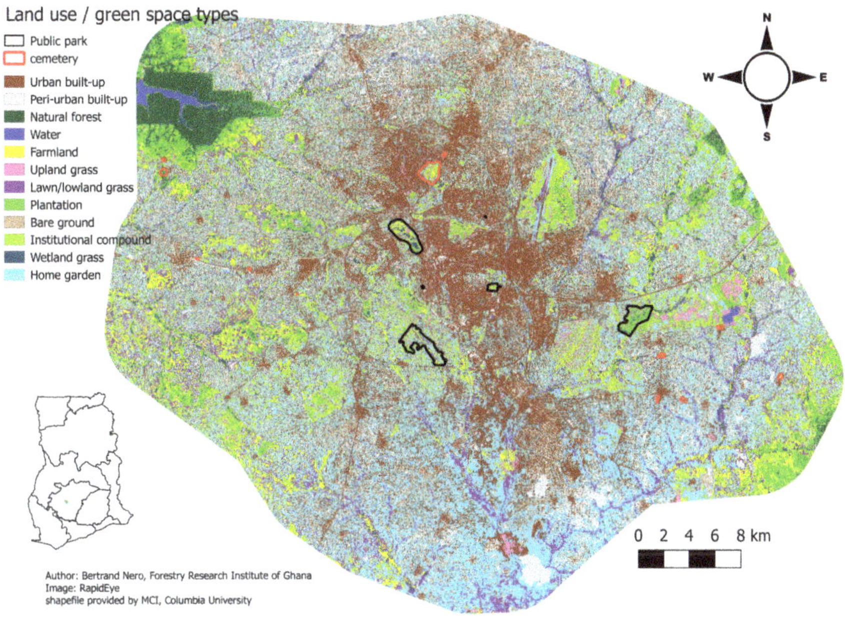

FIGURE 11.2 Distribution of green spaces in Kumasi, Ghana

development beyond the town's erstwhile boundaries. By 1950, Kumasi's land area had expanded to 25 square kilometres. The city has since swelled to its present 254 square kilometres (Figure 11.2) and a population size of approximately 2.5 million residents.

Once dubbed the 'Garden City' of West Africa because of its lush vegetation cover (which comprised ninety per cent of its total land area prior to the 1980s), Kumasi's green cover has since decreased over time to about fifty per cent.[59] The remaining green space is classified into eight dominant UGS types: natural forest, small plantations, farmland along streams, grassland, trees in institutional compounds, public parks, vegetation in cemeteries and sacred groves, and domestic gardens (Figure 11.2).

The changing climatic conditions in the city are quite apparent, with an increase of at least 1°C in daily minimum and maximum temperatures between 1970 and 2000[60] and a twenty per cent decline in precipitation over the past

59 Quagraine, "Urban Landscape Depletion," 219–221; Campion, Betey B., "Urban Wetland Ecology and Floods in Kumasi, Ghana," (PhD diss., University of Bremen, 2012), 194.

60 Manu, Andrew, Yaw A. Twumasi, and Tommy L. Coleman, "Is it the Result of Global Warming or Urbanization? The Rise in Air Temperature in Two Cities in Ghana" (presented at the "5th FIG Regional Conference for Africa," Accra, Ghana, March 8–11, 2006).

four to five decades having been reported.[61] An increase in GHG emissions from the city's transport sector from 665,000 to 860,000 tons between 2000 and 2005 was also reported, a rate of 39,000 tons per year with further implications for local climate.[62]

Biomass energy supplies 64 per cent of the total energy consumed in Ghana, with wood biomass consumption having increased by 72 per cent from 2004 to 2008.[63] Together, Kumasi and Accra consume roughly sixty per cent of the total charcoal produced in Ghana. About 83 per cent of the energy consumed in low-income neighbourhoods in urban Ghana is from biomass.[64] Considering the large piles of organic waste generated and its extensive fuelwood consumption, as reflected by the high proportion of its population engaged in charcoal and firewood collection for cooking, Kumasi has relatively high potential GHG emissions compared to other cities in the region.

Maintaining and managing UGS in Kumasi could offset some of its GHG emissions, minimise the effects of UHI, and help to improve the urban microclimate.

4.2 *Urban Green Spaces: Contributions to Coping with Climatic Impacts in Kumasi*

Green spaces contribute to urban climate resilience in diverse ways. Kumasi's green cover is currently at about fifty per cent of its total land area and is believed to be contributing directly to the reduction of UHI. In cities such as Addis Ababa, Dar es Salaam, and Ouagadougou (Burkina Faso), areas with higher vegetation cover have been shown to be cooler.[65] Evidence for the effectiveness of vegetation cover in combating UHI has also been recorded in Manchester (United Kingdom), where it was shown that by increasing vegetation cover by ten per cent, mean summer ambient temperatures decline by

61 Owusu, Kwadwo, "Changing Rainfall Climatology of West Africa: Implications for Rainfed Agriculture in Ghana and Water Sharing in the Volta Basin," (PhD diss., University of Florida, 2009), 85.

62 Agyeman-Bonsu K.W. et al., "Traffic-Data Driven Modelling of Vehicular Emissions Using COPERT III in Ghana," *American Journal of Scientific and Industrial Research* 1.1 (2010): 32–40.

63 Food and Agriculture Organisation (FAO) "State of the World's Forests," *Food and Agricultural Organization of the United Nations* (2009), 152.

64 Brew-Hammond, Abeeku et al., *Energy Access and Productive Uses for the Urban Poor: Final Report on Ghana Scoping Study* (Washington: The International Bank of Reconstruction, 2011).

65 Lindén, Jenny, "Nocturnal Cool Island in the Sahelian City of Ouagadougou, Burkina Faso," *International Journal of Climatology* 31.4 (2011): 605–620; Lindley et al. "Green Infrastructure," 107–152.

3–4°C, reducing the amount of energy used for air conditioning and thus contributing to a decrease in energy demand for fossil fuels.[66] Most residents in Kumasi use trees in green spaces such as domestic gardens and institutional compounds for shade (Figure 11.3) in order to cope with extreme heat during the day throughout the hot and sunny dry season. Shade under trees provides a cool place to relax, to do business, and for children to play. Additionally, the presence of green cover contributes to the cooling of homes through evapotranspiration, increases space between buildings, and alters the albedo of the urban landscape. This saves costs and cuts emissions due to a reduction in energy expended on air conditioning in offices and in some homes.

As well as acting as carbon sinks, green spaces also trap air pollutants (PM, NOx, CO, and short-lived pollutants e.g. CH_4, O_3, and hydrofluorocarbons) believed to be implicated in urban/global warming.[67] On average, fifty to eighty per cent of fine particle ($PM_{2.5}$) mass in Ghanaian cities is the result of biomass combustion, road dust, and vehicle emissions and, together with other air pollutants, causes up to 6,500 deaths countrywide annually. Low-income communities are more vulnerable than wealthier neighbourhoods. It is noteworthy that the magnitude of the overall effects of such particles on climate remains highly uncertain.[68] Nevertheless, green spaces act as carpets preventing particles from being dislodged from bare surfaces, while trees trap and filter particles circulating in the air. Several studies have shown that urban neighbourhoods with adequate vegetation cover tend to have lower air pollutant concentrations indoors and lower rates of health burdens overall.[69] In Kumasi, sixty per cent of green cover is composed of trees. This large amount of tree (green) cover has a significant air particle absorption and filtering capacity that leads to improved air quality and lowered risks of urban climate-related challenges.

Flooding is another urban climate challenge, for which residents of Kumasi have adopted various coping measures, including building embankments

---

66 Gill, Susannah E. et al., "Adapting Cities for Climate Change: The Role of the Green Infrastructure," *Built Environment* 33.1 (2007): 115–133; Elmqvist, Thomas et al., "Benefits of Restoring Ecosystem Services in Urban Areas," *Current Opinion in Environmental Sustainability* 14 (2015): 101–108.

67 IPCC 2007, "Summary for Policy Makers," 4–5; Cross, J. M., and R. Pierson, *Short-lived Climate Pollutants: Why Are They Important?* (EESI Factsheet, 2013).

68 Fuzzi, Sandro, et al., "Particulate Matter, Air Quality and Climate: Lessons Learned and Future Needs," *Atmospheric Chemistry and Physics* 15.0 (2015): 8217–8299.

69 Zupancic, Tara, Claire Westmacott, and Mike Bulthuis, *The Impact of Green Space on Heat and Air Pollution in Urban Communities: A Meta-Narrative Systematic Review* (Vancouver: Suzuki Foundation, 2015), 1–68; Lindén, "Nocturnal Cool Island," 610–612.

FIGURE 11.3 People in Kumasi under a *Terminalia catappa* tree (coping with heat)
SOURCE: BERTRAND NERO (AUTHOR)

around houses, erecting buildings on stilts, using raised walkways, constructing drainage systems, and relocating to upland areas.[70] However, like in most African cities, institutional (national and local government) capacities and resources to cope with climatic hazards are limited.[71] A relatively simple approach

70 Campion, "Wetland Ecology and Floods," 194.
71 Wilson, Robert H., and Todd G. Smith, "Urban Resilience to Climate Change in Africa," 6.

FIGURE 11.4 Building too close to the watercourse at Atonsu, Kumasi, Ghana
SOURCE: BERTRAND NERO (AUTHOR)

such as appropriate integration of UGS into urban planning and incorporating necessary engineering measures could boost the flood-coping capacities of both local government and individual residents. Urban landscapes with fifty to ninety per cent impervious cover can yield as much as forty to eighty per cent surface run-off from incoming rainfall, whereas a forest landscape yields only 13 per cent surface run-off from similar precipitation events.[72] Impervious surfaces decelerate infiltration rates and hence increase the likelihood of floods. Kumasi is traversed by several streams and wetlands and has about fifty to seventy per cent bare (built and non-built) land cover. Protecting and creating more wetlands, or where necessary dams, for flood mitigation, prohibiting human habitation and environmentally malignant activities near streams and wetlands (Figure 11.4), creating green spaces on non-built bare areas, and designating flood-prone regions and lowlands as green spaces for conservation should fundamentally enhance coping capacity for floods and boost urban resilience to climate change. Economic gains could also arise through the use of green wetlands for tourism and urban agriculture.

72 Pataki, Diane E. et al., "Coupling Biogeochemical Cycles in Urban Environments: Ecosystem Services, Green Solutions, and Misconceptions," *Frontiers in Ecology and the Environment* 9.1 (2011): 27–36.

4.3 *Carbon Sequestration, Culture, and Urban Resilience in Kumasi*

The functions of UGS described above involve reacting to urban climate hazards and thus conform to the stable, preservative view of equilibrist resilience.[73] In this light, the city's systems, agents, and institutions merely strive to cope with an adverse urban environment through, among other measures, the promotion of green spaces. However, UGS may assume a proactive, flexible, and transformative dimension geared at aiding the city to recover from shocks and establishing a new normality over the long term in line with the principles of evolutionary resilience.

In Kumasi as in other cities, active carbon sequestration in urban vegetation and soils, although relatively small in total amount, aligns well with efforts to keep the rise in global mean temperature to below 2°C relative to preindustrial times and thereby achieve lower urban temperatures as well as combat climate change.

It is estimated that the existing UGS cover in Kumasi stores up to 1,934,000 tons of carbon in aboveground trees, equivalent to 211 tons of carbon per hectare. This is similar to the regional average of 202 tons of carbon per hectare for African tropical forests.[74] Remnants of natural forest as well as trees in public parks, and cemeteries have the highest mean carbon stocks, which are significantly higher than those of trees found on grassland, in domestic gardens, or on farmland. Differences due to species composition, stocking density, tree size, and age explain why UGS differ in their respective carbon storage capacities.[75] Native tree species with high wood specific gravities and a diameter greater than one hundred centimetres dominate natural forest as well as vegetation in public parks, cemeteries, and to a lesser extent institutional compounds. Similarly, plantations and domestic gardens are characterised by small-sized trees of varied species, origins, and uses. Low carbon storage in grasslands is attributable to low tree density in such areas.

The benefits of UGS in relation to the offsetting of CO_2 emissions are threefold:

73 White, and O'Hare, "From Rhetoric to Reality," 943–945.

74 Lewis, Simon L. et al., "Increasing Carbon Storage in Intact African Tropical Forests," *Nature* 457.19 (2009): 1003–1006; Adu-Bredu S. et al., *Carbon Stocks Under Four Land Use Systems in Three Varied Ecological Zones in Ghana* (Rome: FAO, 2008), 105–114.

75 Nowak, David J., "Atmospheric CO2 Reduction by Chicago's Urban Forest," in *Chicago's Urban Forest Ecosystems: Results of the Chicago Urban Forest Climate Project*, ed. McPherson E. Gregory, David J. Nowak, and Rowan A. Rowntree (General Technical Report NE 186, Radnor, PA, USDA, Forest Services, Northern Research Station, 1994), 201.

(1) the CO_2 sequestered;
(2) the emissions avoided due to shade and evaporative cooling, and
(3) the emissions avoided due to climate regulation.[76]

In Kumasi, the CO_2 equivalent (CO_2e) of carbon storage in trees is currently estimated to be 7.1 million tons. Assuming that it takes fifty years to build up this level of carbon storage, the city's average annual sequestration rate would be 142,000 tons CO_2e. A rough estimate of CO_2e emissions reveals that 500,000 tons CO_2e are emitted in the city per year, including emissions from residential and non-residential areas, the transport system,[77] and waste deposits.[78] Accordingly, the share of CO_2e sequestered annually by urban tree vegetation represents 25 to 30 per cent of Kumasi's total emissions.

The benefits of tree carbon sequestration in Kumasi in terms of its monetary or CO_2e value were estimated based on the social cost of carbon for 2010 using the procedures outlined in Nowak et al. The social cost of aboveground tree carbon storage was estimated to be USD 106.8 million for the city. This represents the amount of damage avoided by removing CO_2 from the atmosphere and storing it in the city's aboveground vegetation. This avoided damage further strengthens urban resilience to climate change.

Finally, UGS provide social and cultural benefits and may thus strengthen local communities in Kumasi. Trees in cemeteries, especially those emerging directly from gravesites, are believed to symbolise that the departed are now contently resting in heaven. Certain trees christened designated as 'spirits' and those around shrines (although few in number) may be preserved as a mark of respect for the relevant deity. Home gardening in urban areas can in some cases be interpreted as residents seeking to preserve their tradition of cultivation.[79] To some, trees are symbolic representations of people, depicting parental care through the shelter they provide and their communal life when growing together in a forest.[80] As meeting places for community

76 Escobedo, Francisco et al., "Analyzing the Efficacy of Subtropical Urban Forests in Offsetting Carbon Emissions from Cities," *Environmental Science and Policy* 13.5 (2010): 362–372.

77 Agyeman-Bonsu et al., "Traffic-Data Driven Modelling of Vehicular Emissions," 32–40.

78 KMA WMD, *Data for Purposes of Planning Waste Management Intervention Programmes* (Kumasi: Kumasi Waste Management Department, 2010).

79 Eichemberg, Mayra Teruya, "Species Composition and Plant Use in Old Urban Homegardens in Rio Claro, Southeast of Brazil," *Acta Botanica Brasilica* 23.4 (2009): 1057–1075; Aworinde, D. O., S. M. Erinoso, B. O. Ogundairo, and A. O. Olanloye, "Assessment of Plants Grown and Maintained in Home Gardens in Odeda Area Southwestern Nigeria," *Journal of Horticulture and Forestry* 5.2 (2013): 29–36.

80 Dwyer, John F., Herbert W. Schroeder, and Paul H. Gobster, "The Significance of Urban Trees and Forests: Toward a Deeper Understanding of Values," *Journal of Arboriculture* 17.10 (1991): 276–284.

members, UGS foster harmony and provide a platform for greater interaction where community challenges may be discussed. Therefore, UGS contribute in ways that go beyond making communities more resilient to environmental risks and change.

4.4 *Managerial Issues: Stakeholders' Interplay and Urban Green Spaces*

Community participation is fundamental to the establishment of UGS, while a general consensus is crucial to make them effective. In developed countries, top-down mechanisms by which governments and local authorities plan, decide, and invest in promoting urban greening can be a seemingly hegemonic process, whereas in developing countries actions are determined more by bottom-up processes. Hence, policy instruments that are able to ensure social and environmental performance are key.

In this context, the resilience premise was applied to identify factors that favour or negatively influence the existence and maintenance of UGS.[81] This implies that increasing urban resilience to climate change requires improving populations' adaptation and mitigation capabilities as described above. Resilience may include societal and ecological subsystems in mutual interaction[82] and is circumstance- and time-dependent, relying on constant adjustments within the system in response to external changes, thus leading to adaptive processes.[83]

The performance of each UGS and its potential to contribute to urban resilience to changing climatic conditions were assessed against the following contrasting criteria:

(1) the strengthening decisions and actions that promote UGS and;
(2) the weakening decisions and actions that undermine UGS.

In both cases, three aspects were taken into account: the managerial actions (what?), the actors or persons responsible (who?), and the underlying reasons (why?). Responses were recorded and their tabulation carried out in an elicitation workshop involving local academics and the authors, complemented by first-hand information gathered from stakeholders in the field and secondary data from literature (Table 11.2). In the Kumasi case study, the measures that favour or disfavour UGS are a blend of top-down and bottom-up mechanisms.

81 Carpenter, Steve R. et al., "From Metaphor to Measurement: Resilience of What to What?" *Ecosystems* 4 (2001): 765–781.
82 Gallopin, Gilberto C., "Linkages between Vulnerability, Resilience, and Adaptive Capacity," *Global Environmental Change* 16 (2006): 293–303.
83 Smit, Barry, and Johanna Wandel, "Adaptation, Adaptive Capacity and Vulnerability," *Global Environmental Change* 16 (2006): 282–292.

TABLE 11.2 Managerial actions/decisions, actors, and underlying reasons that favour and weaken UGS existence and maintenance in Kumasi, Ghana

Urban green spaces (UGS)	Favouring UGS	Weakening UGS
Plantation	What? Planting trees	What? Tree felling, tree cutting
	Who? Private landowners, administrators of public institutions, chiefs	Who? Fuelwood gatherers, land developers, chiefs, city authorities (Kumasi metropolitan assembly), town and country planners
	Why? Aesthetics, shade, firewood, enhance social interaction, mitigate climate change, regulate biogeochemical cycles, carbon sinks	Why? Impediments to 'development', expansion in urban land, damage to public infrastructure (building foundations, electricity/telephone cables), public health and safety (habitats for dangerous animals, criminal hideouts), need for fuelwood
Home garden	What? Food/fruit crop cultivation, lawns	What? Housing without home gardens
	Who? Private residential heads, tenants in governmental residential areas	Who? Owner
	Why? Food security, augment household income, medicines, beautification, pleasure, improved air quality, inherited practice, shade, love of vegetation, provision of environmental services	Why? Fear of hazards like snakes, invasion by criminals or intruders; destruction of buildings/walls through roots and branches, cultural reasons, urbanisation (converting gardens into more profitable structures)

TABLE 11.2 (*cont.*)

Urban green spaces (UGS)	**Favouring** UGS	**Weakening** UGS
Institutional compounds	What? Tree plantings Who? Heads/activists in institutions/public offices Why? Shade, fruits, beautification, windbreaks, erosion checks, boundaries, influence of management interest and background	What? Bare compounds, land use change Who? Institutional authorities/heads Why? Public hazard, destroys buildings, habitats for dangerous animals, hideouts for criminals, fallen branches, generates waste, litter; lack of management know-how and tools, allocation to other uses e.g. building construction
Farmlands	What? Cultivation of marginal lands Who? Tenant farmers, (unskilled) urban dwellers, labourers Why? Food production, income generation, pleasure	What? Uncultivated fields, use of black waters Who? City authorities, land owners Why? High demand for land, urbanisation (land use change), flood prone, pollution source (fertilisers & pesticides)
Cemeteries/sacred grooves	What? Tree cultivation, tree maintenance Who? Traditional heads (chiefs), local/city authorities Why? Revere the dead, respect, traditional/cultural beliefs, fear of spirits beautification, shade	What? Bare cemeteries, logging Who? Traditional heads (chiefs), local/city authorities, tree thieves Why? Create space for more burial grounds, clear encroachment, demand for land

Street trees	What? Plan(t) rows of trees along streets	What? Bare streets, logging trees, poor or no maintenance
	Who? City authorities, local stakeholders, individuals	Who? City authorities (government), individuals
	Why? Beautification, provide shade and cool areas (private use also), improve visibility and reduce accidents	Why? Interfere with electricity cables, smooth roadways, reduce hazards e.g. falling branches, protruding stems, etc., change of land use e.g. into shops
Public parks/gardens	What? Plant trees and grasses	What? Land use change
	Who? Government, city authorities, investors	Who? City authorities, investors
	Why? Shade, recreation, entertainment, public health, beautification, tourism, conservation of genetic resources	Why? Poor foresight, high maintenance costs, more profitable alternative use (business opportunities), neglect
Grasslands	What? Marginal lands	What? Conversion of grasslands
	Who? Private land owners, city authorities (Kumasi metropolitan assembly)	Who? Private owners, city authorities
	Why? Flood mitigation, grazing, biodiversity conservation	Why? More profitable alternative use, better alternative land uses, urbanisation (pressure for land use change)

Urban green spaces are maintained by different interest groups for varied specific purposes. Because of this, these interest groups may convert UGS to other preferred uses under the influence of different socio-economic, cultural, and/or political factors.

The relevant actors in UGS management are the city authorities, private owners, chiefs, and to a limited extent non-governmental and community-based organisations. In the case of the city authorities, their competencies and roles are regulated by local and national legislation and conventional property rights. These regulations, defined in the most current policy documents, mostly go unapplied despite being comprehensively and theoretically well framed. The authority of the local and central governments over land is limited and lacks enforcement capacity. Hence, the scope of influence in defining and administering UGS is frequently restricted to jurisdiction of, for example, public parks (botanical gardens, zoos, and protected areas) and vegetation in institutional compounds.

Moreover, it is frequently observed that a site originally designated as an UGS is in practice utilised differently or even converted into grey infrastructure, not only by individuals with an interest in the benefits of such a change but also by public authorities with ulterior motives. This is not a new phenomenon. Kumasi has been subject to several city plans since its founding. The 1945 plan, conceived as a quintessential 'Garden City' plan in accordance with Howard's concept, advocated for the creation of a three hundred-metre green belt along stream channels and the establishment of urban parks within Kumasi. However, land use redesignations as a result of population growth and urban sprawl have since rendered these early plans obselete. Most of the designated green belt zone is currently occupied by dense grey/brown infrastructure, such as buildings, roads, and other land uses detrimental to UGS.[84] Nevertheless, although Kumasi's UGS may have declined drastically, its label as a 'Garden City' is still widely used with pride by its inhabitants.

The role of chiefs in the Asante region is unique. With the exception of small areas of state land, all land in the region is held in trust for the Asante people by their king, the *Asantehene*. He allocates land through a network of local chiefs in conjunction with the office of the administrator of stool lands.[85]

84 Schmidt, Stephan, "Cultural Influences and the Built Environment: An Examination of Kumasi Ghana," *Journal of Urban Design* 10.3 (2005): 353–370.

85 Stool land refers to land held in allodial title (i.e. land ownership that is outright and absolute) by a traditional head on behalf of a community or group of people; Devas, Nick, and David Korboe, "City Governance and Poverty: The Case of Kumasi," *Environment and Urbanization* 12.1 (2000): 123–136.

These rights are acknowledged by society and recognised by central government and local authorities.

Some chiefs and their people, however, regard UGS as wasteland; hence protection thereof can be contentious. Even recognised public parks and gardens as well as sensitive wetlands that require mandatory protection may be threatened if land values appreciate and demand for land is high. The government's policy of non-interference in the chieftaincy weakens its ability to promote and enhance the development of UGS. On the other hand, the power of traditional authorities could be harnessed for the purposes of UGS conservation, environmental protection, and climate change mitigation/adaptation. Citizens swear allegiance to their chiefs; thus urban environmental policies instituted and administered through chiefs can promote a green, climate-resilient city. Indeed, the involvement of chiefs in environmental management is gaining momentum in cities throughout the country.

Although officially regulated mainly by local but also by national authorities, the management of UGS in Kumasi tends to be the de facto responsibility of private land users. Hence, individual behaviour still appears to be the major determinant of both favourable and unfavourable decisions in relation to the greening of the city (Table 11.2). Individual behaviour is generally complex and unpredictable, but a pattern can be observed whereby decisions taken by land users are practical and incentivised by short-term interests. For instance, a substantial number of tree-owning residents in Kumasi acknowledged the direct benefits—shade, air purification, food, and so on—derived from UGS, but stated categorically that they would transform these spaces and erect more profitable structures such as buildings and shops if the need arose. Several household heads resent trees and green spaces in their compounds because of the hazards that they pose. Among other disincentives, trees and green spaces regularly interfere with the roofs and foundations of buildings, litter compounds with leaves and debris, increase fuel loadings and thus the risk of fire, provide habitats for dangerous animals such as snakes and scorpions, increase the risk of incurring damage through falling branches, and may serve as hideouts for criminals. Yet there were also others who wished that they had green spaces in their compounds but who lacked the necessary space. In general, reasons favouring the establishment and maintenance of UGS include:

(1) beautification—independent of socioeconomic status, property stand of the UGS, and position of authority;
(2) provision of resources such as food, fodder, fuelwood, etc. (tangible benefits);
(3) provision of services, e.g. air quality improvement, shade, windbreaks, erosion checks, flood mitigation, etc. (intangible benefits) (Table 11.2).

On the other hand, UGS are threatened by:

(1) land use conversion towards more profitable uses e.g. houses, public buildings, roads, channels, etc.;
(2) neglect of landscape plans and ad hoc legislation;
(3) lack of awareness among urbanites of the value of UGS.

Typical insitutional problems in the Global South also characterise the management of UGS, such as the lack of appropriate legislation or planning and enforcement capacity. Consequently, the importance of UGS may be downplayed in favour of more profitable short-term interests and activities. Although often justified by a lack of financial means, these incentivies are related to systemic malfunctions such as corruption or neglect for the rule of law.

At the level of the individual, decisions concerning management and promotion of UGS can make a considerable positive or negative impact. These decisions appear to be influenced by a person's background (connection with nature, rural origins), education (knowledge of the importance and roles of UGS), and short-term interests. Although a question that is yet to be thoroughly examined, it is likely that people who live on the outskirts of cities are more likely to have vegetation in their compounds than those who live in the city centre. People living in suburban areas typically belong to the relatively wealthy class, most of whom are well educated and have learnt to appreciate and conserve nature. However, the permanence of such UGS, mainly domestic gardens, cannot be guaranteed as they are predominantly temporary sites awaiting conversion into more lucrative structural use when the economy becomes favourable.

5 Conclusions

Considering the growing awareness of urbanisation, understanding that urban resilience to the impacts of climate change is an important, multidimensional issue accomplished via various means is crucial. Bolstering climate resilience in cities involves several types of activities, and the role of green spaces in this effort must not be peripheral.

This chapter has discussed urban population growth and urban land expansion in Africa in the context of climate change and in relation to the role of green spaces, presenting Kumasi as a case study to illustrate the potential of African cities for climate resilience. Urbanisation unequivocally drives depletion of natural land cover and exacerbates anthropogenic environmental impacts. However, urban green spaces provide opportunities for both reactive (adaptive) and proactive (mitigative) measures to improve a city's resilience to climate change and its impacts. Through alteration of urban albedo, shading, and evapotranspirational cooling, UGS offer a cost-effective and adaptable means of coping with warming. Urban vegetation interferes with water and air movements on and above land surfaces, lowering the risk of flooding and air pollution. Absorption of GHGs by vegetation (especially trees) plays an important role in air quality improvement and in mitigating both local and global warming. To meaningfully improve urban resilience, UGS must be well planned, adequately integrated into urban space, and underpinned by strong institutions. Under these conditions, abandoned roads, buildings, and other compacted bare surfaces in cities that usually lie idle and unproductive could be converted into green spaces. Vegetation builds up biodiversity, breaks up the compacted concretes in soil, and over time restores ecological equilibrium to such spaces, thus boosting a city's capacity to cope with climatic stress.

Inadequate implementation of existing legislation and a lack of ad hoc policies leave the initiative to customary and private interests. As such, de facto bottom-up processes determine the existence and management of UGS. Here, the roles of chieftancies (traditional leadership) and individual landowners are particularly crucial. Involvement of the government, traditional leaders, and civil society in defining priorities, streamlining actions, and enforcing policy are essential requisites for sustaining and enhancing green cover in African cities.

Bibliography

Abd'Razack, Nelson T.A., and Ahmad Nazri bin Muhamad L. "Wood Fuel Consumption and Ecological Footprint of African Cities." *International Journal of Education and Research* 1.2 (2013): 1–18.

Ackbarally, Nasseem. "Environment–Mauritius: Hold Your Fire." Published by Inter Press Service, July 15, 2009. Accessed December 16, 2014. http://www.ipsnews.net/2009/07/environment-mauritius-hold-your-fire/.

ActionAid. "Climate Change, Urban Flooding and the Rights of the Urban Poor in Africa." ActionAid, 2006.

Adams, Bert N. and Edward Mburugu. "Kikuyu Bridewealth and Polygyny Today." *Journal of Comparative Family Studies* 25.2 (1994): 159–166.

Adano, Wario R., Ton Dietz, and Karen Witsenburg. "Climate Change, Violent Conflict and Local Institutions in Kenya's Drylands." *Journal of Peace Research* 49.1 (2012): 65–80.

Adarkwa, Kwafo Kwasi. "The Role of Kumasi in National Development–Kumasi as a Central Place." In *Future of the Tree: Towards Growth and Development of Kumasi*, edited by Kwasi K. Adarkwa. Kumasi: KNUST University Press, 2011.

Adepoju, M.O., A.C. Mellington, and K.T. Tansey. "Land Use/Land Cover Change Detection in Metropolitan Lagos (Nigeria): 1984–2002." ARPS Annual Conference, Reno Nevada, May 1–5, 2006.

Adger, W. Neil, Saleemul Huq, Katrina Brown, Declan Conway, and Mike Hulme. "Adaptation to climate change in the developing world." *Progress in Development Studies* 3.3 (2003): 179–195.

Aduah, M. and P. Baffoe. "Remote Sensing for Mapping Land-use/cover Changes and Urban Sprawl in Sekondi-Takoradi, Western Region of Ghana." *The International Journal of Enginering and Science (IJES)* (2013): 66–73.

Adu-Bredu Stephen, Mark Abekoe, E. Tachie-Obeng, and P. Tschakert. *Carbon Stocks under Four Land Use Systems in Three Varied Ecological Zones in Ghana.* Rome: FAO, 2008.

Agrawal, Arun. "Dismantling the Divide between Indigenous and Scientific Knowledge." *Development and Change* 26.3 (1995): 413–39.

Agrawala, Shardul, Annett Moehner, Andreas Hemp, Maarten van Aalst, Sam Hitz, Joel Smith, Hubert Meena, Stephen M. Mwakifwamba, Tharsis Hyera and Obeth U. Mwaipopo. *Development and Climate Change in Tanzania: Focus on Mount Kilimanjaro.* Paris: Organisation for Economic Cooperation and Development, 2003.

Agwu, Julia and Appollonia A. Okhimamhe. *Gender and Climate Change in Nigeria: A Study of Four Communities in North-Central and South-Eastern Nigeria.* Ikoyi, Lagos: Heinrich Böll Foundation, Nigeria, 2009.

Agyeman-Bonsu, K.W., I.K. Dontwi, D. Tutu-Benefoh, D.E. Bentil, O.G. Boateng, K. Asuobonteng, and W. Agyemang. "Traffic-data Driven Modelling of Vehicular Emissions Using COPERT III in Ghana." *American Journal of Scientific and Industrial Research* 1.1 (2010): 32–40.

Ahrends, Antje., Neil D. Burgess, Simon A.H. Milledge, Mark T. Bulling, Brendan Fisher, James C.R. Smart, G. Philip Clarke, Boniface E. Mhoro, and Simon L. Lewis. "Predictable Waves of Sequential Forest Degradation and Biodiversity Loss Spreading from an African City." *Proceedings of the National Academy of Sciences of the United States of America* 107.33 (2010): 14556–14561.

Akeampong, Emmanuel. *Between the Sea and the Lagoon: An Eco-social History of the Anlo in Southeastern Ghana, c. 1850 to Recent Times.* Athens: Ohio University Press, 2001.

Akinyoade, Akinyinka, Wijnand Klaver, Sebastiaan Soeters, and Dick Foeken, ed. *Digging Deeper. Inside Africa's Agricultural, Food and Nutrition Dynamics.* Leiden, Boston: Brill, 2014.

Alexander, Catherine and Josuha O. Reno. "From Biopower to Energopolitics in England's Modern Waste Technology." *Anthropological Quarterly* 87.2 (2014): 335–358.

Alexander, Jocelyn. *The Unsettled Land: State-Making and the Politics of Land in Zimbabwe 1893–2003*. Harare, Oxford: Weaver Press, and James Currey, 2006.

Allegretti, Antonio. "Mainstreaming Climate Change Adaptation in Drylands Development Planning in Tanzania. Traditional and Government Planning Processes." Research Report Tanzania Natural Resources Forum (TNRF) and International Institute for Environment and Development (IIED), 2012.

Allen, Richard B. "The Mascarene Slave-trade and Labour Migration in the Indian Ocean during the Eighteenth and Nineteenth Centuries." In *The Structure of Slavery in Indian Ocean Africa and Asia*, edited by G. Campbell, 33–50. London: Frank Cass, 2004.

Allen, Richard B. *Slaves, Freedmen, and Indentured Laborers in Colonial Mauritius.* Cambridge: Cambridge University Press, 1999.

Al-Sharif, Abubakr A.A., Biswajeet Pradhan, Helmi Zulhaidi Modh Shafri, and Shattri Mansor. "Spatio-temporal Analysis of Urban and Population Growths in Tripoli Using Remotely Sensed Data and GIS." *Indian Journal of Science and Technology* 6.8 (2013): 5134–5142.

Altman, Irit, Roel Boumanns, Joe Roman, Suchi Gopal, and Les Kaufman. "An Ecosystem Framework for Marine Ecosystem-Based Management." In *The Sea:Marine Ecosystem-Based Management*, volume. 16, edited by Michael Fogarty, 245–276. Harvard: Harvard University Press, 2014.

Ambler, Charles H. *Kenyan Communities in the Age of Imperialism: The Central Region in the Late Nineteenth Century*. New Haven, 1988.

Amoako, Clifford and David Korboe. "Historical Development, Population Growth and Present Structure of Kumasi." In *Future of the Tree: Towards Growth and Development of Kumasi* ed. Kwasi K. Adarkwa. Kumasi: KNUST University Press, 2011.

Anderson, Alison. "Rethinking Climate Change Communication." In *Culture, Politics and Climate Change. How Information Shapes Our Common Future*, edited by Deserai A. Crow and Maxwell T. Boykoff. Abingdon, New York: earthscan from Routledge, 2014.

Anderson, David. *Eroding the Commons: The Politics of Ecology in Baringo, Kenya 1890–1963*. Oxford: James Currey, 2002.

Anderson, David. "Depression, Dust Bowl, Demography, and Drought: The Colonial State and Soil Conservation in East Africa during the 1930s." Un *Colonialism and Nationalism in Africa, vol. 2: The Colonial Epoch in Africa*, ed., G. Maddox, 209–231. New York: Garland, 1993.

Anderson, David and Richard Grove, ed. *Conservation in Africa: People, Policies and Practice*. Cambridge: Cambridge University Press, 1987.

Anderson, David and David Throup. "Africans and Agricultural Production in Colonial Kenya: The Myth of the War as a Watershed." *The Journal of African History* 26.4 (1985).

Angel, Shlomo Jason Parent, Daniel L. Civco, Alexander Bleic, David Potered. "The Dimensions of Global Urban Expansion: Estimates and Projections for All Countries, 2000–2050." *Progress in Planning* (2011): 53–107.

Angula, Margaret. *Gender and Climate Change: Namibia Case Study*. Cape Town: Heinrich Böll Foundation Southern Africa, 2010.

Annez, P., R. Buckley, and J. Kalarickal. "African Urbanization as Flight? Some Policy Implications of Geography." *Urban Forum* 21 (2010): 221–234.

Araujo, Julio, John Marsham, Dave Rowell, Nkulumo Zinyengere, Andrew Ainslie, Andrew Clenaghan, Rosalind Cornforth, Giovanni DeGiusti, Barbara Evans, Declan Finney, Dan Lapworth, David Macdonald, Celia Petty, John Seaman, Fredrick Semazzi, and Celia Way. "East Africa's Climate: Planning for an Uncertain Future." In *Africa's Climate*, ed. Future Climate for Africa: 11–16.

Ash, Neville and Martin Jenkins, *Biodiversity and Poverty Reduction: The Importance of Biodiversity for Ecosystem Services*. Cambridge: UNEP-WCMC, 2007.

Aworinde, D.O., S.M. Erinoso, B.O. Ogundairo, and A.O. Olanloye. "Assessment of Plants Grown and Maintained in Home Gardens in Odeda Area Southwestern Nigeria." *Journal of Horticulture and Forestry* 5.2 (2013): 29–36.

Ayila, Adzandeh E., Fabiyi O. Oluseyi, and Bello Yakasai Anas. "Statistical Analysis of Urban Growth in Kano Metropolis, Nigeria." *International Journal of Environmental Monitoring and Analysis* 2.1 (2014): 50–56.

Babugura, Agnes. *Gender and Climate Change: South Africa Case Study*. Cape Town, South Africa: Heinrich Böll Foundation Southern Africa, 2010.

Baer, Hans, and Merrill Singer. *The Anthropology of Climate Change*. London: Routledge, 2014.

Baker, Judy L. *Urban Poverty: A Global Overview (No. UP 5)*. Washington: The World Bank, 2008.

Banana, Canaan. *Towards a Socialist Ethos: Socialism without Socialists is Capitalism*. Harare: College Press, 1987.

Banuri, Tariq. "Climate Change 2001: Mitigation. A Report of Working Group III of the Intergovernmental Panel on Climate Change (IPCC)." Technical Summary, Third Assessment Report, 2001.

Barbier, Bruno, Hamma Yacouba, Harouna Karambiri, Malick Zoromé, and Blaise Somé. "Human Vulnerability to Climate Variability in the Sahel: Farmers' Adaptation Strategies in Northern Burkina Faso." *Environmental Management* 43.5 (2009): 790–803.

Barnwell, Patrick J. and Auguste Toussaint. *A Short History of Mauritius*. London: Longmans, Green & Co, 1949.

Bassett, T.J., and D. Crummey, ed. *African Savannas: Global Narratives and Local Knowledge of Environmental Change*. Oxford: James Currey, 2003.

Bäthge, Sandra. *Climate Change and Gender: Economic Empowerment of Women Through Climate Mitigation and Adaptation?* Eschborn: Deutsche Gesellschaft für Technische Zusammenarbeit (GTZ) Gmb, 2010.

Baer, Drake. " 'The Biggest Change of Our Time' is Happening Right Now in Africa. Africa's Population Explosion Will Change Humanity." *Tech Insider Innovation*, August 18, 2015. Accessed February 28. 2016. http://www.techinsider.io/africas-population-explosion-will-change-humanity-2015–8.

BBC. "Africa–China Exports Fall by 40% after China Slowdown." *BBC News*, January 13, 2016. Accessed November 20, 2016. www.bbc.com/news/world-africa-35303981.

Beauchemin, Cris. "Rural–Urban Migration in West Africa: Towards a Reversal? Migration Trends and Economic Situation in Burkina Faso and Cote d'Ivoire." *Population Space and Place* 17 (2011): 47–72.

Beauchemin, Cris and Philippe Bocquier. "Migration and Urbanization in Francophone West Africa: An Overview of the Recent Empirical Evidence." *Urban Studies* 41 (2004): 2245–2272.

Beck, Coby. "Warming is Due to the Urban Heat Island Effect'—No, it isn't." *GRIST*, October 29, 2006. Accessed February 2015. http://grist.org/article/warming-is-due-to-the-urban-heat-island-effect/.

Beinart, William and Colin Bundy. *Hidden Struggles in Rural South Africa: Politics and Popular Movements in the Transkei and Eastern Cape, 1890—1930*. London: James Currey, 1987.

Beinart, William, Peter Delius, and Stanley Trapido, ed. *Putting a Plough to the Ground: Accumulation and Dispossession in Rural South Africa, 1850—1930*. Johannesburg: Ravan Press, 1986.

Beinart, William. “Soil Erosion Conservationism and Ideas about Development: A Southern African Exploration, 1900–1960.” *Journal of Southern African Studies* 11.1 (1984): 52–83.

Belliveau, Suzanne, Barry Smit, and Ben Bradshaw. “Multiple Exposures and Dynamic Vulnerability: Evidence from the Grape Industry in the Okanagan Valley, Canada.” *Global Environmental Change* 16 (2006): 364–378.

Béné, Christophe, Rachel Godfrey-Wood, Andrew Newsham, and Mark Davies. *Resilience: New Utopia or New Tyranny? Reflection about the Potentials and Limits of the Concept of Resilience in Relation to Vulnerability-Reduction Programmes*. IDS Working Paper 405. Brighton, UK: Institute of Development Studies, 2012.

Béné, Christophe, Stephen Devereux, and Rachel Sabates-Wheeler. *Shocks and Social Protection in the Horn of Africa: Analysis from the Productive Safety Net Programme in Ethiopia*. IDS Working Paper 395. Brighton, UK: Institute of Development Studies, 2012.

Berman, Bruce. *Control and Crisis in Colonial Kenya: The Dialetic of Domination*. London: James Currey, 1990.

Bernard, H.R. *Research Methods in Anthropology: Qualitative and Quantitative Approaches*. Walnut Creek: Altamira Press, 2002.

Birkmann, Jörn. “Measuring Vulnerability to Promote Disaster-Resilient Societies: Conceptual Frameworks and Definitions.” In *Measuring Vulnerability to Natural Hazards: Towards Disaster Resilient Societies*, edited by Jörn Birkmann. Tokyo: UNU Press, 2006.

Bjarnesen, Jesper. “Diaspora at Home? Wartime Mobilities in the Burkina Faso-Cote d’Ivoire Transnational Space.” PhD diss., University of Uppsala, Sweden, 2013.

Blackden, C.M. *Literature Review on Intra-Household Resource Allocation*. Washington, DC, September, 2009.

Blaikie, Piers, Terry Cannon, Ian Davis, and Ben Wisner. *At Risk: Natural Hazards, People’s Vulnerability and Disasters*. London: Routledge, 1994.

Blaikie, Piers and H. Brookfield. “Defining and Debating the Problem.” In *Land Degradation and Society*, edited by P. Blaikie and H. Brookfield, 1–26. London: Methuen, 1987.

Bloomberg. “The Impact on Africa from Falling Oil Prices.” *Bloomberg Business*, October 6, 2015. Accessed February 28, 2016. http://www.bloomberg.com/news/videos/2015-10-06/the-impact-on-africa-from-falling-oil-prices.

Bocquier, Philippe, and S. Traore. *Urbanisation et dynamique migratoire en Afrique de l’Ouest: la croissance urbaine en panne*. Paris: Editions L’Harmattan, 2000.

Boersema, Jan J. *Beelden van Paaseiland. Over de Duurzaamheid van een Cultuur*. Amsterdam, Antwerpen: Atlas, 2011.

Bollig, Michael and Sara de Wit. “Commentary on: Orlove, Ben, Heather Lazrus, Grete K. Hovelsrud, and Alessandra Giannini, Recognitions and Responsibilities: On the Origins and Consequences of the Uneven Attention to Climate Change around the World.” *Current Anthropology* 55.3 (2014): 262–264.

Bolwig, Simon. "Livelihood Practices and Land Use in the Sahel. Labour Allocation and Adaptive Capability among the Fulani Rimaybe in Northern Burkina Faso." PhD diss., University of Copenhagen, 1999.

Boserup, Ester. *The Conditions of Agricultural Growth: The Economics of Agrarian Change under Population Pressure*. New York: Aldine Pub. Co., 1965.

Bossard, Laurent. "West Africa: Security Crisis and Food Crisis." *OECD Insights*, January 6, 2016. Accessed February 28, 2016. http://oecdinsights.org/2016/01/06/west-africa-security-crisis-and-food-crisis/.

Boswell, K.V. "Migration, War, and Repatriation from Cote D'ivoire." PhD diss., Indiana University, 2010.

Bowyer, Paul, Guy P. Brasseur, and Daniela Jacob. "The Role of Climate Services in Adapting to Climate Variability and Change." In *Handbook of Climate Change Adaptation*. Edited by Walter Leal Filho. Berlin: Springer, 2015.

Bradley, P.N. *Woodfuel, Women and Woodlots: The Foundations of a Woodfuel Development Strategy for East Africa*. London: Macmillan, 1991.

Bradshaw, Sarah and Maureen Fordham. *Women, Girls and Disasters*. London: DFID, 2013.

Breusers, M. "Responses to Climate Variability in the Kaya Region, Burkina Faso." In *The Impact of Climate Change on Drylands: With a Focus on West-Africa*, edited by A. Dietz, R. Ruben, A. Verhagen, 207–241. Dordrecht: Kluwer Academic Publishers, 2004.

Brew-Hammond, Abeeku, Fred Ohene Akuffo, Rudith King, Ato Quansah, Seth Ohemeng Dapaah, Owusu Amponsah, Dominic Eshun, and Alex Antobere. *Energy Access and Productive Uses for the Urban Poor: Final Report on Ghana Scoping Study*. Washington: The international Bank of Reconstruction, 2011.

Brody, Alyson, Justina Demetriades, and Emily Esplen, *Gender and Climate Change: Mapping the Linkages—A Scoping Study on Knowledge and Gaps*. Brighton: BRIDGE, Institute of Development Studies, 2008.

Brooks, Georg E. *Landlords and Strangers: Ecology, Society, and Trade in West Africa, 1000—1630*. Boulder, Col.: Westview Press, 1993.

Brooks, George E. "A Provisional Historical Schema for Western Africa Based on Seven Climatic Periods." *Cahiers d'Etudes Africaines* 26.101 (1986): 43–62.

Bryant, Raymond L. and Sinead Bailey. *Third World Political Ecology*. London: Routledge, 1997.

Bryant, Raymond L. "The Rise and Fall of Taungya Forestry: Social Forestry in Defence of the Empire," *The Ecologist* 24.1 (1994).

Bryant, Raymond L. "Political Ecology: An Emerging Research Agenda in Third-World Studies." *Political Geography* 11.1 (1992): 12–36.

Bryceson, Deborah. "The Scramble in Africa: Reorienting Rural Livelihoods." *World Development* 30.2 (2002): 725–739.

Bundy, Colin. *Rise and Fall of the South African Peasantry*. London: Heinemann, 1979.

Buttel, Frederick H., Ann P. Hawkins, and Alison G. Power. "From Limits to Growth to Global Change. Constraints and Contradictions in the Evolution of Environmental Science and Ideology." *Global Environmental Change* 1.1 (1990): 57–66.

Caillods, Françoise. "Regional Divides in Global Environmental Change Research Capacity." In *World Social Science Report 2013: Changing Global Environments*, ed. ISSC and UNESCO. Accessed 3 August 2017. doi:10.1787/9789264203419-en.

Caldwell, C. John. *African Rural–Urban Migration: The Movement to Ghana's Towns.* Australian National University Press (1969): 257.

Campbell, Horace. "Mamdani, Mugabe and the African Scholarly Community." *Pambazuka News*, December 18, 2008. Accessed June 8, 2011. http:// www.pambazuka.org/en/category/features/52845.

Campion, Betey B. "Urban Wetland Ecology and Floods in Kumasi, Ghana." PhD Thesis, University of Bremen, 2012.

Cardona, Omar D. "The Need for Rethinking the Concepts of Vulnerability and Risk from a Holistic Perspective: A Necessary Review and Criticism for Effective Risk Management." In *Mapping Vulnerability: Disasters, Development and People*, edited by Greg Bankoff, Georg Frerks, and Dorothea Hilhorst. London: Earthscan, 2004.

Carpenter, Steve R., Brian Walker, J. Marty Anderies, and Nick Abel. "From Metaphor to Measurement: Resilience of What to What?." *Ecosystems* 4 (2001): 765–781.

Carruthers, Jane. "Environmental History in Africa." In *A Companion to Global Environmental History*, ed. J.R. McNeill. Chichester: Wiley-Blackwell, 2014.

Carruthers, Jane. *The Kruger National Park: A Social and Political History*. Pietermaritzburg: University of Natal Press, 1995.

Carswell, Grace. "Multiple Historical Geographies: Responses and Resistance to Colonial Conservation Schemes in East Africa." *Journal of Historical Geography* 32.2 (2006): 398–421.

Carter, Marina. *Servants, Sirdars and Settlers: Indians in Mauritius, 1834–1874*. Delhi: Oxford University Press, 1995.

Castles, Stephen and Mark J. Miller. *The Age of Migration: International Population Movements in the Modern World*. New York: Palgrave Macmillan, 2009.

Central Electricity Board. "Annual Report." http://ceb.intnet.mu/CorporateInfo/ar2011.pdf.

Chant, Sylvia. "Re-thinking the 'Feminization of Poverty' in Relation to Aggregate Gender Indices." *Journal of Human Development* 7.2 (2006): 201–220.

Chen, Shaohua, and Ravallion, Martin. "Absolute Poverty Measures for the Developing World, 1981–2004." *Proceedings of the National Academy of Sciences of the United States of America* 104.43 (2007): 16757–16762.

Chitiga, Gibson, and Percyslage Chigora. "An Analysis of the Implications of the Fast Track Land Reform Program on Climate Change and Disaster Management in

Zimbabwe: A Case of Chegutu District." *Journal of Sustainable Development in Africa* 12.2 (2010): 124–143.

Christensen, J.H., B. Hewitson, A. Busuioc, A. Chen, I. Held, R. Jones, R.K. Kolli, W.-T. Kwon, R. Laprise, V. Magana Rueda, L. Mearns, C.G. Menendez, J. Raisanen, A. Rinke, A. Sarr, and P. Whetton. "Regional Climate Projections." In *Climate Change 2007: The Physical Science Basis. Contribution of Working Group 1 to the Fourth Assessment Report of the Intergovernmental Panel on Climate Change*, edited by S. Solomon. Cambridge: Cambridge University Press, 2007.

Chummun, Jaykumar. "The Potential for Using Renewable Sources of Energy in Mauritius." In *Climate-Smart Technologies: Integrating Renewable Energy and Energy Efficiency in Mitigation and Adaptation Responses*, edited by Walter L. Filho, Franziska Mannke, Romeela Mohee, Veronika Schulte and Dinesh Surroop, 207–218. Berlin: Springer, 2013.

Clancy, Joy S., Tanja Winther, Margaret N. Matinga, and Sheila Oparaocha. *Gender Equity in Access to and Benefits from Modern Energy and Improved Energy Technologies.* Leusden: ENERGIA/Norad/World Bank, 2011.

Clark, J.F.M. " 'The Incineration of Refuse is Beautiful': Torquay and the Introduction of Municipal Refuse Destructors." *Urban History* 34.4 (2007): 255–277.

Clarke, Duncan. "Africa 2050 Economic Revolution." *Global Briefing*, April 2013. Accessed February 2015. http://www.global-briefing.org/2013/04/africa-2050-economic-revolution/.

Climate and Development Knowledge Network (CDKN). "Helping Developing Countries to Design and Deliver Climate Compatible Development." Last modified January 22, 2016. Accessed February 28, 2016. http://r4d.dfid.gov.uk/PDF/Outputs/CDKN/CDKN-english-brochure-lores-sept2011.pdf.

Climate and Development Knowledge Network (CDKN). *The IPCC's Fifth Assessment Report: What's in it for Africa?* 2014.

Climate Change Secretariat: UNFCCC, ed. *The Nairobi Work Programme on Impacts, Vulnerability and Adaptation to Climate Change. Action on the Ground. A Synthesis in the Areas of Education, Training and Awareness-rising for Adaptation.* Bonn: UNFCCC, 2010.

CoCT. *Showcasing Excellence. Mayor's Portfolio of Urban Sustainability*. City of Cape Town/Isixeko Sasekapa/Stad Kaapstad. Last modified 2014. Accessed February 24, 2016. https://www.capetown.gov.za/en/EnvironmentalResourceManagement/publications/Documents/CCT_Mayors_Portfolio_of_Urban_Sustainability_2014-06.pdf.

Cohen, A.P. "Participant Observation." In *Ethnographic Research*, edited by Roy Ellen, 216–229. London: Academic Press, 1984.

Colson, Elizabeth. *The Social Consequences of Resettlement. The Impact of the Kariba Resettlement upon the Gwembe Tonga.* Manchester: Manchester University Press, 1971.

Connell, John. *Islands at Risk? Environments, Economies and Contemporary Change.* Cheltenham: Edward Elgar, 2013.

Coquery-Vidrovitch, Catherine. *The History of African Cities South of The Sahara: From the Origins to Colonization*, trans. Mary Baker. Princeton: Markus Wiener Publication, 2005.

Cordell, Dennis D., Joel W. Gregory, and Victor Piché. *Hoe and Wage: A Social History of Circular Migration System in West Africa.* Boulder: WestviewPress, 1996.

Crate, Susan A. and Mark Nuttall. *Anthropology and Climate Change: From Encounters to Actions.* Walnut Creek: Left Coast Press, 2009.

Creese, A. and W. Pokam. "Central Africa's Climate System." in *Africa's Climate*, edited by Future Climate for Africa, 4–10.

Croll, Elisabeth. and David Parkin. "Anthropology, the Environment and Development." In *Bush Base: Forest Farm—Culture, Environment and Development*, edited by Elisabeh Croll & David Parkin, 3–10. London: Routledge, 1992.

Crosby, Alfred W. *The Columbian Exchange: The Biological and Cultural Consequences of 1492.* Westport, Conn.: Greenwood, 1972.

Cruikshank, Julie. "Melting Glaciers and Emerging Histories in the Saint Elias Mountains." In *The Anthropology of Climate Change: An Historical Reader,* edited by Michael Dove, 261–275. Boston: Wiley Blackwell, 2014. Originally published in Marisol de la Cadena and Orin Starn, *Indigenous Experience Today.* Oxford: Berg, 2007.

Curtin, Philip D. *Disease and Empire: The Health of European Troops in the Conquest of Empire.* Cambridge: Cambridge University Press, 1998.

Dankelman, Irene. "Climate Change: Learning from Gender Analysis and Women's Experiences of Organising for Sustainable Development." *Gender & Development* 10.2 (2002): 21–29.

Davies, Zoe G., Jill L. Edmondson, Andreas Heinemeyer, Jonathan R. Leake, and Kevin J. Gaston. "Mapping an Urban Ecosystem Service: Quantifying Above-ground Carbon Storage at a City-wide Scale." *Journal of Applied Ecology* 48.5 (2011): 1125–1134.

de Bruijn, Mirjam, Han van Dijk, Mayke Kaag, Kiky van Til. *Sahelian Pathways: Climate and Society in Central and South Mali.* Leiden: African Studies Centre, 2005.

Demeritt, David. "The Construction of Global Warming and the Politics of Science." *Annals of the Association of American Geographers* 91.2 (2001): 307–337.

Demetriades, Justina and Emily Esplen. "The Gender Dimensions of Poverty and Climate Change Adaptation." *IDS Bulletin* 39.4 (2008): 24–31.

Demographia. *World Urban Areas. Built-Up Urban Areas or World Agglomerations*, 10th edition, 2014-05 revision. Belleville, Illinois: Demographia, 2014. .

Denzin, Norman K., and Yvonna S. Lincoln. *The Sage Handbook of Qualitative Research.* Los Angeles: Sage, 2011.

Descola, Philippe. "Constructing Natures: Symbolic Ecology and Social Practice." In *Nature and Society: Anthropological Perspectives*, edited by Philippe Descola and Gísil Pálsson, 82–102. London: Routledge, 1996.

Dessai, Suraje, Mike Hulme, Robert Lempert, and Roger Pielke Jr. "Climate Prediction: A Limit to Adaptation?." In *Adapting to Climate Change: Thresholds, Values, Governance*, edited by W. Neil Adger, Irene Lorenzoni, and Karen L. O'Brien. Cambridge: Cambridge University Press, 2009.

Devas, Nick and David Korboe. "City Governance and Poverty: The Case of Kumasi." *Environment and Urbanization* 12.1 (2000): 123–136.

de Waal, Alexander. *Famine that Kills: Darfur, Sudan.* Oxford: Oxford University Press, 2005.

Dewees, Peter A. "Trees and Farm Boundaries: Farm Forestry, Land Tenure and Reform in Kenya." *Africa: Journal of the International African Institute* 65.2 (1995). Accessed October 3, 2014. doi:10.2307/1161191.

De Wit, Sara. "Changing Patterns of Rain or Power? How an Idea of Adaptation to Climate Change Travels Up and Down to a Village in Simanjiro, Maasailand Northern Tanzania." *SPP 1448 Working Papers Series,* no. 11 (2015): 1–25. Accessed, October 10, 2015. http://www.spp1448.de/fileadmin/media/galleries/SPP_Administration/Working_Paper_Series/SPP1448_WP11_de_Wit.pdf.

De Wit, Sara. *Global Warning. An Ethnography of the Encounter Between Global and Local Climate-Change Discourses in the Bamenda Grassfields, Cameroon.* Bamenda/Leiden: Langaa & African Studies Centre, 2015.

De Wit, Sara. "Denaturalizing Adaptation, Resocializing the Climate: Theoretical and Methodological Considerations on How to Follow a Travelling Idea of Climate Change." In *Denaturalizing Climate Change: Migration, Mobilities and Space,* edited by Friederike Gesing, Johannes Herbeck, and Silja Klepp, 56–64. Artec-paper no. 200, 2014. Accessed November 3, 2015. http://www.uni-bremen.de/fileadmin/user_upload/single_sites/artec/artec_Dokumente/artec-paper/200_paper.pdf.

D'haen, Sarah A.L., Jonas Ø. Nielsen, and Eric F. Lambin. "Beyond Local Climate: Rainfall Variability as a Determinant of Household Nonfarm Activities in Contemporary Rural Burkina Faso." *Climate and Development* 6.2 (2014): 144–165.

Diallo, Boubacar A. and Bao Zhengyu. "Land Cover Change Assessment Using Remote Sensing: Case Study of Bamako, Mali." *Researcher* 2.4 (2010): 7–17.

Dietz, Ton. "Comparing the Agricultural Performance of Africa and Southeast Asia over the Last Fifty Years." In *Asian Tigers, African Lions. Comparing the Development Performance of Southeast Asia and Africa,* edited by. Bernard Berendsen, Ton Dietz, Henk Schulte Nordholt, Roel van der Veen, 85–128. Leiden, Boston: Brill, 2013.

Dietz, Ton, Kjell Havnevik, Mayke Kaag, Terje Oestigaard, ed. *African Engagements. Africa Negotiating an Emerging Multipolar World.* Leiden, Boston: Brill, 2011.

Dietz, Ton, and Els Veldhuizen. "Population Dynamics. An Important Intervening Variable." In *The Impact of Climate Change on Drylands, with a Focus on West Africa,* edited by Antonius J. Dietz, Ruerd Ruben, and Jan Verhagen. Dordrecht, Boston, London: Kluwer, 2004.

Dietz, Ton, Ruerd Ruben, Jan Verhagen, Saa Dittoh, M. Konaté, David Millar, Edward Ofori-Sarpong, Hassane Saley, and Ndiaye Cheikh Sylla. "Climate Change Preparedness in West Africa." In *The Impact of Climate Change on Drylands, with a Focus on West Africa,* edited by Ton Dietz, Ruerd Ruben, and Jan Verhagen, 403–408. Dordrecht, Boston, London: Kluwer, 2004.

Dietz, Ton, and Fred Zaal. "The Provisioning of African Cities, with Ouagadougou as a Case." In *Re-aligning Actors in an Urbanizing World: Institutions and Governance in an International Perspective,* edited by Isa Baud, and Johan Post, 263–284. Ashgate: Aldershot, 2002.

Djoudi, Houria and M. Brockhaus. "Is Adaptation to Climate Change Gender Neutral? Lessons from Communities Dependent on Livestock and Forests in Northern Mali." *International Forestry Review* 13.2 (2011): 123–135.

Dove, Michael R. "Introduction." In *The Anthropology of Climate Change. An Historical Reader,* edited by Michael Dove, 1–36. West Sussex: Wiley-Blackwell, 2014.

Durham University. "The Politics of Climate Change" (talk by Anthony Giddens, October 31, 2012). Last modified November 7, 2012. Accessed August 9, 2015. http://www.youtube.com/watch?v=yCgzxnRKhcU&list=PL1zMD_kTXdjzTG8CnFyh1abezH-FmkP8V.

Dwyer, John F., Herbert W. Schroeder, and Paul H. Gobster. "The Significance of Urban Trees and Forests: Toward a Deeper Understanding of Values." *Journal of Arboriculture* 17.10 (1991): 276–284.

East Africa Protectorate. *Report on the Forests of British East Africa by D. E. Hutchins. With a map and 25 photographs*. Published 1909. Cd. 4723. LX.

Edmondson, Jill L., Zoe G. Davies, Sarah A. McCormack, Kevin J. Gastonad, and Jonathan R. Leake. "Land-cover Effects on Soil Organic Carbon Stocks in a European City." *Science of the Total Environment* 472 (2014): 444–453.

Eguavoen, Irit, Karsten Schulz, Sara de Wit, Florian Weisser, Detlef Müller-Mahn. "Political Dimensions of Climate Change Adaptation. Conceptual Reflections and African Examples." In *Handbook of Climate Change Adaptation,* edited by Walter Leal Filho, 1183–1198. Berlin, Heidelberg: Springer, 2015.

Ehrlich, Paul. *The Population Bomb*. New York: Ballantine, 1968.

Eichemberg, Mayra T. "Species Composition and Plant Use in Old Urban Homegardens in Rio Claro, Southeast of Brazil." *Acta Botanica Brasilica* 23.4 (2009): 1057–1075.

Elahee, Khalil. The Challenges and Potential Options to Meet the Peak Electricity Demand in Mauritius." *Journal of Energy in Southern Africa* 22.3 (2011): 8–15.

Elahee, Khalil. "Energy Management Challenges in a Small Island Economy: The Case of Mauritius." *Energy and Environment* 21.7 (2010): 803–813.

Elmqvist, Thomas, Heikki Setälä, Steven N. Handel, Sander van der Ploeg James C. Aronson, James N. Blignaut, Erik Gómez-Baggethun, David J. Nowak, Jakub

Kronenberg, Rudolf S. de Groot. "Benefits of Restoring Ecosystem Services in Urban Areas." *Current Opinion in Environmental Sustainability* 14 (2015): 101–108.

Endfield, Georgina. "Reculturing and Particularizing Climate Discourses. Weather, Identity, and the Work of George Manley." *Osiris* 26.1 (2011): 145.

Epule, Terence, Changhui Peng, Laurent Lepage, and Zhi Chen. "The Causes, Effects and Challenges of Sahalian Droughts: A Critical Review." *Regional Environmental Change* 14 (2014): 145–156.

Escobar, Arturo. "Whose Knowledge, Whose Nature? Biodiversity, Conservation, and the Political Ecology of Social Movements". *Journal of Political Ecology* 5 (1998): 53–82.

Escobar, Arturo. "Constructing Nature: Elements for a Poststructural Political Ecology." In *Liberation Ecologies: Environment, Development, Social Movements*, edited by Richard Peet and Michael Watts, 46–68. London: Routledge, 1996.

Escobedo, Francisc, Sebastian Varela, Min Zhao, John E. Wagner, and Wayne Zipperer. "Analyzing the Efficacy of Subtropical Urban Forests in Offsetting Carbon Emissions from Cities." *Environmental Science and Policy* 13.5 (2010): 362–372.

Evans, Julian. *Plantation Forestry in the Tropics.* Oxford: Clarndon Press, 1992.

Fairhead, James and Melissa Leach. *Misreading the African Landscape: Society and Ecology in a Forest-Savanna Mosaic.* Cambridge: Cambridge University Press, 1996.

Fanan, Ujoh, Dlama Kwabe Isa, and Oluseyi Ifatimehin Olarewaju. "Urban Expansion and Vegetal Cover Loss in and around Nigeria's Federal Capital City." *Journal of Ecology and Natural Environment* 3.1 (2010): 1–10.

FAO. *Gender and Development Plan of Action 2002–2007.* Rome: Food and Agriculture Organisation of the United Nations, 2003.

Farbotko, Carol, and Heather Lazrus. "The First Climate Refugees? Contesting Global Narratives of Climate Change in Tuvalu." *Global Environmental Change* 22.2 (2012): 382–390.

Farley, John. *Bilharzia: A History of Imperial Tropical Medicine.* Cambridge: Cambridge University Press, 1991.

Fenn, Elizabeth A. *Smallpox Americana: The Great Smallpox Epidemic of 1775–1782.* New York: Hill and Wang, 2002.

Ferguson, James. *The Anti-Politics Machine. 'Development,' Depoliticization, and Bureaucratic Power in Lesotho.* Cambridge: Cambridge University Press, 1990.

Field, Christopher B. *Managing the Risks of Extreme Events and Disasters to Advance Climate Change Adaptation—A Special Report of Working Groups I and II of the Intergovernmental Panel on Climate Change.* Cambridge: Cambridge University Press, 2012.

Financial Times. "African Growth Feels the Strain from China's Slowdown." *Financial Times*, October 27, 2015. Accessed February 28, 2016. http://www.ft.com/cms/s/0/ef4d851a-7bc5-11e5-98fb-5a6d4728f74e.html#axzz3ydMKJhCE.

Fischer, Amy. *Hope Dries Up? Women and Girls Coping with Drought and Climate Change in Mozambique*. Report prepared for CARE International, accessed October 20, 2017. http://careclimatechange.org/wp-content/uploads/2016/11/El_Nino_Mozambique_Report_final.pdf.

Foeken, Dick, and Sam O. Owuor. "Multi-spatial Livelihoods in Sub-Saharan Africa: Rural Farming by Urban Households. The Case of Nakuru Town, Kenya." In *Mobile Africa: Changing Patterns of Movement in Africa and Beyond*, edited by Mirjam de Bruijn, Rijk van Dijk, and Dick Foeken, 125–140. Leiden: Brill, 2001.

Fohlmeister, Sandr, Stephan Pauleit, Adrien Coly, Hamidou Touré, Kumelachew Yeshitela. "The Way Forward: Climate Resilient Cities for Africa's Future." In *Urban Vulnerability and Climate Change in Africa: A Multidiscplinary Approach*. Heidelberg, New York, Dordrecht, London: Springer, 2015.

Fong, Wee Kean. *Global Report on Human Settlements 2011: Cities and Climate Change*. London, Washington DC: UN Human Settlement Program, 2011.

Food and Agriculture Organisation (FAO). "State of the World's Forests." *Food and Agricultural Organization of the United Nations* (2009).

Foolmaun, Rajendra K., Dinkar S. Chamilall, and G. Munhurrun. "Overview of Non-hazardous Solid Waste in the Small Island State of Mauritius." *Resources, Conservation and Recycling* 55 (2011): 966–972.

Forkuor, Gerald and Cofie Olufunke. "Dynamics of Land-use and Land-cover Change in Freetown, Sierra Leone and Its Effects on Urban and Peri-Urban Agriculture—A Remote Sensing Approach." *International Journal of Remote Sensing* 32.4 (2011): 1017–1037.

Frontline. "Frontline SMS Africa." Accessed August 9, 2015. http://www.frontlinesms.com/tag/africa/.

Fuzzi, Sandro, Urs Baltensperger, Kenneth S. Carslaw, S. Decesari, Hugo A.C. Denier van der Gon, Maria C. Facchini, David Fowler, Ilan Koren, Ben Langford, U. Lohmann, Eiko Nemitz, Spyros N. Pandis, Ilona Riipinen, Yinon Rudich, Martijn Schaap, J.G. Slowik, Dominick Spracklen, E. Vignati, Martin Wild, Martin Williams, S. Gilardoni. "Particulate Matter, Air Quality and Climate: Lessons Learned and Future Needs." *Atmospheric Chemistry and Physics* 15.0 (2015): 8217–8299.

GAIA. "Keep Mauritius Incinerator-free." Accessed December 16, 2014. http://www.no-burn.org/article.php?id=374.

GAIA. "Mauritius Says No to MSW Incinerator." Published June 22, 2009. Accessed December 16, 2014. http://www.no-burn.org/article.php?id=734.

Gaidzanwa, Rudo B. "Women and Land in Zimbabwe." Paper presented at the Conference "Why Women Matter in Agriculture," Sweden, April 4–8, 2011.

Gallopin, Gilberto C. "Linkages between Vulnerability, Resilience, and Adaptive Capacity." *Global Environmental Change* 16 (2006): 293–303.

Gardner, H.M. "Re-Afforestation in Kenya Colony by Means of Shifting Cultivation" *Empire Forestry Journal* 2.1 (1923).

Gemenne, François and Alexandre Magnan. *The Other Migrants Preparing for Change: Environmental Changes and Migration in the Republic of Mauritius*. Geneva: International Organisation for Migration, 2011. Accessed February 2, 2019. http://publications.iom.int/bookstore/free/the_other_migrants.pdf.

Gergel, Diana. "Water Resources Development: Engineering the Future of Global Health." *Issues in Brief* 27 (2013): 2–3.

Gesing, Friederike, Johannes Herbeck, and Silja Klepp, ed. *Denaturalizing Climate Change: Migration, Mobilities and Space.* Artec Paper 200, 2014. Accessed November 3, 2015. http://www.uni-bremen.de/fileadmin/user_upload/single_sites/artec/artec_Dokumente/artec-paper/200_paper.pdf.

Geteria, Wamugunda. "Management of the Mau Ecosystem." In *Forest Landscape and Kenya's Vision 2030: Proceedings of the 3rd Annual Forestry Society of Kenya (FSK) Conference and Annual General Meeting Held at the Sunset Hotel, Kisumu. 30th September–3rd October, 2008*, edited by D.O. Ogweno, P.S. Opanga, and A.O. Obara. Kisumu: Forestry Society of Kenya, 2009.

GGDC. "Growth of World Population, GDP and GDP Per Capita before 1820." Last modified September 3, 2008. Accessed February 2016. http://www.ggdc.net/MADDISON/other_books/appendix_B.pdf.

Ghana Statistical Service (GSS). *2010 Population and Housing Census Final Results Ghana Statistical Service*. Accra: Sakoa Press Ltd, 2012.

Giblin, James L. *The Politics of Environmental Control in Northeastern Tanzania, 1840–1940*. Philadelphia: University of Pennsylvania Press, 1992.

Giddens, Anthony. *The* Politics *of Climate Change*. Cambridge: Polity, 2009.

Giles-Vernick, Tamara and James L. Webb, ed. *Global Health in Africa: Historical Perspectives on Disease Control.* Athens: Ohio University Press, 2013.

Gill, Susannah E., J.F. Handley, Roland Ennos, and Stephan Pauleit. "Adapting Cities for Climate Change: The Role of the Green Infrastructure." *Built Environment* 33.1 (2007): 115–133.

Gißibl, Bernhard. *The Nature of German Imperialism. Conservation and the Politics of Wildlife in Colonial East Africa.* New York, Oxford: Berghahn Books, 2016.

Gleditsch, Nils P. "Special Issue: Climate Change and Conflict." *Journal of Peace Research*, 49.1 (2012).

Goh, Amelia H.X. *A Literature Review of the Gender-Differentiated Impacts of Climate Change on Women's and Men's Assets and Well-Being in Developing Countries*. CAPRi Working Paper No. 106. Washington, D.C.: International Food Policy Research Institute, 2012. http://dx.doi.org/10.2499/CAPRiWP106.

Goldman, Mara and Fernando Riosmena. "Adaptive Capacity in Tanzanian Maasailand: Changing Strategies to Cope With Drought in Fragmented Landscapes." *Global Environmental Change* 23 (2013): 588–597.

Goody, Jack. *Technology, Tradition, and the State in Africa.* London: Oxford University Press, 1971.

Gordon David and Shepard Krech III. "Indigenous Knowledge and the Environment." In *Indigenous Knowledge and the Environment in Africa and North America*, edited by David Gordon and Shepard Krech III, 1–24. Athens: Ohio University Press, 2012.

Goudou, D., J. Gué-Traoré, Mathieu Ouédraogo, Zénabou Segda, Badiane Ndour, Ndèye Yacine, Moussa Sall, F. Gueye, Keffing Sissoko, Zougmoré, Robert B. Moussa, S. Abdoulaye. *Village Baseline Study: Site Analysis Report for Kaffrine—Kaffrine, Senegal (SE0112).* Copenhagen: CGIAR Research Program on Climate Change, Agriculture and Food Security (CCAFS), 2012.

Gray, Leslie C. and William G. Moseley. "A Geographical Perspective on Poverty–Environmental Interactions." *The Geographical Journal* 171.1 (2004): 9–23.

Greschke, Heike. "The Social Facts of Climate Change: An Ethnographic Approach." In *Grounding Global Climate Change. Contributions from the Social and Cultural Sciences,* edited by Heike Greschke and Julia Tischler, 121–138. Dordrecht: Springer, 2015.

Grimm, Nancy B. Stanley H. Faeth, Nancy E. Golubiewski, Charles L. Redman, Jianguo Wu, Xuemei Bai, and John M. Briggs, "Global Change and the Ecology of Cities." *Science* 319 (2008): 756–760.

Grove, Richard H. *Green Imperialism: Colonial Expansion, Tropical Island Edens and the Origins of Environmentalism 1600—1860.* Cambridge: Cambridge University press, 1997.

Gugler, Josef. "The Son of the Hawk Does Not Remain Abroad: The Urban–Rural Connection in Africa." *African Studies Review* 45 (2002): 21–41.

Habtezion, Senaya, *Gender, Agriculture and Food Security.* New York: UNDP, 2012.

Hagberg, Sten and Jasper Bjarnesen. " 'Good Guys' and 'Bad Guys': The Burkinabe Public Debate on the Ivorian Crisis." In *Une anthropologie entre pouvoirs et historie: Conversations autour de l'aeuvre oeuvre??? de Jean-Pierre Chauveau,* edited by E. Jul-Larsen, P.J. Laurent, P.Y. Le Meurand and E. Leonard, 509–534, Paris: Karthala, 2011.

Hakikazi Catalyst. *A Future for Pastoralism. A Guide to the African Union's Policy Framework for Pastoralism in Africa: Securing, Protecting and Improving the Lives, Livelihoods and Rights of Pastoral Communities.* Arusha, 2011.

Hampshire, Kate. "Flexibility in Domestic Organization and Seasonal Migration among the Fulani of Northern Burkina Faso." *Africa* 3 (2006): 402–426.

Hampshire, Kate and Sara C. Randall. "Seasonal Labour Migration Strategies in the Sahel: Coping with Poverty or Optimizing Security?" *International Journal of Population Geography* 5 (1999): 367–385.

Hardin, Garret. "Tragedy of the Commons." *Science* 126.3859 (1968): 1243–1248.

Harlan, Sharon L. and Darren M. Ruddell. "Climate Change and Health in Cities: Impacts of Heat and Air Pollution and Potential Co-benefits from Mitigation and Adaptation." *Current Opinion in Environmental Sustainability* 3.3 (2011): 126–134.

Harris, John R. and Michael P. Todaro. "Migration, Unemployment and Development: A Two Sector Analysis." *American Economic Review* 60 (1970): 126–142.

Harsch, Ernest. " 'Arab Spring' Stirs African Hopes and Anxieties. Will Popular Upsurge in the North Help Spur Reform South of the Sahara?" *Africa Renewal*, August, 2011. Accessed November 20, 2016. http://www.un.org/africarenewal/magazine/august-2011/%E2%80%98arab-spring%E2%80%99-stirs-african-hopes-and-anxieties.

Hartley, Andrew, Dave Rowell, Serge Janicot, Françoise Guichard, Ian Macadam, Chris Taylor, and Douglas J. Parker. "A Century of Climate Change: 1950–2050." In *Africa's Climate,* edited by Future Climate for Africa: 33–38.

Hartwig, G.W., and K.D. Patterson, ed. *Disease in African History: An Introductory.*

Hassan, Ahmed Abdelhalim M. "Changes in the Urban Spatial Structure of the Greater Cairo Metropolitan Area." *ISPRS—International Archives of the Photogrammetry, Remote Sensing and Spatial Information Sciences* XXXVIII-4/(C21–133) (2011): 133–136.

Hastrup, Kirsten, ed. *The Social life of Climate Change Models: Anticipating Nature.* New York, London: Routledge, 2014.

Hastrup, Kirsten. "Nature: Anthropology on the Edge." In *Anthropology and Nature*, edited by Kirsten Hastrup, 1–26. New York: Routledge, 2014. *Survey and Case Studies.* Durham: Duke University Press, 1978.

Hastrup, Kirsten. "Comparing Climate Worlds: Theorising Across Ethnographic Fields." In *Grounding Global Climate Change. Contributions from the Social and Cultural Sciences,* edited by Heike Greschke and Julia Tischler, 139–154. Dordrecht: Springer, 2015.

Hayashi, Reiko. "Long Term Population Dynamics in Africa: A Perspective from the Urban Structure." Paper presented at the conference "Fifth African Population Conference," Union for African Population Studies–Arusha, December 10–14, 2007. Accessed February 2015. Last modified 2007. http://uaps2007.princeton.edu/papers/70296.

Hayes. P. "The 'Famine of the Dams:' Gender, Labour and Politics in Colonial Ovamboland, 1929–1930." In *Namibia under South African Rule: Mobility and Containment, 1915–1945*, P. Hayes et al., ed. Oxford: James Currey, 1998.

Headrick, Daniel R. *The Tools of Empire: Technology and European Imperialism in the Nineteenth Century*. New York: Oxford University Press, 1981.

Healy, Stephen. "Post-normal Science in Postnormal Times." *Futures* 43.2 (2011): 204. Accessed June 8, 2015. doi:10.1016/j.futures.2010.10.009.

Heeralall, Nirmal. "The World's Fastest Growing Middle Class." *UHY The Network for Doing Business*. Last modified probably 2014. Accessed February 2015. http://www.uhy.com/the-worlds-fastest-growing-middle-class/.

Heilig, Gerhard. *Demographics*. Accessed February 28, 2016. http://www.demographics.at/Growth-Decline/images/PPP2015_Africa_Tot-Pop_1950-2100.gif, and http://www.demographics.at/Growth-Decline/fig_ppp2015_africa-europe_tot-pop.htm.

Helliker, Kirk. “Dancing on the same Spot: NGOs.” In *Contested Terrain: Land Reform and Civil Society in Contemporary Zimbabwe*, edited by Sam Moyo, Kirk Helliker, and Tendai Murisa, 239–274. Pietermaritzburg: S&S Publishers, 2008.

Heltberg, Rasmus, Naomi Hossain, and Anna Reva, ed. *Living through Crises: How the Food, Fuel, and Financial Shocks Affect the Poor*. Washington DC: World Bank, 2012.

Henderson, J. Vernon, Adam Storeygard, and Uwe Deichmann. “Is Climate Change Driving Urbanization in Africa?.” *The World Bank* (2014): 50.

Henderson, Jeffrey, Peter Dicken, Martin Hess, Neil Coe and Henry Wai-Chung Yeung. “Global Production Networks and the Analysis of Economic Development.” *Review of International Political Economy* 9(2002): 4436–4464.

Henley, David. *Asia-Africa. Development Divergence. A Question of Intent*. London: Zed Books, 2015.

Henry, Sabine, Bruno Schoumaker, and Cris Beauchemin. “The Impact of Rainfall on the First Out-Migration: A Multi-level Event-History Analysis in Burkina Faso.” *Population and Environment* 25 (2004): 423–460.

Herskovits, Melville J. “The Cattle Complex in East Africa.” PhD diss., Columbia University Press, 1926.

Hodgson, Dorothy L. *The Church of Women. Gendered Encounters Between Maasai and Missionaries.* Indiana: Indiana University Press, 2005.

Hodgson, Dorothy L. “ ‘Once Intrepid Warriors’: Modernity and the Production of Maasai Masculinities,” in *Gendered Modernities,* edited by Dorothy L. Hodgson, 105–145. New York: Palgrave, 2001.

Hodgson, Dorothy L. “ ‘These Are not Our Priorities’: Maasai Women, Human Rights, and the Problem of Culture.” In *Gender and Culture at the Limits of Rights,* edited by Dorothy L. Hodgson, 138–157. Philadelphia: University of Pennsylvania Press, 2011.

Holman, Dennis. *Inside Safari Hunting, with Eric Rundgren.* London: W.H. Allen, 1969.

Homewood, Katherine M., and W.A. Rodgers. *Maasailand Ecology. Pastoralist Development and Wildlife Conservation in Ngorongoro, Tanzania.* Cambridge University Press, 1991.

Hoogeveen, Johannes, Bas van der Klaauw, and Gijsbert van Lomwel. *On the Timing of Marriage, Cattle and Weather Shocks*. Netherlands: Tinbergen Institute Discussion Paper TI 2004–073/3, 2004.

Hornborg, Alf. “The Fossil Interlude: Euro-American Power and the Return of the Physiocrats.” In *Cultures of Energy: Power, Practices, Technology*, edited by Sarah Strauss, Stephanie Rupp and T. Love, 41–59. Walnut Creek: Left Coast Press, 2013.

Horta, Ana, Harold Wilhite, Luísa Schmidt, and Françoise Bartiaux. “Socio-technical and Cultural Approaches to Energy Consumption: An Introduction.” *Nature and Culture* 9.2 (2014): 115–121.

Hovorka, Alice J. “The No. 1 Ladies’ Poultry Farm: A Feminist Political Ecology of Urban Agriculture in Botswana, Gender, Place & Culture,” *Journal of Feminist Geography* 13.3. (2006): 207–225.

Hughes McDermott, David. *From Enslavement to Environmentalism: Politics on a Southern African Frontier.* Seattle: University of Washington Press, 2006.

Hulme, Mike. *Why We Disagree about Climate Change: Understanding Controversy. Inaction and Opportunity.* Cambridge: Cambridge University Press, 2009.

Hulme, Mike. "Problems with Making and Governing Global Kinds of Knowledge." *Global Environmental Change* 20 (2010): 558–564.

Hulme, Mike. " 'Meet the Humanities.' " *Nature Climate Change* 1 (2011): 177–179.

Hulme, Mike. "Reducing the Future to Climate: A Story of Climate Determinism and Reductionism." *Osiris* 26.1 (2011): 245–266.

Hulme, Mike. "Climate and its Changes: A Cultural Appraisal." *Geo* 2.5 (2015): 1–11.

Hulme, Mike. "The Idea of Climate Change. Exploring Complexity, Plurality and Opportunity." *GAIA* 19.3 (2010): 171–174.

Hulme, Mike. "The Conquering of Climate: Discourses of Fear and Their Dissolution." *The Geographical Journal* 174.1 (2008): 5–16.

Hulme, Mike. "Geographical Work at the Boundaries of Climate Change." *Transactions of the Institute of British Geographers* 33 (2008): 5–11.

Hulme, Mike. "Climatic Perspectives on Sahelian Desiccation: 1973–1998." *Global Environmental Change* 11 (2001): 19–29.

Hulme, Mike, Ruth Doherty, Ngara,Todd, New, Mark, and David Lister. "African Climate Change: 1900–2100." *Climate Research* 17 (2001): 145–168.

Ibrahim Index of African Governance. Accessed December 16, 2014. http://www.moibrahimfoundation.org/iiag/.

Igoe, Jim. "National Parks and Human Ecosystems. The Challenge to Community Conservation. A Case Study from Simanjiro, Tanzania." In *Conservation and Mobile Indigenous Peoples: Displacement, Forced Settlement and Sustainable Development,* edited by Dawn Chatty and Marcus Colchester, 77–96. Oxford: Berghahn Books, 2002.

Igoe, Jim, and Dan Brockington. "Pastoral Land Tenure and Community Conservation: A Case Study From North-East Tanzania." *IIED Drylands Programme: Pastoral Land Tenure Series 11*, 1999.

Iliffe, John. *Africa: The History of a Continent.* Cambridge: Cambridge University Press, 2007.

Imo, Moses. "Interactions amongst Trees and Crops in Taungya Systems of Western Kenya." *Agroforestry Systems* 76.2 (2009). Accessed October 3, 2014. doi:10.1007/s10457-008-9164-z.

Imo, Moses. "Professional and Societal Mismatch in Kenyan Forestry: Is There a Right Way to Manage Our Forests?" In *Forest Landscape and Kenya's Vision 2030: Proceedings of the 3rd Annual Forestry Society of Kenya (FSK) Conference and Annual General Meeting Held at the Sunset Hotel, Kisumu. 30th September–3rd October, 2008,* edited by D.O. Ogweno, P.S. Opanga, and A.O. Obara. Kisumu: Forestry Society of Kenya, 2009.

Ingold, Tim. *The Perception of the Environment. Essays on Livelihood, Dwelling and Skill.* London: Routledge, 2000.

Institute for Advanced Sustainability Studies (IASS). *The Future of Africa's Energy Supply: Potentials and Development Options for Renewable Energy*, edited by Rainer Quitzow, Sybille Roehrkasten, David Jacobs, Benjamin Bayer, El Mostafa Jamea, Yvonne Waweru, Patrick Matschoss. Potsdam, 2016.

Institute for Environmental and Legal Studies. "Solid Waste Management in Mauritius." Accessed December 16, 2014. http://iels.intnet.mu/solidwaste_mau.htm.

Integrated Regional Information Networks (IRIN). "Uganda Charcoal Boom a Bust for Forests." *IRIN Humanitarian News and Analysis*, February 7, 2010. Accessed February 24, 2016. http://www.irinnews.org/report/94810/uganda-charcoal-boom-a-bust-for-forests.

Intergovernmental Panel on Climate Change (IPCC). "Working Group II: Impacts, Adaptation and Vulnerability," last modified June 6, 2017, accessed July 11, 2017. http://www.ipcc.ch/ipccreports/tar/wg2/index.php?idp=384.

Intergovernmental Panel on Climate Change (IPCC). *Climate Change 2007—Impacts, Adaptation and Vulnerability.* Contribution of Working Group II to the Fourth Assessment Report of the IPCC. Cambridge: Cambridge University Press, 2007.

International Energy Agency (IEA). *Africa Energy Outlook: A Focus on Energy Prospects in Sub-Saharan Africa, World Energy Outlook Special Report.* Paris, 2014.

International Fund for Agriculture Development (IFAD). Community—Based Integrated Natural Resources Management Project in Lake Tana Watershed-Ethiopia (Third Draft), 2007.International Rivers. Accessed August 2, 2017, https://www.internationalrivers.org/programs/africa.

International Organisation Migration (IOM). *Migration, Environment and Climate Change: Assesing the Evidence.* Geneva: IOM, 2009.

International Renewable Energy Agency (IRENA). *Africa 2030: Roadmap for a Renewable Energy Future.* Abu Dhabi, 2015.

IPCC, *Climate Change 2013: The Physical Science Basis. Contribution of Working Group I to the Fifth Assessment Report of the Intergovernmental Panel on Climate Change,* edited by T.F. Stocker, D. Qin, G.K. Plattner, M. Tignor, S.K. Allen, J. Boschung, A. Nauels, Y. Xia, V. Bex and P.M. Midgley. Cambridge, New York: Cambridge University Press, 2013.

IPCC. *Managing the Risks of Extreme Events and Disasters to Advance Climate Change Adaptation: A Special Report of Working Groups I and II of the Intergovernmental Panel on Climate Change,* edited by C.B. Field, V. Barros, T.F. Stocker, D. Qin, D.J. Dokken, K.L. Ebi, M.D. Mastrandrea, K.J. Mach, G.-K. Plattner, S.K. Allen, M. Tignor, and P.M. Midgley. Cambridge: Cambridge University Press, 2012.

IPCC. "Summary for Policymakers." In *Climate Change 2007: The Physical Science Basis. Contriution of Working Group 1 to the Fourth Assessment Report of the*

Intergovernmental Panel on Climate Change, edited by Dahe Qin et al. Cambridge, New York: Cambridge University Press, 2007.

Isaacman, Allan F. and Barbara S. Isaacman. *Dams, Displacement, and the Delusion of Development: Cahora Bassa and Its Legacies in Mozambique, 1965–2007*. Athens: Ohio University Press, 2013.

Isaacman, Allen and Richard Roberts, ed. *Cotton, Colonialism, and Social History in Sub-Saharan Africa*. Portsmouth, N.H.: Heinemann, 1995.

Ismael, F.M., S. Seeruttun, C. Barbe, and A. Gaungoo. "A Review of Changing Cultural Practices to Improve Productivity of Sugar Cane in Mauritius." *Proceedings of the South African Sugar Technologists' Association* 81 (2008): 538–544.

Jacobs, Nancy J. *Environment, Power, and Injustice: A South African History*. Cambridge: Cambridge University Press, 2003.

Jacobs, Nancy J. "Grasslands and Thickets: Bush Encroachment and Herding in the Kalahari Thornveld." *Environment and History* 6.3 (2000): 289–316.

Jasanoff, Sheila. "A New Climate for Society." *Theory, Culture & Society* 27.2–3 (2010): 233–253.

Jedwab, Remi. *Why is African Urbanization Different? Evidence from Resource Exports in Ghana and Ivory Coast*. Manuscript, Paris School of Economics, 2012.

Jenkins, Paul. "Urban Management, Urban Poverty and Urban Governance: Plan-ning and Land Management in Maputo." *Environment and Urbanization* 12.1 (2000): 137–152.

Jogoo, Vasantt K. "The Political Economy of Transitioning to a Green Economy in Mauritius." In *Transitioning to a Green Economy: Political Economy of Approaches in Small States*, edited by Nadine Smith, Anna Halton and Janet Strachan, 128–153. London: Commonwealth Secretariat, 2014.

Johnson, David. "Mamdani, Moyo and 'Deep Thinkers' on Zimbabwe." *Pambazuka News*, February 12, 2009. Accessed June 8, 2011. http://www.pambazuka.org/en/category/comment/54039.

Kajima, Saeko. "Institutional Interactions and the Implementation of Climate Change and Sustainable Development Programmes: A Case Study of Mauritius' 'Maurice Ile Durable'." Unpublished MSc dissertation, University of Edinburgh, 2012.

Kalame, Fobissie B., Robert Aidoob, Johnson Nkem, Oluyede C. Ajayied, Markku Kanninena, Olavi Luukkanena, and Monica Idinobae. "Modified Taungya System in Ghana: A Win–win Practice for Forestry and Adaptation to Climate Change?" *Environmental Science & Policy* 14.5 (2011): 519–530. Accessed October 3, 2014. doi:10.1016/j.envsci.2011.03.011.

Kamusoko, Courage, Jonah Gamba, and Hitomi Murakami. "Monitoring Urban Spatial Growth in Harare Metropolitan Province, Zimbabwe." *Advances in Remote Sensing* 2.0 (2013): 322–331.

Kanogo, Tabitha?. *Squatters and the Roots of Mau Mau, 1905–63*. London: James Currey, 1987.

Karekezi, Stephen, John Kimani, and Oscar Onguru. *Climate Change and Energy Security in East Africa*. Nairobi: AFREPREN/FWD, 2009.

Kasenally, Swalay. "Coal Power: The Politics of Indecision." *Le Mauricien*, February 28, 2012. Accessed December 16, 2014. http://www.lexpress.mu/article/coal-power-politics-indecision.

Kasirye, Birikadde G., Magi Matinga, and Joy Clancy. "Fuel Security and Supply Dynamics in Internally Displaced Persons' Camps of Northern Uganda." *Journal of Humanitarian Assistance*, accessed October 20, 2017. http://sites.tufts.edu/jha/archives/462.

Kaufman, Frederick. "The Man Who Stole the Nile: An Ethiopian Billionaire's Outrageous Land Grad." *Harper's Magazine*, July 2014, 36–42.

Kayembe wa kayembe, Matthieu, Mathieu de Maeyer, and Eleonore Wolff. "The Mapping of the Urban Growth of Kinshasa (DRC) through High Resolution Remote Sensing between 1995 and 2005." In *Remote Sensing—Applications*, edited by Escalante Boris. Rijeka: InTech, 2008.

Kennedy, Dane K. *Islands of White: Settler Society and Culture in Kenya and Southern Rhodesia, 1890–1939*. Duke University Press, 1987.

Kenya Forest Department. *Squatter and Shamba Management*. Bulletin No. 33. Published 20 May 1932.

Kenya Forest Department. "Annual Report of the Forest Department, 1932." Nairobi Government Printer, 1932. The National Archives of the UK (TNA), CO 544/38.

Kenya Forest Department. "Annual Report of the Forest Department, 1933." Nairobi Government Printer, 1933. The National Archives of the UK (TNA), CO 544/40.

Kenya Forest Department. "Annual Report of the Forest Department, 1934." Nairobi Government Printer, 1934. The National Archives of the UK (TNA), CO 544/44.

Kenya Forest Department. "Annual Report of the Forest Department, 1938." Nairobi Government Printer, 1938. The National Archives of the UK (TNA), CO 544/55.

Kenya Forest Department. *Plantation Management Costs & Practical Hints*. Bulletin No. 34, September 28, 1939.

Kenya Forest Department. "Annual Report of the Forest Department, 1951–1953." Nairobi Government Printer, 1953. The National Archives of the UK (TNA), CO 544/78.

Kenya Forest Department. "Annual Report of the Forest Department, 1954–1955." Nairobi Government Printer, 1955. The National Archives of the UK (TNA), CO 544/80.

Kenya Forest Department. "KFD Annual Report," published 1953.

Kenya Forest Department. "KFD Annual Report," published.

Kenya Forest Department. "Annual Report of the Forest Department, 1960." Nairobi Government Printer, 1960. The National Archives of the UK (TNA), CO 544/97.

Kenya Forest Department. "Annual Report of the Forest Department, 1961." Nairobi Government Printer, 1961. The National Archives of the UK (TNA), CO 544/100.

Kenya Forest Department. "Annual Report of the Forest Department, 1963." Nairobi Government Printer, 1963. The National Archives of the UK (TNA), CO 544/106.

Kenya Forest Department. "KFD Annual Report," published 1963.

Kenya Forest Department. "Annual Reports of the Forest Department, 1920–1963." Nairobi Government Printer, 1920–1963. The National Archives of the UK (TNA), CO 544/12–CO 544/106.

Kenya Forest Department. "Taungya in Kenya: The 'Shamba System'." Paper presented at the World Symposium on Man-Made Forests and their Industrial Importance, Canberra: Food and Agriculture Organization of the United Nations, 14–25 April 1967.

Kenya Forest Service. *PELIS Now Halted In Kimondi 2c Forest*. Published August 2011. Accessed October 3, 2014. http://www.kenyaforestservice.org/index.php?option=com_content&view=category&layout=blog&id=228&Itemid=168&limitstart=5.

Kenyatta, Jomo. *Facing Mount Kenya: The Tribal Life of the Gikuyu.* London: Secker and Warburg, 1938.

Kessides, Christine. *The Urban Transition in Sub-Saharan Africa.* Wahington: The Cities Alliance, 2006.

Keulemans, Maarten. "Klimaatdebat is in zekere zin een voortzetting van Bijbels Eindtijdverhaal," *Volkskrantkatern Vonk*, March 24, 2014.

Khan, Fareida. "Soil Wars: The Role of the African National Soil Conservation Association in South Africa, 1953–59." *Environmental History* 2.4 (1997): 439–459.

Khan, Fareida. "Rewriting South Africa's Conservation History: The Role of the Native Farmers Association." *Journal of Southern African Studies* 20.4 (1994): 499–516.

Kiple, Kenneth F. *The Caribbean Slave: A Biological History*. Cambridge: Cambridge University Press, 2002.

Kiriinya, C.K. "The Rise and Fall of Taungya: Lessons from Kenya." *Agroforestry Today (ICRAF)* 6.3 (1994).

Kitching, Gavin. *Class and Economic Change in Kenya: The Making of an African Petite Bourgeoisie 1905–1970.* London: Yale University Press, 1980.

Kitto, Elizabeth. "Before European Colonialism was Africa Essentially Rural?." *Retrospective* online (2012).

Kjekhus, H. *Ecology Control and Economic Development in East African History: The Case of Tanganyika, 1850–1950.* London: Heinemann, 1977.

Klopp, Jacqueline M. and Jeffrey W. Paller. "How Africa Can Build Inclusive, Safe and Sustainable Cities." *The Conversation*, October 9, 2015. Accessed February 28, 2016. http://theconversation.com/how-africa-can-build-inclusive-safe-and-sustainable-cities-48445.

KMA. *The Composite Budget of the Kumasi Metropolitan Assembly for the 2013 Fiscal Year*. Kumasi Metropolitan Assembly, 2013.

KMA WMD. *Data for Purposes of Planning Waste Management Intervention Programmes.*Kumasi: Kumasi Waste Management Department, 2010.

Knutti, Reto. “Should we Believe Model Predictions of Future Climate Change?.” *Phil. Trans. R. Soc. A* 366 (2008): 4647–4664. doi:10.1098/rsta.2008.0169.

Koech, Eric, Eliud Kireger, and Joel Laigong. “Forest Resources as a Source of Energy and Food Security: Two Contradictory Issues or Two Sides of the Same Challenge?” In *Forest Landscape and Kenya's Vision 2030: Proceedings of the 3rd Annual Forestry Society of Kenya (FSK) Conference and Annual General Meeting Held at the Sunset Hotel, Kisumu. 30th September–3rd October, 2008*, ed. D.O. Ogweno, P.S. Opanga and A.O. Obara. Kisumu: Forestry Society of Kenya, 2009.

Kolkman, Harold. “Inequity and Strife in Community Based Natural Resource Management.” In *Culture, Organization and Management in South Africa: In Search of Equity*, edited by Marja Spierenburg and Harry Wels. New York: Nova Science Publishers, 2004.

Konseiga, Adama. “Household Migration Decisions as Survival Strategy: The Case of Burkina Faso.” *Journal of African Economies* 16 (2007): 198–233.

Kothari, Uma and Rorden Wilkinson. “Global Change, Small Island State Response: Restructuring and the Perpetuation of Uncertainty in Mauritius and Seychelles.” *Journal of International Development* 25 (2013): 92–107.

KPMG. “Mining in Africa. Towards 2020.” Last modified 2013. Accessed February 2015. https://www.kpmg.com/Africa/en/IssuesAndInsights/Articles-Publications/Documents/Mining%20Indaba%20brochure.pdf.

Kreike, Emmanuel. *Environmental Infrastructure in African History: Examining the Myth of Natural Resource Management in Namibia*. Cambridge: Cambridge University Press, 2013.

Kreike, Emmanuel. *Deforestation and Reforestation in Namibia: The Global Consequences of Local Contradictions*. Leiden: Brill, 2010.

Kreike, Emmanuel. *Re-Creating Eden: Land Use, Environment, and Society in Southern Angola and Northern Namibia*. Portsmouth, NH: Heinemann, 2004.

Kundzewicz, Zbigniew W., Shinjiro Kanae, Sonia I. Seneviratne, John Handmer, Neville Nicholls, Pascal Peduzzi, Reinhard Mechler, Laurens M. Bouwer, Nigel Arnell, Katharine Mach, Robert Muir-Wood, G. Robert Brakenridge, Wolfgang Kron, Gerardo Benito, Yasushi Honda, Kiyoshi Takahashi and Boris Sherstyukov. “Flood Risk and Climate Change: Hydrological Perspectives,” *Hydrological Sciences Journal* 59.1 (2012): 1–28.

Kvale, Steinar. *InterViews: An Introduction to Qualitative Research Interviewing*. Thousands Oaks: Sage Publications, 1996.

Kwashirai, Vimbai C. “Ecological and Poverty Impacts of Zimbabwe's Land Struggles 1980 to Present.” *Global Environment: A Journal of History and Natural and Social Sciences* 2.3 (2009): 222–253.

Lanjouw, Peter. “Nonfarm Employment and Poverty in Rural El Salvador.” *World Development* 29 (2001): 529–547.

La Rédaction. "Not in My Backyard," L'express, May 22, 2007, accessed December 16, 2014. https://www.lexpress.mu/article/not-my-backyard.

Lasco, Rodel D. Rafaela J.P. Delfino, Delia C. Catacutan, Elisabeth S. Simelton, David M. Wilson. "Climate Risk Adaptation by Smallholder Farmers: The Roles of Trees and Agroforestry." *Current Opinion in Environmental Sustainability* 6 (2014): 83–88.

Law, Robin. *The Horse in West African History: The Role of the Horse in the Societies of Pre-Colonial West Africa*. Oxford: Oxford University Press, 1980.

Leach, Melissa and Robin Mearns, ed. *The Lie of the Land: Challenging Received Wisdom on the African Environment.* Portsmouth: Heinemann, 1996.

Leach, Melisssa and Robin Mearns. *Beyond the Fuelwood Crisis: People, Land and Trees in Africa*. London: Earthscan, 1988.

Leakey, L.S.B. *Mau Mau and the Kikuyu.* London: Methuen & Co. Ltd., 1952.

Leal Filho, Walter. "Climate Change at Universities: Results of a World Survey." In *Universities and Climate Change. Introducing Climate Change to University Programmes.* edited by Walter Leal Filho. Berlin, Heidelberg: Springer, 2010.

Lebel, Thierry and Abdou Ali. "Recent Trends in the Central and Western Sahel Rainfall Regime (1990–2007)." *Journal of Hydrology* 375 (2009): 52–64.

Leggewie, Claus and Harald Welzer. *Das Ende der Welt, wie wir sie kannten: Klima, Zukunft und die Chancen der Demokratie.* Frankfurt: Fischer, 2009.

Leichenko, Robin M. and Karen O'Brien. *Environmental Change and Globalization: Double Exposures.* Oxford: Oxford University Press, 2008.

Le Jeune, Gael, Jean Poirier, and Victor Piche. "Toward a Reconsideration of Female Migration Patterns in Burkina Faso." *Canadian Studies in Population* 31 (2005): 145–177.

Leslie, Paul W. and Terrence J. McCabe. "Response Diversity and Resilience in Social-Ecological Systems." *Current Anthropology* 54.2 (2013): 114–129.

Lewis, Justin, and Tammy Boyce, ed. *Climate Change and the Media.* New York, Bern, Berlin, Bruxelles, Frankfurt a.M., Oxford, Wien: Peter Lang, 2009.

Lewis, Simon L., Gabriela Lopez-Gonzalez, and Hannsjörg Wöll. "Increasing Carbon Storage in Intact African Tropical Forests." *Nature* 457.19 (2009): 1003–1006.

Lim Tung, Odile. "Development and the Balancing of Interests in Environmental Policy and Law in Mauritius." In *The Balancing of Interests in Environmental Law in Africa*, edited by Michael Faure and Willemien du Plessis, 261–294. Pretoria: Pretoria University Law Press, 2011.

Lindén, Jenny. "Nocturnal Cool Island in the Sahelian City of Ouagadougou, Burkina Faso." *International Journal of Climatology* 31.4 (2011): 605–620.

Lindley, Sarah J., Susannah E. Gill, Gina Cavan, Kumelachew Yeshitela, Alemu Nebebe, Tekle Woldegerima, Deusdedit Kibassa, Riziki Shemdoe, Florian Renner, Katja Buchta, Hany Abo-El-Wafa, Andreas Printz, Fatimatou Sall, Adrien Coly, and Ndèye Marème Ndour. "Green Infrastructure for Climate Adaptation in African Cities." In *Urban Vulnerability and Climate Change in Africa*, edited by Stephan Pauleit, Adrien

Coly, Sandra Fohlmeister, Paolo Gasparini, Gertrud Jørgensen, Sigrun Kabisch, Wilbard J. Kombe, Sarah Lindley, Ingo Simonis, and Kumelachew Yeshitela. Heidelberg, New York: Springer, 2015.

Lockwood, Matthew. "What Can Climate-Adaptation Policy in Sub-Saharan Africa Learn from Research on Governance and Politics?." *Development Policy Review* 31.6 (2013): 647–676. Accessed June 8, 2015. doi: 10.1111/dpr.12029.

Loeb, Edwin M. *In Feudal Africa*. Bloomington, Indiana University, 1962.

Logie, J.P., and W.G. Dyson. *Forestry in Kenya: A Historical Account of the Development of Forest Management in the Colony.* Nairobi: Colony and Protectorate of Kenya, 1962.

Luig, Ute and Achim von Oppen. "Einleitung: Zur Vergesellschaftung von Natur in Afrika." In *Naturaneignung in Afrika als sozialer und symbolischer Prozess*, edited by Ute Luig und Achim von Oppen, 5–27. Berlin: Das Arabische Buch, 1995.

Luximon, Y., and M.D. Nowbuth. "A Status of Food Security in Mauritius in Face of Climate Change." *European Water* 32 (2010): 3–14.

Lwando, Catherine. "Climate Variability and Gender: Emerging Experiences from Western Zambia." *Environment and Natural Resources Research* 3.4 (2013): 133–146.

Lwasa, Shuaib. "Adapting Urban Areas in Africa to Climate Change: The Case of Kampala." *Current Opinion in Environmental Sustainability* 2.3 (2010): 166–171.

Lyons, Maryinez. *The Colonial Disease: A Social History of Sleeping Sickness in Northern Zaire, 1900–1940*. Cambridge: Cambridge University Press, 1992.

Mabogunje, Akin L. "Global Urban Poverty Research Agenda: The African Case." Seminar Presented at the Global Urban Poverty: Setting the Research Agenda, (2005).

Mackenzie, A. Fiona D. *Land, Ecology and Resistance in Kenya, 1880–1952*. Edinburgh: Edinburgh University Press, 1998.

Maddox, Gregory, James Giblin, and Isaria N. Kimambo, ed. *Custodians of the Land: Ecology and Culture in the History of Tanzania*. London: James Currey, 1996.

Mago, Stephen. "Urban Youth Unemployment in Africa: Whither Socio-Economic Problems." *Mediterranean Journal of Social Sciences* 5.9 (2014): 33–40.

Malo, Meshack, "Visioning the Energy Dimension of Kenya: Changing Forest Landscape under Vision 2030." In *Forest Landscape and Kenya's Vision 2030: Proceedings of the 3rd Annual Forestry Society of Kenya (FSK) Conference and Annual General Meeting Held at the Sunset Hotel, Kisumu. 30th September–3rd October, 2008*, edited by D.O. Ogweno, P.S. Opanga and A.O. Obara. Kisumu: Forestry Society of Kenya, 2009.

Mamdani, Mahmood. "Lessons of Zimbabwe." *London Review of Books* 30.23 (2008): 17–21.

Manu, Andrew, Yaw A. Twumasi, and Tommy L. Coleman. "Is it the Result of Global Warming or Urbanization? The Rise in Air Temperature in Two Cities in Ghana." Presented at the 5th FIG Regional Conference, 2006.

Marin-Burgos, Victoria. "Social Environmental Conflicts and Agrofuel Crops: The Case of Oil Palm Expansion in Colombia," in *The Political Ecology of Agrofuels*, edited by Kristina Dietz, Bettina Engels, Oliver Pye, and Achim Brunnengräber. Abingdon: Routledge, 2015.

Masera, Omar R., Barbara D. Saatkamp, and Daniel D. Kammen. "From Linear Fuel Switching to Multiple Cooking Strategies: A Critique and Alternative to the Energy Ladder Model." *World Development* 28.12 (2000): 2083–2103.

Mathu, Winston. "Forest Plantations and Woodlots in Kenya." *African Forest Forum Working Paper Series* 1.13 (2011).

Mauritius Meteorological Services. Accessed December 16, 2014. http://metservice.intnet.mu/climate-services/climate-change.php.

Mazingi, Lucy, and Richard Kamidza. "Inequality in Zimbabwe." In *Tearing us Apart: Inequality in Southern Africa*, edited by Herbert Jauch, and Deprose Muchena, 322–383. OSISA, 2010. Accessed July 05, 2017. http://www.osisa.org/sites/default/files/sup_files/chapter_5_-_zimbabwe.pdf.

MacKenzie, John M. *The Empire of Nature: Hunting, Conservation and British Imperialism*. Manchester: Manchester University Press, 1988.

Mbow, Cheikh H. Neufeldt, M. van Noordwijk, P.A. Minang, G. Kowero, and E. Luedeling. "Agroforestry Solutions to Address Food Security and Climate Change Challenges in Africa." *Current Opinion in Environmental Sustainability, Sustainability Challenges* 6 (February 2014): 61–67. Accessed October 3, 2014. doi:10.1016/j.cosust.2013.10.014.

Mbugua, D.K. "Deepening Forestry Sector Reforms in Kenya." In *Forest Landscape and Kenya's Vision 2030: Proceedings of the 3rd Annual Forestry Society of Kenya (FSK) Conference and Annual General Meeting Held at the Sunset Hotel, Kisumu. 30th September–3rd October, 2008*, edited by D.O. Ogweno, P.S. Opanga and A.O. Obara. Kisumu: Forestry Society of Kenya, 2009.

McCabe, Terrence J. "Giving Conservation a Human Face? Lessons from Forty Years of Combining Conservation and Development in the Ngorongoro Conservation Area, Tanzania." In *Conservation and Mobile Indigenous People. Displacement, Forced Settlement, And Sustainable Development,* edited by Dawn Chatty and Marcus Colchester, 61–76. New York, Oxford: Berghahn Books, 2002.

McCabe, Terrence J. "Sustainability and Livelihood Diversification among the Maasai of Northern Tanzania." *Human Organization* 62.2 (2003): 100–111.

McCabe, Terrence J., and Paul W. Leslie, and Laura DeLuca. "Adopting Cultivation to Remain Pastoralists: The Diversification of Maasai Livelihoods in Northern Tanzania." *Human Ecology* 38.3 (2010): 321–334.

McCann, James C. *Maize and Grace: Africa's Encounter with a New World Crop, 1500–2000*. Cambridge, Mass.: Harvard University Press, 2005.

McCann, James C. "Climate and Causation in African History," *International Journal of African Historical Studies* 32.2–3 (1999): 261–280.

McCann, James C. *Green Land, Brown Land, Black Land: An Environmental History of Africa, 1800—1990*. Portsmouth: Heinemann, 1999.

McCann, James C. "The Plow and the Forest: Narratives of Deforestation in Ethiopia, 1840–1992." *Environmental History* 2.2 (1997): 138–159.

McCann, James C. "Britain, Ethiopia, and Negotiations for the Lake Tana Dam Project, 1922–1935." *International Journal of African Historical Studies* 14.4 (1981): 667–99.

McCracken, John. "Conservation and Resistance in Colonial Malawi: The 'Dead North' Revisited." In *Social History and African Environments*, edited by William Beinart and JoAnn MacGregor. Athens: Ohio University Press, 2003.

Meerow, Sara, Joshua P. Newell, and Melissa Stults. "Defining Urban Resilience: A Review." *Landscape and Urban Planning* 147 (2016).

Meinert, Carmen. "Introduction: Climate and Culture in East-Asia," in *Nature, Environment and Culture in East Asia: The Challenge of Climate Change*, edited by Carmen Meinert, 1–20. Leiden: Brill, 2013.

Menzies, Nicholas. "Three Hundred Years of Taungya: A Sustainable System of Forestry in South China." *Human Ecology* 16.4 (1988): 361–376.

Mertz, Ole, Cheikh Mbow, Anette Reenberg, Lorenzo Genesio, Eric F. Lambin, Sarah D'haen, Malicki Zorom, Kjeld Rasmussen, Drissa Diallo, Bruno Barbier, Ibrahim B. Moussa, Awa Diouf, Jonas Østergaard Nielsen, and Inge Sandholt. "Adaptation Strategies and Climate Vulnerability in the Sudano-Sahelian Region of West Africa." *Atmospheric Science Letters* 12 (2011): 104–108.

Mertz, Ole, Cheikh Mbow, Jonas Østergaard Nielsen, Abdou Maiga, Drissa Diallo, Anette Reenberg, Awa Diouf, Bruno Barbier, Ibrahim Bouzou Moussa, Malicki Zorom, Ibrahim Ouattara, and Daniel Dabi. "Climate Factors Play a Limited Role for Past Adaptation Strategies in West Africa." *Ecology and Society* 15 (2010).

MEWNR. *Analysis of the Charcoal Value Chain in Kenya. Final Report.* Nairobi: Ministry of Environment, Water and Natural Resources, Government of Kenya, 2013. Last modified 2013. Accessed February 25, 2016. http://www.kenyaforestservice.org/documents/redd/Charcoal%20Value%20Chain%20Analysis.pdf.

Middleton, John. *The Central Tribes of the North-Eastern Bantu: The Kikuyu, Including Embu, Meru, Mbere, Chuka, Mwimbi, Tharaka, and the Kamba of Kenya, Ethnographic Survey of Africa—East Central Africa 5.* London: International African Institute, 1953.

Ministry of Environment and Sustainable Development. *Maurice Ile Durable Green Paper: Towards a National Policy for a Sustainable Mauritius.* Port Louis: Government of Mauritius, 2011.

Misilu, Mia N.E., Earl Bailey, Tushiminine Martin Mpinda, Abraham Carleen, Jiangfeng Li, and Liqin Zhang. "Ongoing Informal Settlements in Democratic Republic of Congo: Implementing New Urban Policy for Creating Sustainable Neighborhoods." *Journal of Sustainable Development* 7.5 (2014): 254–265.

Mohammed, Eiman A., Hanan M. Alawad, Khalid A. Elsayed Zeinelabdein, and Abdelhafi G. Ali. "Urban Expansion and Population Growth in Omdurman City, Sudan

Using Geospatial Technologies and Statistical Approaches." *American Journal of Earth Sciences* 2.1 (2015): 1–7.

Mohee, Romeela and Ackmez Mudhoo. "Energy from Biomass in Mauritius: Overview of Research and Applications." In *Waste to Energy: Green Energy and Technology*, edited by Avraam Karagiannidis, 297–321. London: Springer, 2012.

Moore, David. "Marxism and Marxist Intellectuals in Schizophrenic Zimbabwe: How many Rights for Zimbabwe's left? A Comment." *Historical Materialism* 12.4 (2004): 405–425.

Moore, Henrietta L. and Todd Sanders. "Anthropology and Epistemology." In *Anthropology in Theory. Issues in Epistemology*, edited by Henrietta L. Moore and Todd Sanders, 1–18. West Sussex: Wiley-Blackwell, 2014.

Morgan, Gareth. *Images of Organization*. London, Delhi: Sage, 2006.

Morrissey, James W. "Understanding the Relationship between Environmental Change and Migration: The Development of an Effects Framework Based on the Case of Northern Ethiopia." *Global Environmental Change* 23 (2013): 1501–1510.

Mortimore, Michael. "Adapting to Drought in the Sahel: Lessons for Climate Change." *Wiley Interdisciplinary Reviews-Climate Change* 1 (2010): 134–143.

Mortimore, Michael, and William M. Adams. "Farmer Adaptation, Change and 'Crisis' in the Sahel." *Global Environmental Change* 11 (2001): 49–57.

Morton John F. "The Impact of Climate Change on Smallholder and Subsistence Agriculture." *PNAS* 104.50 (2007): 19680–19685.

Mosley, Paul. *The Settler Economies: Studies in the Economic History of Kenya and Southern Rhodesia, 1900–1963*. Cambridge: Cambridge University Press, 1983.

Moyo, Sam, and Paris Yeros, eds. *Reclaiming the Land: The Resurgence of Rural Movements in Africa, Asia and Latin America*. London: Zed Books, 2005.

Mubila, Maurice, Mohamed Safouane Ben Aissa, and Charles Leyeka Lufumpa. "The Middle of the Pyramid. Dynamics of the Middle Class in Africa." *African Development Bank Market Brief*, April 20, 2011. Accessed November 20, 2016. http://www.afdb.org/fileadmin/uploads/afdb/Documents/Publications/The%20Middle%20of%20the%20Pyramid_The%20Middle%20of%20the%20Pyramid.pdf.

Mulwafu, Wapulumuka. *Conservation Song: A History of Peasant-State Relations and the Environment in Malawi, 1860–2000*. Cambridge: The White Horse Press, 2011.

Mundia, C.N. and M Aniya. "Analysis of Land Use/Cover Changes and Urban Expansion of Nairobi City Using Remote Sensing and GIS." *International Journal of Remote Sensing* 26.13 (2005): 2831–2849.

Muriuki, Godfrey. *A History of the Kikuyu, 1500–1900*. London: Oxford University Press, 1974.

Muys, Bart, Lindsey Norgrove, Tena Alamirew, Rhoda Birech, Elisa Chirinian, Yoseph Delelegn, Albrecht Ehrensperger, Carol A. Ellison, Adem Feto, Bernhard Freyer, Jan Gevaert, Simon Gmünder, Raymond E. E. Jongschaap, Manfred Kaufmann, Jodie

Keane, Marc Kenis, Boniface Kiteme, Jackson Langat, Reginald Lyimo, Violet Moraa, Jacinta Muchugu, Aklilu Negussie, Caroline Ouko, Mathurin W. Rouamba, Iria Soto, Manfred Wörgetter, Rainer Zah, and Rigoberto Zetina. "Integrating Mitigation and Adaptation into Development. The Case of Jatropha Curcas in Sub-Saharan Africa." *Global Change Biology, Bioenergy* 6 (2014): 169–171.

Mwangi, Samson W. "Partnerships in Urban Environmental Management. An Approach to Solving Environmental Problems in Nakuru, Kenya." *Environment and Development*, 12.2 (2000): 77–92.

Mz coop. "Managing Your Agricultural Cooperative." Accessed August 9, 2015. http://moodle.itcilo.org/mycoop/.

Nabikolo, D., B. Bashaasha, M.N. Mangheni, and J.G.M. Majaliwa. "Determinants of Climate Change Adaptation among Male and Female Headed Farm Households in Eastern Uganda." *African Crop Science Journal*, 20.2 (2012): 203–212.

Nalepa, Rachel. "The Gobal Land Rush: Implications for Food Fuel, and the Future of Development." *Pardee Papers* 13 (2011): 1–36.

Narayan, Deepa. *Can Anyone Hear Us? Voices from 47 Countries*. Washington DC: The World Bank, 1999.

National Archives of Namibia (NAN), NAO 41. *Native Commissioner Ovamboland (henceforth NCO), Famine Reports*, September–October and December 1930.

National Archives of Namibia (NAN), NAO 41. *Report on Famine Relief Works: Oukwanyama*. September 1930.

National Archives of Namibia (NAN), NAO 18. *Monthly Reports Oukwanyama*. June, July and September–November 1930.

National Archives of Namibia (NAN), NAO 18. *Monthly Report Oukwanyama*. August 1930.

National Archives of Namibia (NAN), NAO 18. *Monthly Reports Oukwanyama*. June–August 1930.

National Archives of Namibia (NAN), NAO 18. *Monthly Report for Ovamboland*. September 1928.

National Archives of Namibia (NAN), NAO 18. *Monthly Report Oukwanyama*. October 1927.

National Archives of Namibia (NAN), NAO 40, Officer Commanding (henceforth O/C) Oshikango to O/C NAO, Oshikango, 11 March 1929, "Stock Losses: Ukuanyama;" and "Particulars Obtained from Headmen in Western Ukuanyama Area re. Famine etc.;" "Particulars in Regard to Famine Ukuanyama Areas received from headmen east of the main road from Ondonga to Oshikango," and Appendixes to O/C Oshikango to O/C NAO, Oshikango 31 July and 8 August 1929.

National Archives of Namibia (NAN), NAO 40–41. *Reports on Famine Relief Works: Oukwanyama*. August 1930–February 1931 and Trust Fund Ovamboland, Statement for February 1930.

National Archives of Namibia (NAN), NAO 41. *Famine Relief Ukuanyama, Works & Issue Schedules*. October–December 1929.

National Archives of Namibia (NAN), NAO 18. *Monthly Reports Oukwanyama*. September–December 1929.

National Archives of Namibia (NAN), NAO 18. *Monthly Reports Oukwanyama*. September–October 1929.

National Archives of Namibia (NAN), NAO 18. *Monthly Report Oukwanyama*. September 1929.

National Archives of Namibia (NAN), NAO 9. O/C NAO to Secretary SWA, 6 September 1929.

National Archives of Namibia (NAN), NAO 40. Director of Works to O/C NAO. Windhoek 17 June 1929.

National Archives of Namibia (NAN), NAO 18. *Monthly Reports Oukwanyama*. May–July 1929.

National Archives of Namibia (NAN), NAO 40, O/C Oshikango to O/C NAO. *Oshikango*, 8 July 1929.

National Archives of Namibia (NAN), NAO 18. *Monthly Reports Oukwanyama*. January–February 1929.

National Archives of Namibia (NAN), NAO 18. *Monthly Report Oukwanyama*, January 1929; NAO 40.

National Archives of Namibia (NAN), NAO 18. *Monthly Report for Ovamboland*, January 1929.

National Archives of Namibia (NAN), NAO 18. *Monthly Reports for Ovamboland*, July–August 1928.

National Archives of Namibia (NAN), Native Affairs Ovamboland (NAO) 18, Monthly Reports for Ovamboland July–September 1928 and NAO 40, "Note of interview with his honour the administrator [of SWA] 9/2/29 in regard to prospective famine conditions in Ovamboland."

National Archives of Namibia (NAN), Native Affairs Ovamboland NAO 40, O/C NAO to Secretary SWA. *Ondangwa*. 18 March 1929.

National Archives of Namibia (NAN), NAO 40, O/C NAO to Secretary South West Africa (SWA). Ondangwa, 15 January 1929 and Assistant Priest St. Mary's Mission to O/C NAO, [Odibo]. 10 January 1929.

Ndagala, D.K. "Operation Imparnati: The Sedentarization of the Pastoral Maasai in Tanzania." *Nomadic People* 10 (1982): 28–39.

Nelson, Fred, Rugemeleza Nshala, W.A. Rodgers. "The Evolution and Reform of Tanzanian Wildlife Management." *Conservation and Society* 5.2 (2007): 232–61.

Nelson, Valerie and Tanya Stathers. "Resilience, Power, Culture, and Climate: A Case Study from Semi-arid Tanzania, and New Research Directions." *Gender & Development* 17.1 (2009): 81–94.

Nelson, Valerie, Kate Meadows, Terry Cannon, John Morton and Adrienne Martin, "Uncertain Predictions, Invisible Impacts, and the Need to Mainstream Gender in Climate Change Adaptations." *Gender & Development* 10.2 (2002): 51–59.

NEPAD. *Agriculture in Africa. Transformation and Outlook.* Addis Ababa: NEPAD, 2013.

Neumann, Roderick P. *Imposing Wilderness: Struggles over Livelihood and Nature Preservation in Africa.* Berkeley: University of California Press, 2000.

Neumayer, Eric, and Thomas Plümper, "The Gendered Nature of Natural Disasters: The Impact of Catastrophic Events on the Gender Gap in Life Expectancy, 1981–2002." *Annals of the Association of American Geographers*, 97 (2007): 551–566.

Nguh, Balgah S. "Land Tenure and Land Use Dynamics in Limbe City, South West Region of Cameroon." *Agriculture Science Development* 2 (2013): 14–24.

Nhemachena, Charles and Rashid Hassan. *Microlevel Analysis of Farmers' Adaptation to Climate Change in South Africa.* Washington, DC: International Food Policy Research Institute, 2008.

Nicholson, Sharon. "On the Question of the 'Recovery' of the Rain in the West African Sahel." *Journal of Arid Environments* 63 (2005): 615–641.

Nicholson, Sharon E. "The Methodology of Historical Climate Reconstruction and its Application to Africa." *Journal of African History* 20.1 (1979): 31–49.

Nicholson, Sharon E. "Climatic Variation in the Sahel and Other African Regions during the Past Five Centuries." *Journal of Arid Environments* 1 (1978): 3–24.

Nielsen, Jonas Ø. and Anette Reenberg. "Exploring Causal Relations: The Societal Effects of Climate Change." *Geografisk Tidsskrift-Danish Journal of Geography* 112 (2012): 89–92.

Nielsen, Jonas Ø. and Anette Reenberg. "Cultural Barriers to Climate Change Adaptation: A Case Study from Northern Burkina Faso." *Global Environmental Change* 20 (2010): 142–152.

Nielsen, Jonas Ø. and Anette Reenberg. "Temporality and the Problem with Singling out Climate as a Current Driver of Change in a Small West African Village." *Journal of Arid Environments* 74 (2010): 464–474.

Nielsen, Jonas Ø. "Drought and Marriage: Exploring the Interconnection between Climate Variability and Social Change through a Livelihood Perspective." In *The Question of Resilience: Social Responses to Climate Change*, edited by Kirsten Hastrup, 159–177. Copenhagen: The Royal Danish Society of Arts and Letters, 2009.

Niemeijer, David and Vaientina Mazzucato. "Soil Degradation in the West African Sahel: How Serious is it?" *Environment: Science and Policy for Sustainable Development* 44 (2002): 20–31.

NIRAS. "Ethiopia Tana Beles Integrated Water Resource Development Programme." Accessed August 10, 2014. http://www.niras.com/business-areas/development-consulting/references/water/tbwme.aspx.

Nowak David J., Eric J. Greenfield, Robert E. Hoehn, and Elizabeth Lapoint. "Carbon Storage and Sequestration by Trees in Urban and Community Areas of the United States" *Environmental Pollution* (2013): 229–236.

Nowak, David J. and Daniel E. Crane. "Carbon Storage and Sequestration by Urban Trees in the USA." *Environmental Pollution* 116.3 (2002): 381–389.

Nowak, David J. "Atmospheric CO2 Rreduction by Chicago's Urban Forest." In *Chicago's Urban Forest Ecosystems: Results of the Chicago Urban Forest Climate Project*, edited by McPherson E. Gregory, David J. Nowak, and Rowan A. Rowntree. General Technical Report NE 186, Radnor, PA, USDA, Forest Services, Northern Research Station, 1994.

Nwaka, Geoffrey I. "Planning Sustainable Cities in Africa." *Canadian Journal of Urban Research* 5.1 (1996).

Nyamnjoh, Francis B. "Blinded by Sight: Divining the Future of Anthropology in Africa." *Africa Spectrum* 47.2–3 (2012): 63–92.

Nyandiga, Charles O. "Strengthening Forest Governance: How Far to Go in Decentralization and Participation?" In *Forest Landscape and Kenya's Vision 2030: Proceedings of the 3rd Annual Forestry Society of Kenya (FSK) Conference and Annual General Meeting Held at the Sunset Hotel, Kisumu. 30th September–3rd October, 2008*, edited by D.O. Ogweno, P.S. Opanga, and A.O. Obara. Kisumu: Forestry Society of Kenya, 2009.

O'Brien, Karen, Jonathan Reams, Anne Caspari, Andrew Dugmore, Maryam Faghihimane, Ioan Fazey, Heide Hackmann, David Manuel-Navarrete, John Marks, Riel Miller, Kari Raivio, Patricia Romeo-Lankao, Hassan Virji, Coleen Vogel, and Verena Winiwarter. "You Say You Want a Revolution? Transforming Education and Capacity Building in Response to Global Change." *Environmental Science & Policy* 28 (2013): 49. Accessed June 8, 2015. doi:10.1016/j.envsci.2012.11.011.

OECD/IEA. *World Energy Outlook 2016*. Paris: International Energy Agency, 2016.

O/C Oshikango Station to O/C NAO. *Oshikango*. 11 March 1929.

Oduol, Peter A. "The Shamba System: An Indigenous System of Food Production from Forest Areas in Kenya." *Agroforestry Systems* 4.4 (1986): 365–373. Accessed October 10, 2014. doi:10.1007/BF00048108.

Odwori, Paul O., Philip M. Nyangweso, and Mark O. Odhiambo. "Alleviating Food Insecurity and Landlessness through Pelis in Kenya." Paper presented at the 4th International Conference of the African Association of Agricultural Economists, Hammamet, Tunisia, September 22–25, 2013.

Ogbazi, Joy U. "Alternative Planning Approaches and the Sustainable Cities Programme in Nigeria." *Habitat International* 40 (2013): 109–118.

Omolo, Nancy A. *Gender and Climate Change-induced Conflict in Pastoral Communities: Case Study of Turkana in North-Western Kenya*, accessed January 1, 2017. http://www.ajol.info/index.php/ajcr/article/viewFile/63312/51195.

Orlove, Ben, Heather Lazarus, Grete K. Hovelsrud, and Alessandra Giannini. "Recognitions and Responsibilities: On the Origins and Consequences of the Uneven Attention to Climate Change around the World." *Current Anthropology* 55.3 (2014): 249–275.

Osman, Mahomed. "Maurice Ile Durable." Published by The United Nations, November 1, 2013. Accessed December 16, 2014. http://sustainabledevelopment.un.org/content/documents/4074durable.pdf.

Otunga, Charles, John Odindi, and Onisimo Mutanga. "Land Use and Land Cover Change in the Fringe of Ethekwini Municipality: Implications for Urban Green Spaces Using Remote Sensing." *South African Journal of Geomatics* 3.2 (2014): 145–162.

Otzelberger, Agnes. *Gender-Responsive Strategies on Climate Change: Recent Progress and Ways Forward for Donors*. Brighton: BRIDGE, Institute of Development Studies, 2011.

Owusu, Kwadwo. "Changing Rainfall Climatology of West Africa: Implications for Rainfed Agriculture in Ghana and Water Sharing in the Volta Basin." PhD Thesis, University of Florida, 2009.

Özden, Kemal and Enwere Chigoyie. "Urbanisation and its Political Challenges in Developing Countries." *Eurasian Journal of Business and Economics* 5.10 (2012): 99–120.

Ozor, Frederick U. "Research Governance and Scientific Knowledge Production in The Gambia." *South African Journal of Science* 110.9/10 (2014): 6. Accessed June 8, 2015. doi:10.1590/sajs.2014/20130185.

Padgham, Jon, Hassan Virji, and Clark Seipt. "Promoting Climate Change Curricular Development at African Universities." *Environmental Development* 5 (2013): 169. Accessed June 8, 2015. doi:10.1016/j.envdev.2012.08.001.

Palmer, Robin and Neil Parsons, ed. *The Roots of Rural Poverty in Central and Southern Africa*. Berkeley: University of California Press, 1977.

Parker-Flynn, James E. "The Intersection of Mitigation and Adaptation in Climate Law and Policy." *University of California, Davis* 38.1 (2014).

Parker, Helen, Nathaniel Mason, Naomi Oates, Roger Calow, William Chadza, and Eva Ludi. *Gender, Agriculture and Water Insecurity*. London: Overseas Development Institute (ODI), 2016.

Parnell, Susan and Ruwani Walawege. "Sub-Saharan African Urbanisation and Global Environmental Change." *Global Environmental Change* 21 (2011): 12–20.

Pataki, Diane E, Margaret M. Carreiro, Jennifer Cherrier, Nancy E. Grulke, Viniece Jennings, Stephanie Pincetl, Richard V. Pouyat, Thomas H. Whitlow, and Wayne C. Zipperer. "Coupling Biogeochemical Cycles in Urban Environments: Ecosystem Services, Green Solutions, and Misconceptions." *Frontiers in Ecology and the Environment* 9.1 (2011): 27–36.

Pearce, Fred. "Will Huge New Hydro Projects Bring Power to African People?." *Yale Environment 360*, May 30, 2013. http://e360.yale.edu/features/will_huge_new_hydro_projects_bring_power_to_africas_people.

Peet, Richard and Michael Watts. "Liberation Ecology: Development, Sustainability, and Environment in an Age of Market Triumphalism." In *Liberation Ecologies: Environment, Development, Social Movements*, edited by Richard Peet and Michael Watts, 1–45. London: Routledge, 1996.

Peires, J.P. *The Dead will Arise: Nongqawuse and the Great Xhosa Cattle Killing Movement of 1856–7*. Johannesburg: Ravan Press, 1989.

Peng, Shushi, Shilong Piao, Philippe Ciais, Pierre Friedlingstein, Catherine Ottle, François-Marie Bréon, Huijuan Nan, Liming Zhou, and Ranga B. Myneni. "Surface Urban Heat Island across 419 Global Big Cities." *Environmental Science & Technology* 46.2 (2012): 696–703.

Perry, Alex. "Africa Rising." *Time Magazine*, December 3, 2012.

Population Reference Bureau (PRB). "Africa and the Demographic Divided." Last modified April 2013. Accessed February 2015. http://www.prb.org/Publications/Reports/2013/africa-demographic-dividend.aspx.

Potts, Deborah. "Challenging the Myths of Urban Dynamics in Sub-Saharan Africa: The Evidence from Nigeria." *World Development* 40 (2012): 1382–1393.

Potts, Deborah. *Whatever Happened to Africa's Rapid Urbanization?*. Africa Research Institute, Counterpoints, 2012.

Pouyat, Richard V., Ian D. Yesilonis, and David J. Nowak. "Carbon Storage by Urban Soils in the United States." *Journal of Environmental Quality* (2006): 1566–1575.

Pratt, Mary L. "Arts of the Contact Zone." *Profession* (1991): 33–40.

Quagraine, Victor K. "Urban Landscape Depletion in the Kumasi Metropolis" In *Future of the Tree: Towards Growth and Development of Kumasi*, ed. Kwasi K. Adarkwa. Kumasi: Knust University Press, 2011.

Rahman Osmani. Siddiqur. *Food Security, Poverty and Women: Lessons from Rural Asia*. IFAD, 2010, accessed October 5. http://www.ifad.org/gender/thematic/rural/rural_2.htm.

Rain, David *Eaters of the Dry Season. Circular Labor Migration in the West African Sahel*. Boulder, Colorado: Westview Press, 1999.

Rakodi, Carole, Rose Gatabaki-Kamau, and Nick Devas. "Poverty and Political Conflict in Mombasa." *Environment and Urbanization* 12.1 (2000): 153–1170.

Ramankutty, Navin, Lisa Graumlich, Frédéric Achard, Diogenes Alves, Abha Chhabra, Ruth S. De Fries, Jonathan A. Foley, Helmut Geist, Richard A. Houghton, Kees K. Goldewijk, Eric F. Lambin, Andrew Millington, Kjeld Rasmussen, Robin S. Reid, and Billie L. Turner II. "Global Land-cover Change: Recent Progress, Remaining Challenges" In *Land Use and Land-cover Change—Local Processes and Global Impacts*, edited by Eric F. Lambin and H. Geist. Berlin: Springer, 2010.

Ramessur, Roshan T., R.P. Gunputh, and Taruna S. Ramessur. "Climate Compatible Development: Legal Implications in the Coastal Zone and Inclusive Development for Mauritius." *Journal of Coastal Development* 16 (2013): 3.

Ramjeawon, Toolseeram. "Life Cycle Assessment of Electricity Generation from Bagasse in Mauritius." *Journal of Cleaner Production* 16 (2008): 1727–1734.

Rasmussen, LauraV. and Anette Reenberg. "Collapse and Recovery in Sahelian Agro-Pastoral Systems: Rethinking Trajectories of Change." *Ecology and Society* 17 (2012): 14.

Ravi, Rajan S. *Modernizing Nature: Forestry and Imperial Eco-Development 1800–1950.* Oxford: OUP, 2006.

Read, Hannah, Muyeye Chambwera, and Laurel Murray. *Tried and Tested: Learning from Farmers on Adaptation to Climate Change.* London: IIED, 2013.

Reenberg, Anertte, Laura V. Rasmussen, and Jonas Ø. Nielsen. "Causal Relations and Land Use Transformation in the Sahel: Conceptual Lenses for Processes, Temporal Totality and Inertia." *Geografisk Tidsskrift-Danish Journal of Geography* 112 (2012): 159–173.

Richards, Michael. *Social and Environmental Impacts of Agricultural Large-Scale Land Acquisitions in Africa—With a Focus on West and Central Africa.* Washington, DC: Rights and Resources Initiative, 2013.

Richey, Rita C. "Reflections on the 2008 AECT Definitions of the Field." *TechTrends* 52.1 (2008): 24–25. Last accessed June 8, 2015. doi:10.1007/s11528-008-0108-2.

Ripplinger, Pascal. "Renewable Energies in Sub-Saharan Africa: A Case Study of the Gambia." MSc. Thesis, Department of Geography, University of Bonn, 2014.

Robbins, Paul. *Political Ecology: A Critical Introduction.* Oxford: Blackwell, 2004.

Robertson, Maria L.B. and Cecilie Rubow. "Engaged World-Making: Movements of Sand, Sea, and People at two Pacific Islands." In *Anthropology and Nature,* edited by Kirsten Hastrup, 63–78. New York: Routledge, 2014.

Rocheleau, Dianne, Barbara Thomas-Slayter, and Esther Wangari. *Feminist Political Ecology.* London: Routledge, 1996.

Rodenberg, Birte. *Climate Change Adaptation from a Gender Perspective.* Bonn: German Development Institute, 2009.

Roncoli, Carla, Keith Ingram, Christine Jost, and Paul Kirshen. "Meteorological Meanings: Farmers' Interpretations of Seasonal Rainfall Forecasts in Burkina Faso." In *Weather, Culture, Climate,* edited by Sarah Strauss and Benjamin S. Orlove. Oxford: Berg, 2003.

Roth, Susan and Cindy Trumbore. *The Mangrove Tree. Planting Trees to Feed Families.* New York: Lee and Low Books, 2011.

Roussos, Peter. *Zimbabwe: An Introduction to the Economics of Transformation.* Harare: Baobab Books, 1988.

Rudiak-Gould, Peter. "The Influence of Science Communication on Indigenous Climate Change Perception: Theoretical and Practical Implications." *Human Ecology* 42.1 (2014): 75–86.

Rudiak-Gould, Peter. *Climate Change and Tradition in a Small Island State: The Rising Tide.* New York: Routledge, 2013.

Rudiak-Gould, Peter. " 'We Have Seen It with Our Own Eyes': Why We Disagree about Climate Change Visibility." *Weather, Climate and Society* 5 (2013): 120–132.

Rudiak-Gould, Peter. "Promiscuous Corroboration and Climate Change Translation: A Case Study from the Marshall Islands." *Global Environmental Change* 22.1 (2012): 46–54.

Rudiak-Gould, Peter. "Climate Change and Anthropology. The Importance of Reception Studies." *Anthropology Today* 27.2 (2011): 9–12.

Sachedina, Hassanali. T. "Wildlife is Our Oil: Conservation, Livelihoods and NGOs in the Tarangire Ecosystem,Tanzania." PhD diss., University of Oxford, 2008.

Sachikonye, Lloyd M. *Labour Legislation in Zimbabwe: Historical and Contemporary Perspectives*. Harare: Zimbabwe Institute of Development Studies, 1985.

Sahalu, Atalel G. *Analysis of Urban Land Use and Land Cover Changes: A Case Study in Bahir Dar, Ethiopia*. MSc diss., Universidade Nova de Lisboa, Lisbon, Portugal, 2014.

Saringe, Emanuel. "Impact of Climate Change to Pastoralist and Hunter-Gatherer Communities of Tanzania. A Case Study of Kiteto and Simanjiro Districts." *Pastoralist Indigenous Non Governmental Organizations (PINGO's) Forum*, 2011.

Satterthwaite, David. "The Transition to a Predominantly Urban World and its Underpinnings." *Human Settlements Discussion Paper Series: Urban Change* 4 (2007): 99.

Satterthwaite, David. "Reducing Urban Poverty: Constraints on the Effectiveness of Aid Agencies and Development Banks and Some Suggestions for Change." *Environment and Urbanization* 13.1 (2001): 137–157.

Schaffnit-Chatterjee, Claire. "High Population Growth in Africa: A Mixed Blessing." *DHL. Insight in Business in Africa*. Last modified September 9, 2013. Accessed February 2015. http://www.howwemadeitinafrica.com/high-population-growth-in-africa-a-mixed-blessing/30043/.

Schmidt, Stephan. "Cultural Influences and the Built Environment. An Examination of Kumasi Ghana." *Journal of Urban Design* 10.3 (2005): 353–370.

Schneider, Birgit, and Thomas Nocke. "Image Politics of Climate Change. Introductio." In *Image Politics of Climate Change. Visualizations, Imaginations, Documentations*. Edited by Birgit Schneider, and Thomas Nocke. Bielefeld: transcript, 2014.

Schneidewind, Uwe and Angelika Zahrnt. *The Politics of Sufficiency. Making It Easier to Live the Good Life*. Munich: oekom, 2014.

Schreiber, Dominik. "Der Klimawandel: Aufstieg eines globalen Narrativs." In *Figuren des Globalen: Weltbezug und Welterzeugung in Literatur, Kunst und Medien*, edited by Christian Moser and Linda Simonis. Göttingen: Bonn University Press at V&R unipress, 2014.

Scoones, Ian, Nelson Marongwe, Blasio Mavedzenge, Felix Murimbarimba, Jacob Mahenehene, and, Chrispen Sukume. "Livelihoods after Land Reform in Zimbabwe: Understanding Processes of Rural Differentiation." *Journal of Agrarian Change* 12.4 (2012): 503–527.

Scott, James C. *The Moral Economy of the Peasant: Rebellion and Subsistence in Southeast Asia*. New Haven, Conn.: Yale University Press, 1976.

Scudder, Thayer. *The Ecology of the Gwembe Tonga*. Manchester: Manchester University Press, 1962.

Sebego, Reuben J., and Thando D. Gwebu. “Patterns, Determinants, Impacts and Policy Implications of the Spatial Expansion of an African Capital City: The Greater Gaborone Example.” *International Journal of Sustainable Built Environment* 2.2 (2013): 193–208.

Seegobin, Ram, and Lindsey Collen. “The La Chaumière Waste-to-Energy Project Criticized.” *Lalit*, April 23, 2009. Accessed December 16, 2014. https://www.lalitmauritius.org/en/newsarticle/843/the-la-chaumire-waste-to-energy-project-criticized/.

Senior, Catherine, Andrew Turner, Simon Vosper, Richard Washington, and Richard Graham. “Improving Climate Modelling for Africa,” in *Africa's Climate*, edited by Future Climate for Africa, 38–43. Cape Town, 2016.

Seto, Karen C., Burak Güneralp, and Lucy R. Hutyra. “Global Forecasts of Urban Expansion to 2030 and Direct Impacts on Biodiversity and Carbon Pools.” *Proceedings of the American Academy of Sciences* 109.40 (2012): 16083–16088.

Seto, Karen C., Michail Fragkias, Burak Güneralp, and Michael K Reilly. “A Meta-analysis of Global Urban Land Expansion.” *Plos One* 6.8 (2011): 1–9.

Shanahan, Mike, Willie Shubert, Cameron Scherer, Teresa Corcoran. *Climate Change in Africa: A Guidebook for Journalists (Paris: UNESCO, 2013); Ampomah, Gifty, and Tahia Devisscher, Adaptation Toolkit. Guidebook for Researchers and Adaptation Practitioners Working with Local Communities*. Dakar: ENDA, 2013.

Shemdoe, Riziki, Gabriel Kassenga, and Stephen Mbuligwe. “Implementing Climate Change Adaptation and Mitigation Interventions at the Local Government Levels in Tanzania: Where Do We Start ?.” *Current Opinion in Environmental Sustainability* 13 (2015): 32–41.

Shepherd, Andrew. *The Geography of Poverty, Disasters and Cimate Extremes in 2030*. London: Overseas Development Institute (ODI), 2013.

Showers, Kate B. *Imperial Gullies: Soil Erosion and Conservation in Lesotho*. Athens, Ohio University Press, 2005.

Siemens. “African Green City Index.” Last modified 2011. Accessed February 26, 2016. http://www.siemens.co.za/sustainable-development/pdf/African-Green-City-Index.pdf.

Simone, Abdou M. “The Urbanity of Movement: Dynamic Frontiers in Contemporary Africa.” *Journal of Planning Education and Research* 31.4 (2011): 379–391.

Smit, Barry, and Johanna Wandel. “Adaptation, Adaptive Capacity and Vulnerability.” *Global Environmental Change* 16 (2006): 282–292.

Smucker, Thomas A., Ben Wisner, Adolfo Mascarenhas, Pantaleo Munishi, Elizabeth E. Wangui, Gaurav Sinha, Daniel Weiner, Charles Bwenge, Eric Lovell. “Differentiated

Livelihoods, Local Institutions, and the Adaptation Imperative: Assessing Climate Change Adaptation Policy in Tanzania." *Geoforum* 59 (2015): 39–50.

Social Development Department. *Gender-Responsive Social Analysis: A Guidance Note. Incorporating Social Dimensions into Bank-Supported Projects*. Washington, DC: The World Bank, 2005.

Spear, Thomas. "Introduction." In: *Being Maasai,* edited by Thomas Spear and Richard Waller, 1–24. London: Villiers Publications, 1993.

Spear, Thomas, and Richard Waller, edited by *Being Maasai. Ethnicity & Identity in East Africa.* London: Villiers Publications, 1993.

Spradley, James P. *The Ethnographic Interview*. Fort Worth: Harcourt Brace Jovanovich College Publishers, 1979.

Stainforth, D.A., Allen, M.R., Tredger, E.R. and Smith, L.A. "Confidence, Uncertainty and Decision-Support Relevance in Climate Predictions." *Phil. Trans. R. Soc. A* 365 (2007): 2145–2161. doi:10.1098/rsta.2007.2074.

St. Antoine, Pamela de. "Engineer Felix Ah-Kee: Mauritius Should Look at Alternatives to Proposed Coal Plant." *Le Mauricien*, April 7, 2013. Accessed December 16, 2014. http://www.lemauricien.com/article/engineer-felix-ah-kee-mauritius-should-look-alternatives-proposed-coal-plant.

Statistics Mauritius. *Energy and Water Statistics.* Government of Mauritius, 2013. February 16, 2019. http://statsmauritius.govmu.org/English/StatsbySubj/Pages/Energy-and-Water.aspx.

Stehr, Nico and Hans von Storch. *Über Klima, Wetter, Mensch.* München: Beck, 1999.

Stitcher, Sharon. *Migrant Labour in Kenya: Capitalism and African Response 1895–1975.* Harlow: Longman, 1982.

Strauss, Sarah, Stephanie Rupp, and Thomas Love. "Powerlines: Cultures of Energy in the Twenty-First Century." In *Cultures of Energy: Power, Practices, Technology*, edited by Sarah Strauss, Stephanie Rupp and Thomas Love, 10–40. Walnut Creek: Left Coast Press, 2013.

Strauss, Sarah and Ben Orlove. "Up in the Air: The Anthropology of Weather and Climate." In *Weather, Climate, Culture*, edited by Sarah Strauss and Ben Orlove. Oxford: Berg, 2003.

Sullivan, Leann M. "Five Challenges to the Future of Transboundary Water Governance." *Issues in Brief* 29 (2014).

Sutcliffe, John.V. and Y.P. Parks. *The Hydrology of the Nile*. Wallingford, Oxfordshire: International Association of Hydrological Sciences (IAHS), 1999.

Swai, Okuli W., Jonathan S. Mbwambo, and Flavianus T. Magayane. "Gender and Adaptation Practices to the Effects of Climate Change in Bahi and Kondoa Districts Dodoma Region, Tanzania." *Journal of Sustainable Development* 5.12 (2012).

Sylla, L., D. Xiong, H. Y. Zhang, and S. T. Bangoura. "A GIS Technology and Method to Assess Environmental Problems from Land Use/Cover Changes: Conakry, Coyah and Dubreka Region Case Study." *Egyptian Journal of Remote Sensing and Space Science* 15.1 (2012): 31–38.

Tanzania Natural Resources Forum (TNRF). "Integrating Pastoralist Livelihoods and Wildlife Conservation? Options for Land Use and Conflict Resolution in Loliondo Division, Ngorongoro District." *Policy Paper,* 2011.

Tanzania Natural Resources Forum (TNRF) and International Institute for Environment and Development (IIED). "Implications of Climate Change for Drylands Planning in Tanzania at District and National Levels: Opportunities and Challenges." Workshop Report held in Arusha, May 29–31, 2012.

Taylor, Anna, and Camaren Peter. "Strengthening Climate Resilience in African Cities: A Framework for Working with Informality." *Climate & Development Knowledge Network (CDKN) Working Paper* (2014).

TEEB. *The Economics of Ecosystems and Biodiversity: Mainstreaming the Economics of Nature: A Synthesis of the Approach, Conclusions and Recommendations of TEEB.* London: Earthscan, 2010.

Tiffen, Michael and Francis Gichuki. *More People, Less Erosion: Environmental Recovery in Kenya.* Chichester: Chichester: John Wiley and Sons, 1994.

Tilley, Helen. *Africa as a Living Laboratory: Empire, Development, and the Problem of Scientific Knowledge, 1870–1950.* Chicago: University of Chicago Press, 2011.

Tilley, Helen. "Global Histories, Vernacular Science, and African Genealogies; or, Is the History of Science Ready for the World?" *Isis* 101.1 (2010): 114–115.

Tischler, Julia. *Light and Power for a Multiracial Nation: The Kariba Dam Scheme in the Central African Federation*. Cambridge: Palgrave McMillian, 2013.

The Economist. "Africa Rising. The Hopeful Continent." *The Economist,* December 1, 2011. Accessed February 2015. http://www.economist.com/node/21541015.

The Economist. "The Hopeless Continent." *The Economist,* May 13, 2000.

Thorlakson, Tannis, Henry Neufeldt, and François Collart Dutilleul. "Reducing Subsistence Farmers' Vulnerability to Climate Change: Evaluating the Potential Contributions of Agroforestry in Western Kenya." *Agric Food Security* 1.15 (2012).

Thorsen, Dorte. "Weaving In and Out of Employment and Self-Employment: Young Rural Migrants in the Informal Economy of Ouagadougou." *International Development Planning Review* 35 (2013): 203–218.

Time Magazine. "Africa Rising." *Time Magazine,* March 30, 1998.

Tompkins, Emma L. and W. Neil Adger. "Does Adaptive Management of Natural Resources Enhance Resilience to Climate Change?." *Ecology and Society* 9.2 (2004).

Toussaint, Auguste. *History of the Indian Ocean.* Translated by June Guicharnaud. London: Routledge & Kegan Paul, 1966.

Tropp, Jacob A. *Natures of Colonial Change: Environmental Relations in the Making of the Transkei.* Athens: Ohio University Press, 2006.

Tschakert, Petra. "Views from the Vulnerable: Understanding Climatic and Other Stressors in the Sahel." *Global Environmental Change* 17 (2007): 381–396.

Turnpenny, John, Marvis Jones, and Irene Lorenzoni. "Where Now for Post-Normal Science?: A Critical Review of its Develop-ment, Definitions, and Uses." *Science, Technology, & Human Values* 36.3 (2011): 287–330. Accessed June 8, 2015. doi:10.1177/0162243910385789.

Tved, Terje. "Hydrology and Empire: The Nile, Water Imperialism and the Partition of Africa." *The Journal of Imperial and Commonwealth History* 39,2, 2011, 173–194.

Twyman, Jennifer, Green, Molly, Bernier, Quinn, Kristjanson, Patricia M., Russo, Sandra, Tall, Arame, Ampaire, Edidah L., Nyasimi, Mary, Mango, Joash, McKune, Sarah, Mwongera, Caroline, Badiane Ndour, Ndèye Yacine. *Adaptation Actions in Africa: Evidence that Gender Matters*. (Copenhagen: CCAFS, 2014).

Uekötter, Frank. "You Ain't Seen Nothing Yet: A Death-Defying Look at the Future of the Climate Debate." In *Grounding Global Climate Change. Contributions from the Social and Cultural Sciences*,175–181. Dordrecht: Springer, 2015.

Ujoh, Fanan, Olarewaju O. Ifatimehin, and Adams N. Baba. "Detecting Changes in Land Use/Cover of Umuahia, South-Eastern Nigeria, Using Remote Sensing and GIS Techniques." *Confluence Journal of Environmental Studies* 6 (2011): 72–80.

Unger, Alon and Lee W. Riley. "Slum Health: From Understanding to Action." *PLoS Medicine* 4.10 (2007): 1561–1566.

UN-Habitat. *Planning Sustainable Cities. UN-Habitat Practices and Perspectives*. Nairobi: United Nations Human Settlements Programme, 2010. Accessed February 25, 2016. https://portals.iucn.org/library/sites/library/files/documents/Man-Dev-143.pdf.

UN-Habitat. "The UN-Habitat Urban Population Database: UN-Habitat for a Better Future." Accessed August 27, 2015. http://urbandata.unhabitat.org/download-data/#/countries/1.

UN International Year of Small Island Developing States 2014. Accessed December 16, 2014. http://www.un.org/en/events/islands2014/smallislands.shtml.

Union (of South Africa) Government Representative Namacunde (henceforth UNG). *Union Administration 2 folder 1922–1946*. UGR Oshikango, *Monthly Report* December 1929.

United Nations. "World Urbanization Prospects: 2014 Revisions, Highlights." (ST/ESA/SER.A/352) (2014).

United Nations Department of Economic and Social Affairs (UN DESA). *The World's Women 2010, Trends and Statistics*. New York: UN DESA, 2010.

United Nations Development Programme (UNDP). *The Rise of the South: Human Progress in a Diverse World*. New York: United Nations Development Programme, 2013.

United Nations Development Programme (UNDP). *Gender, Climate Change and Community-Based Adaptation*. New York: UNDP, 2010.

United Nations Development Programme (UNDP) and World Health Organisation WHO, *The Energy Access Situation in Developing Countries*. New York: UNDP and WHO, 2009.

United Nations Development Programme (UNDP). *Fighting Climate Change: Human Solidarity in a Divided World* Human Development Report 2007/2008. New York: UNDP, 2007.

United Nations Development Programme (UNDP). "UNDP Human Development Index." Accessed December 16, 2014. http://data.undp.org/dataset/Table-1-Human-Development-Index-and-its-components/wxub-qc5k.

United Republic of Tanzania (URT). *National Adaptation Program of Action* (NAPA for the UNFCCC), 2007.

United States Environment Protection Agency. "Water Topics." Last modified April 12, 2017. Accessed 8 August 2014. http://water.epa.gov/type/watersheds/whatis.cfm https://www.epa.gov/environmental-topics/water-topics.

University of Reading. "Climate Change. Impact and Responses," developed in cooperation with seven East African universities and launched in 2015. Accessed August 28, 2015. http://www.reading.ac.uk/ssc/resource-packs/ccir/.

UNU-IEH. "Experts." Last modified in 2016. Accessed February 28, 2016. http://ehs.unu.edu/experts/researchexperts.

van Aelst, Katrien. *Gender, Households and Climate Change: Adaptation Decision-Making in the Morogoro Region of Tanzania*. PhD Thesis. University of Antwerp: Institute for Development Policy and Management (IOB), 2016.

van Beek, Walter E.A. "Echoes of the End. Myth, Ritual and Degradation." *Focaal* 35 (2000): 29–51.

van Boxel, John. "Uncertainties in Modelling Climate Change." In *The Impact of Climate Change on Drylands, with a Focus on West Africa*, edited by Antonius J. Dietz, Ruerd Ruben, and Jan Verhagen. Dordrecht, Boston, London: Kluwer, 2004.

van den Born, Gertjan, Rik Leemans, and Michael Schaeffer. "Climate Change Scenarios for Dryland West Africa, 1990–2050." In *The Impact of Climate Change on Drylands, with a Focus on West Africa*, edited by Antonius J. Dietz, Ruerd Ruben, and Jan Verhagen. Dordrecht, Boston, London: Kluwer, 2004.

van der Land, Victoria and Diana Hummel. "Vulnerability and the Role of Education in Environmentally Induced Migration in Mali and Senegal." *Ecology and Society* 18 (2013): 14.

van Sittert, Lance. "The Seed Blows About in Every Breeze': Noxious Weed Eradication in the Cape Colony, 1860—1909." *Journal of Southern African Studies*, 26.4 (2000): 655–674.

Vaughan, Megan. *The Story of an African Famine: Gender and Famine in Twentieth-Century Malawi*. Cambridge: Cambridge University Press, 1987.

Vermeiren, Karolien, Anton Van Rompaey, Maarten Loopmans, and Serwajja Eria. "Urban Growth of Kampala, Uganda: Pattern Analysis and Scenario Development." *Landscape and Urban Planning* 106.2 (2012): 199–206.

Versi, Anver. "Out of Africa. The Rise of the African Economy." *Global Briefing*, January 2013. Accessed February 2015. http://www.global-briefing.org/2013/01/out-of-africa-the-rise-of-the-african-economy/.

Vincent, Katharine, Tracy Cull, and Emma R.M. Archer. "Gendered Vulnerability to Climate Change in Limpopo Province, South Africa." In *Gender and Climate Change: An Introduction*, edited by Irene Dankelman. London, Sterling: Earthscan 2010.

Vogel, Coleen and Karen O'Brien. "Vulnerability and Global Environmental Change: Rhetoric and Reality," *AVISO* 13 (2004). http://www.ihdp.unu.edu/docs/Publications/GECHS/Aviso/Aviso13-2004.pdf.

von Hellermann, Pauline. "Things Fall Apart? Management, Environment and Taungya Farming in Edo State, Southern Nigeria." *Africa* 77.03 (2007). Accessed October 3, 2014. doi:10.3366/afr.2007.0052.

von Storch, Hans, and Werner Krauß. *Die Klimafalle. Die gefährliche Nähe von Politik und Klimaforschung*. München: Hanser, 2013.

Wafula, Munialo E. and Chege Moses Gichuho. "The Pattern of Expansion of Urban Residential in Eldoret Municipality, Kenya." *International Journal of Science and Research* 2.4 (2013): 586–589.

Walker, Brian, Stephen Carpenter, John Anderies, Nick Abel, Graeme S. Cumming, Marco Janssen, Louis Lebel, Jon Norberg, Garry D. Peterson, and Rusty Pritchard. "Resilience Management in Social-Ecological Systems: A Working Hypothesis for a Participatory Approach." *Conservation Ecology* 6 (2002).

Warner, Koko and Tamer Afifi. "Where the Rain Falls: Evidence from 8 Countries on How Vulnerable Households Use Migration to Manage the Risk of Rainfall Variability and Food Insecurity." *Climate and Development* 6 (2014): 1–17.

Washington, Richard, Rachel James, Helen Pearce, Wilfried M. Pokam, and Wilfran Moufouma-Okia. "Congo Basin Rainfall Climatology: Can We Believe the Climate Models?" *Philosophical Transactions B* 368.1625 (2013). Accessed July, 24, 2017. doi: 10.1098/rstb.2012.0296.

Washington, Richard, Mike Harrison, Declan Conway, Emily Black, Andrew Challinor, David Grimes, Richard Jones, Andy Morse, Gillian Kay, and Martin Todd. "African Climate Change: Taking the Shorter Route." *Bulletin of the American Meteorological Society* 87.10 (2006): 1355–1366.

Washington, Richard, Mike Harrison, and Declan Conway. *African Climate Report: A Report Commissioned by the UK Government to Review African Climate Science, Policy and Options for Action*. London and Oxford, 2004.

Watkins, Thayer. *The Direct and Indirect Effects of Increased Carbon Dioxide on Plant Growth*, San José State University. Accessed February 2015. http://www.sjsu.edu/faculty/watkins/CO2plants.htm.

Watson, Vanessa and Agbola Babatunde. "Who will Plan Africa's Cities?." *Africa Research Institute, Counterpoints* (2013).

Watts, Michael. *Silent Violence: Food, Famine and Peasantry in Northern Nigeria*. Berkeley: University of California Press, 1983.

Webb, James. *Desert Frontier: Ecological and Economic Change along the Western Sahel, 1600–1850*. Madison: University of Wisconsin Press, 1995.

Weber, Christiane, and Anne Puissant. "Urbanization Pressure and Modeling of Urban Growth: Example of the Tunis Metropolitan Area." *Remote Sensing of the Environment* 86.3 (2003): 341–352.

WEDO, ABANTU (Ghana), Action Aid Bangladesh and ENDA (Senegal). *Gender, Climate Change and Human Security: Lessons from Bangladesh, Ghana and Senegal*. Report prepared for Hellenic Ministry of Foreign Affairs (MFA) and the Hellenic Foundation for European and Foreign Policy ELIAMEP, 2008.

Weiser, Florian, Michael Bollig, Martin Doevenspeck, and Detlef Müller-Mahn. "Translating the 'Adaptation to Climate Change' Paradigm—The Politics of a Travelling Idea in Africa." *The Geographical Journal* 180.2 (2014): 111–119. Accessed January 23, 2015. doi: 10.1111/geoj.12037.

Welzer, Harald. *Climate Wars: What People Will be Killed for in the 21st Century*, trans. Patrick Camiller. Cambridge: Polity Press, 2012 (Original: *Klimakriege. Wofür im 21. Jahrhundert getötet wird*. Frankfurt a.M.: S. Fischer, 2008.

White, Iain, and Paul O'Hare. "From Rhetoric to Reality: Which Resilience, why Resilience, and Whose Resilience in Spatial Planning?." *Environment and Planning C: Government and Policy* 32.5 (2014): 934–950.

Wikipedia. "Water Supply and Sanitation in Burkina Faso." Last modified 2010. Accessed in February 2015. http://en.wikipedia.org/wiki/Water_supply_and_sanitation_in_Burkina_Faso.

Wisner, Ben, Adolfo Mascarenhas, Charles Bwenge, Thomas A. Smucker, Elizabeth E. Wangui, Daniel Weiner, D., and Pantaleo Munishi. "Let Them Eat (Maize) Cake: Climate Change Discourse, Misinformation and Land Grabbing in Tanzania." Paper presented at "The International Conference on Global Land Grabbing II," New York, October 17–19, 2012.

Witcomb, Mark and Peter Dorward. "An Assessment of the Benefits and Limitations of the Shamba Agroforestry System in Kenya and of Management and Policy Requirements for Its Successful and Sustainable Reintroduction." *Agroforestry Systems* 75.3 (2009). Accessed October 3, 2014. doi:10.1007/s10457-008-9200-z.

World Bank. *State of the Clean Cooking Energy Sector in Sub-Saharan Africa.* Washington DC: World Bank, 2012.

World Bank. *Wood-based Biomass Energy Development for Sub-Saharan Africa.* Washington: The World Bank, 2011.

World Bank, FAO, and IFAD. *Gender in Agriculture Source Book.* Washington:World Bank, 2009.

World Bank. *Engendering Development: Through Gender Equality in Rights, Resources, and Voice. A World Bank Policy Research Report.* Washington DC: World Bank, 2001.

World Bank. *Entering the 21st Century: World Development Report 1999/2000.* New York: Oxford University Press, 2000.

World Commission on Environment and Development (WECD). *WCED Brundtland Report.* Published March 20, 1987. Accessed 16 December 2014. http://www.un-documents.net/our-common-future.pdf.

World Food Programme. "Horn of Africa Crisis." Last modified February 2016. Accessed February 28, 2016. https://www.wfp.org/crisis/horn-of-africa.

Word Health Organization (WHO). "Climate and Health." Accessed May 5, 2015. http://www.who.int/globalchange/news/fsclimandhealth/en/.

World Meteorological Organisation (WMO). "WMO Statement on the State of the Global Climate in 2016." 2017. Accessed August 2, 2017. https://public.wmo.int/en/resources/library/wmo-statement-state-of-global-climate-2016.

Wu, Weicheng, Marie-Francoise Courel, and Jeannine Le Rhun. "Application of Remote Sensing to the Urban Expansion Analysis for Nouakchott, Mauritania." *Geocarto International* 18.1 (2003): 17–24.

Yamin, Farhana, Atiq Rahman, and Saleemul Huq. "Vulnerability, Adaptation and Climate Disasters: A Conceptual Overview." *IDS Bulletin* 36 (2005):1–14.

Zongo, Mahamadou. "Introduction." In *Les enjeux autour de la disapora burkinabé: Burkinabé á l'étrangers au Burkina Faso*, edited by Mahamadou Zongo, 11–14. Paris: L'Harmattan, 2010.

Zupancic, Tara, Claire Westmacott, and Mike Bulthuis. *The Impact of Green Space on Heat and Air Pollution in Urban Communities: A Meta-narrative Systematic Review.* Vancouver: Suzuki Foundation, 2015.

zur Heide, Friedrich. *Feasibility Study for a Lake Biosphere Reserve, Ethiopia. BfN-Skripten 317.* Bonn: Federal Agency for Native Conservation, 2012.

Index

Adama Barrow 210
adaptation xv, 4–6, 16, 19, 124, 149, 159, 202, 211, 245, 248
  individual 3
  gender-specific 149
  governance of 209
  masters of 33
  politics of xvii, 210
  projects 160–161
  strategies 2, 35, 37, 43, 98, 100, 119, 147, 165
Addis Ababa, Ethiopia xix, 269, 273, 276, 283
African Union xx, 177, 181
agency
  environmental 73
  human 75–76, 97
  of women 154, 158, 169, 172
Agalega, Mauritius 182–183
agriculture xv, 35, 42, 49–51, 61–62, 66, 75, 100, 125, 142, 147–148, 159–161, 163, 183–185, 213, 220, 232–234, 237–240, 247, 264
  cut-and-burn 64, 185
  education in 55, 62
  industrial 78, 257
  intensive 243
  monoculture/monocrop a. *See also* biodiversity 68
  rain-fed 65, 124, 144, 150, 164
  subsistence 124, 144, 148
  sustainable 59, 68
  traditional 78
  types of 263
  urban 285
ahoi 106–107, 120
air pollution 168, 185, 240, 295
Albion, Mauritius 189, 190, 195
America
  Anglo- 52
  North xix, 45, 52, 76, 80, 191, 220, 235, 252, 280
  Latin/South xv
Angola 80–97, 256, 260
Anthropocene 7, 200
anthropology 29, 46
  cultural 207
  ecological 177–178
  task of 27
Atlantic Ocean. *See* ocean

bagasse 185–192. *See also* plantation, sugarcane; biofuel
Bamako, Mali 276
biodiversity. *See also* agriculture, monocrop 227, 234, 243–144, 262, 291, 295
  aquatic 255
  Biodiversity Action Journalists Gambia 211
  degredation of 184–185, 206
  preservation of 187–188, 248
biofuel xv, 248–249. *See also* plantation, sugarcane, *and* biofuel
birth rate 138, 270. *See also* child mortality
Bobo-Dioulasso, Burkina Faso 129
Brazzaville, Republic of the Congo 269
Brundtland Report 178, 186
Burkina Faso 271, 282
Burma 99–100, 110, 114
Burundi 273
burning fields. *See* agriculture, cut-and-burn

Cairo, Egypt 231, 269
Cape Flats 79
Cape of Good Hope 182
Cape Verde 164, 181, 184
Cameroon xvi
carbon dioxide (CO_2). *See* greenhouse gases
Caribbean 181
cattle. *See* livestock
Central Electricity Board (CEB) 189
Chagos Archipelago xviii, 182
child mortality 226, 228, 268, 271. *See also* birth rate
China xvii, 99, 107, 117, 181, 129, 133, 135
Christianity 33, 44–45
  conversion to 45

climate
conference 53, 58, 181
global 8
history vii, 253
indicators 56–57, 265
influences/effects 63, 65, 97, 156, 159, 236
policy vii, 9, 16, 18, 199, 209, 240, 245, 248, 264
protection vii, 293
resilience 4–5, 41, 154, 157, 160, 209, 265–282, 293–294. *See also* urban climate resilience
variability vi, 10, 12, 24, 26, 43, 57, 66, 70, 123–126, 136, 140–142, 144–147, 239
climate change
adaptation xv, xvii, 2–5, 16, 31–32, 98, 100, 119, 123, 147, 149, 159, 165, 209, 210–212, 214, 239, 244–245, 248–249, 267, 278–280
anthropogenic 6, 26, 161, 178–179, 200
effects of 70, 162, 174, 179, 181, 184, 205–206, 217–218, 239, 250–252, 256, 258, 260, 263, 267, 278
global xx, 2–3, 28, 30, 73–74, 76–79, 97, 177, 179–181, 227, 233, 245, 250, 252
mitigation 2, 4, 19, 177, 198, 201, 208–209, 211, 214, 234, 239, 245, 248–249, 265, 267, 278–279, 288, 293
narratives of 24, 28, 30–31
climatology 203, 207, 240
coal 116, 130, 168, 170, 180, 186, 188–190, 192, 295, 233–234, 236–237, 261, 277, 282
Cologne, Germany xx, 39
colonialism
British 17, 37, 49, 53, 98, 99, 106, 115, 182
French 129, 182
German 36, 38, 80
Portuguese 80, 82–83, 89–95, 255
Community Forest Associations (CFAs) 119
Comoros 181, 184
coral reefs 182, 184–185
corruption 98, 113, 117, 120, 125, 195–196, 228, 294
Cote d'Ivoire 129, 138–140, 144, 146, 271
Cotton 52, 62, 64, 68–69, 174, 182, 239
CO_2. *See* greenhouse gases
cyclone 1, 182, 242
anticyclone 1
Dakar, Senegal 269, 279
Dar es Salaam, Tanzania 269, 276–277, 282
democracy 183
Democratic Republic of the Congo 169, 254, 269, 272
demographic change 19
diarrhoea 157
diversity
biological. *See* biodiversity
cultural 227
genetic 78
Douala, Cameroon 269
Drought xix, 3–6, 9, 13, 17, 35, 37, 42–45, 49, 51, 53, 56–66, 70–73, 78–90, 95–100, 104–120, 223–128, 139, 142, 147–149, 156, 161–169, 173–176, 206, 221, 238–239, 245, 253, 256, 265, 277–278
Dubai, United Arab Emirates 39
Dutch East India Company (VOC) 182

East Africa 9, 38, 182, 234, 239, 241, 255, 269
earthquake, 161
Ebola 78, 203, 251
ecological
crisis 55, 81
equilibrium 77, 178, 295
ecology
dam 264
human 179, 252
marine 263
of famines 81
of rivers 257
political 74, 96, 149, 152, 154–155, 157, 177, 179, 250–251, 262
ecosystem 42, 63, 80, 149, 153–157, 169, 174–175, 262, 266
degradation of 167, 171, 185. *See also* biodiversity, degradation
marine 170, 184. *See also* ecology, marine
socio- 150. *See also* ecology, human
education
public 207–208, 211, 216–219, 222, 224
transformative learning 199–204, 216, 222, 224
Egypt 236, 255, 258, 260, 269, 273
electricity 52, 130, 137, 169, 187, 189–192, 215, 289, 291. *See also* energy
elephants 57
Elgeyo people 109

El Niño 65
emission 7, 64–65, 180, 233, 265, 276, 277, 286–287
  reduction of 179, 187, 217–218, 246, 279, 283, 287
  rising 4, 179, 193, 209, 282
empiricism 258
energy/power
  autonomy 186–187
  efficiency 177, 187–188, 193, 195, 197–198
  nuclear 200, 233, 272
  production 233, 238
  renewable xv, 7–8, 180–181, 186–192, 195–196, 201, 209, 279
  solar 181, 186–187, 201, 233, 238
  supply 7, 166, 176
  storage (capacity) 190, 256
Eng'ai 45
Environmental Impact Assessment (EIA) 189, 192
Essen, Germany xvii, 1
ethics
  of nature/environmental 14, 51, 62
ethnicity 134, 151, 153, 172
ethnography, xvi, xx, 5, 24–25, 30, 33, 123, 125, 131, 174, 194, 196
EU. *See* European Union
  climate leadership
Europe 5, 18, 20, 45, 52, 77, 110, 113, 140, 181, 183, 199–201, 212, 216, 224, 229–230, 233, 235–236, 252, 256, 259, 268, 280
European Union 183
Ethiopia xvii, 5, 155, 165, 251, 254–260, 273, 275–276

famine
  killer 13, 72, 80
  of the Dams 80, 87–93, 96–97
  of the Insects 82
farmers
  commercial 52, 58
  forest 103–104, 115
  large-scale 51, 54
  white 52, 54–59
Fast Track Land Resettlement Programme (FTLRP) 54
feminist
  political ecology (FPE) 152–153
fertiliser 55, 62, 136, 139, 143, 165, 185, 192. *See also* pesticides
  artificial/chemical 67–68, 238, 256, 290
  organic 53, 272
  tree species 118
fire
  forest 7, 59, 64–65, 79, 109–110, 114
  protection 103, 114
fishery 14, 61, 170, 183–184, 252, 262–263
flood 65, 78–79, 83, 98, 104, 139, 163, 175, 242, 276, 283, 295
  flash 185, 265, 276
  impact of 118, 120, 128, 149, 155–156, 162, 206, 221, 243, 260–261, 266, 270, 278–279, 285, 290–291
  Noah's 74
  victims 172, 279
floodplain 66, 96, 252, 257, 259, 263, 279
food production 67, 98, 232, 290
fossile fuels 161, 179, 181, 186–190, 195, 227, 233–234, 237, 265, 283
forest
  deforestation xviii, 15, 53, 66, 69, 75, 180, 185, 246, 248, 276–277
  fire. *See* fire
  natural 281, 286
  tropical xviii, 286
forestry
  agroforestry xv–xvii, 17, 98–100, 107, 114, 116, 118, 120, 136
  plantation 17, 98–99
  shamba 98, 101, 112, 115–120
  taungya 110
Friends of the Earth and the Global Anti-Incinerator Alliance (GAIA) 190
frost 242–243

Gambia xvii, xx, 199, 204–205, 209–216, 219–223
gender. *See also* feminist
  analysis 152–153, 172
  roles 149, 163–164, 172–175
Germany xvii, xix, xx, 1, 9, 36, 211, 216, 221
Ghana iv, xix, 5, 19, 129, 166, 213, 215, 217–218, 221, 253, 265–266, 271, 280–285
Githaka 106–107
Glaciers 180
Global North 265
Global South xv, 1, 25, 28, 31, 46, 150, 294

global warming 31, 57, 66, 181, 202, 238, 281, 283, 295
God 44–46, 48–49, 62–63
Godzi, Zimbabwe 68
Great Britain 246
Great Depression 111
greenhouse gases 179, 267, 193, 203, 217–218, 265, 277, 280, 286–287
green space. *See* urban
Guinea Bissau 181, 184, 212
Gulf of Guinea 8, 269, 273

Harare, Zimbabwe 53, 269
HIV/AIDS 203, 228
Human Development Index (HDI) 126, 183
Hombwe, Zimbabwe 68
Horn of Africa 5, 232
household
child-headed 158–159
men-headed 158, 165, 172
women-headed 158, 165, 171, 173

Ibrahim Index of African Governance 183
Impoverishment 67, 178–179
Independent Power Producers (IPPs) 189
Indian Ocean. *See* ocean
Industrialisation 179, 188, 200, 235, 269
Intergovernmental Panel on Climate Change (IPCC) 9, 156, 159, 205, 220, 224, 250, 256, 260
Islamism 244
Ivorian crisis 123

Jaluo people 109

Kaffrine, Senegal 175
Kampala, Uganda 234, 279
Kenya 17, 42, 98–120, 168, 171, 234, 236, 244, 255–256, 260, 272–273
Kenya Forest Department (KFD) 101–120
Kenya Forest Service 98, 115–116
Kenya Land Freedom Army (KLFA) 114
Kenya Meat Commission 108
Kenzamba, Zimbabwe 68
Kikuyu Forest Division 113
Kikuyu people 17, 98, 102, 105–109, 120
Kinshasa, DR of the Congo 231, 269, 272
Kiribati 184
Kipsigis people 109
knowledge production 201–202
Kumasi, Ghana 5, 19, 265–268, 280–294

La Chaumière, Mauritius 191–192, 195
Lagos, Nigeria, 231, 269
Lake
Manyara 38
Victoria 101, 269, 273
Land Acquisition Act of 1992 53
Landfill 187, 189, 191–193, 272
Landslides 180, 185
Leicester, United Kingdom 280
lighting. *See* electricity
Little Ice Age 253, 256
livestock 6, 17, 33, 35, 37–41, 47–48, 59, 62, 64, 69, 80–88, 90–92, 94, 96, 103–113, 116–117, 120, 164–167, 171–173, 183, 232, 236, 238–239, 248, 256, 263–264
locust 104, 111–112, 120. See also plague, insect
lotis 130–131, 133
non 130–132
Lusaka, Zambia 268, 269

Maa language 33, 45
Maasai people 23–24, 28–29, 32–41, 45–46
Madagascar xvi, 182
maize xix, 35, 52, 62, 64, 67–69, 77, 92, 99, 116, 165, 167, 174, 239, 256, 260
Makonde District, Zimbabwe v, xi, 48, 52, 57–58, 69
malaria xix, 77–78, 157, 162, 228, 251, 257, 259, 264, 278
Maldives 181, 184
Mali 164–165, 170, 242, 244, 253, 276
Maputo, Mozambique 269, 272
Mare Chicose, Mauritius 191–193
Marshall Islands 184
Marula
tree 87
wine 87
Mau Mau xvii, 107–111, 114, 120
Maurice Ile Durable, 4, 14, 177, 180, 186, 194
Mauritius vi, 177, 180–186, 188–198
Mauritius CT Power (MCTP) 189–190
Mbugua, David
Mbuji-Mayi, DR of the Congo 269

methane. *See* greenhouse gases
migration
(circular) labour *See also* migration, transnational 12, 35, 123–131, 136, 243–147, 182
rural-to-urban 126
transnational 178
millet 61, 64, 80–82, 85–86, 95–96, 142, 165, 239
modernism vi, 250, 252, 254, 260
Mogadishu, Somalia 269
moisture
soil 9, 252
Mombasa, Kenya 272
monocropping. *See* agriculture, monoculture
monsoon. *See* rainfall
Morogoro, Tanzania 173
Mossi people 130
mountain
Mt. Chirorodziva 63
Mt. Elgon 107
Oakland Hills 97
Mozambique 18, 60, 166, 173, 190, 255–256, 260–261, 272
muhoi. See ahoi
Myanmar. *See* Burma

Nairobi, Kenya 103, 105–106, 205, 269
Namibia 17, 73, 80–91, 93–97, 164–165, 170–171
New Orleans 79
Ngong forest 106
Nicole Syndicate 52
Nigeria xvii, 102, 163, 213, 230, 269, 271, 273
nitrogen 68, 238, 260, 264
nitrous oxide. *See* greenhouse gases
Nyanza people 109

ocean
Atlantic 170, 181, 268
Indian vi, xviii, 8, 177, 180–181, 184, 268
level *See* sea level rise
Pacific 65, 181
temperatures 8–9, 252
Ogiek people 107
Ohaingu, Namibia 83
Okalongo, Namibia 83, 86
Omulunga, Angola 86
Onahulu, Namibia 83
Onamine, Namibia 83
Onamunama, Namibia 83
Ondonga, Namibia 80, 84–86, 89–91, 95–96
Oshikango, Namibia 84, 86–89, 91, 95
Ouagadougou, Burkina Faso x, xi, 13, 123, 125–126, 128–148, 212, 247, 282
Oukwanyama, Angola, Namibia 80–97
Ovamboland, Namibia 80–86, 89–90, 93–96
Overfishing 180, 185. *See also* fishery
Overpopulation 18, 55
Overseas Territory 182

Pacific Ocean. *See* ocean
Parks. *See* urban green space
pesticides 99, 185, 290. *See also* fertiliser
Pointe aux Caves, Mauritius 189–190, 195
Port Louis, Mauritius 184, 190, 193, 195
patron–client relationship 35, 87, 93–94, 98
pollution. *See* air pollution
population
bomb 75–76
growth x, 4, 7, 41, 75, 123, 138, 229–230, 232–233, 241, 244, 266, 270, 276, 280, 292, 294
post-colonial era xvi–xvii, 18, 49–50, 52, 58, 60, 129, 183
plague
insect 83, 104, 111, 120
plantation
coconut 183
sugarcane 182, 185. *See also* bagasse; biofuel
tea 52
tobacco 52, 62, 64, 68–69
Plantation Establishment and Livelihood Improvement Scheme (PELIS). *See* forestry, *shamba*
Pointe aux Caves, Mauritius 189–190, 195
Pointe aux Sables, Mauritius 190, 195
power. *See* energy
protection
animal/plant 114
climate vii, 293
environmental/of nature 14, 18, 64, 293
human 170
strategy 114

rainfall
- anomaly x, 127
- changing rainfall patterns 4, 11, 44, 57, 165, 243
- decreasing 9–10, 41, 56, 65, 142, 166, 184, 250, 260
- extreme 9, 124
- increase/recovery 145–146, 239, 244
- irregular 42–43, 78, 95, 100, 147, 177, 181, 240
- monsoonal 8, 10, 242–243
- prayers 35, 45, 61
- rainy season 5, 9, 35–36, 56–57, 82, 90–91, 127, 139, 141, 242–243, 253

Ramgoolam, Navin 186
refugees
- climate/environmental 243
- from famine 80, 84
- from war 80

religion
- religious leaders 49
- religious traditions/practices 7, 45

renewable energy. *See also* bagasse, biofuel *and* electricity
- hydro 169, 186–187, 190, 233, 237–238, 255, 238–239
- fuelwood 100–101, 113, 118, 163, 167–168, 170–171, 186–187, 234, 237, 277, 282, 289, 293
- photovoltaic 186–187, 190, 201, 233
- solar thermal 186
- waste-to-energy (WtE) 186, 190, 192, 195
- wind 133

Republic of the Congo 169, 254
resource management xv, xviii, 67, 151, 177, 180
Réunion 181–182
Revolution
- Green 238, 247
- Industrial 179, 235

Richards Bay 189
river
- Blue Nile vi, 4, 250–252, 255, 258–260
- Rhine 79
- White Nile 250, 255, 258
- Zambezi vi, 4, 250, 252, 255, 257–261

Rodrigues 182–183
Rwanda 169, 273

Sahel x–xi, 3–6, 12, 76, 118, 123–131, 141–146, 167, 240–244, 247, 253
sanitation 13, 110, 130, 166
São Tomé and Príncipe 181, 184
savanna 42
sea level rise 177, 180–181, 184
Second World War. *See* war
Senegal 170, 175, 214, 279
Serengeti National Park 39
sexual harassment 171
Seychelles 181–182, 184
shamba. See forestry, *shamba*
Shona people 48, 61
Simanjiro District, Tanzania 23, 29, 35–36, 42
slavery 182, 268
Small Island Developing State (SIDS) 31, 177, 181
Small Island State (SIS) vi, 184
social stability 178
soil
- erosion 15–17, 53, 66–70, 74, 139, 142, 166, 180, 184–185, 206, 243, 264, 271, 279, 290, 293
- moisture 9, 252

South Africa 12, 52, 80, 83, 89–90, 94–95, 101, 150, 158, 162–163, 174, 189, 246, 256
South America. *See* America
South China Sea 181
Southern Oscillation. *See* El Niño
St. Brandon 182
sub-Saharan Africa (SSA) vi, xiii, 31, 100, 125–126, 137, 143, 146–152, 159–163, 167–169, 173–174, 201, 228, 250, 266, 268, 273–276
Sudano-Sahelian region xi, 123–128, 131, 141, 144–145
Sugarcane 182–185, 189
Sustainable Mauritius (MID). *See* Maurice Ile Durable
Swahili 33–34, 37, 42, 98, 181

Tanganyika (mainland Tanzania) 36–38
Tanzania v, xi, xvi, 12, 18, 23–24, 28–29, 32–43, 164–167, 173–174, 190, 236, 276
Tarangire National Park 34, 40–41
temperature
- data 241
- development 57, 66, 70, 162, 180, 277, 286

fluctuation/variation 5, 177, 181, 278, 281
measurement 53
moisture–temperature interaction 252
of the ocean. *See* ocean
rise 9, 48–49, 65, 184, 241, 243, 265, 276. *See also* climate change
Terrat, Tanzania xi, 23, 29, 33–36, 39–46
textile industry 185
Thelaphi people 150
Tiriki people 109
tourism 183–185, 243, 285, 291
Tromelin 182
tsunami 161
2004 Indian Ocean 184
Tugen people 109
Tuvalu 184

Uganda 165, 234, 255, 256, 260, 273, 279
railway 101
United Kingdom xiii, xv, xix, 182, 280, 282, 188
United Nations xiii, 202, 246
United Nations Framework Convention on Climate Change (UNFCCC) 202, 205, 209–210
United States of America (USA) 13, 280. *See also* America, North
urban
agriculture. *See* agriculture
climate resilience vi, 265–267, 282
green space xix, 19, 163, 217, 238, 265–266, 275, 278–290, 292–295
heat island (UHI) 265, 277–278, 282
urbanisation xi, xv–xvii, 4, 7, 125, 174, 231, 233–236, 244, 263, 265–273, 276–277, 289–291, 294

Vasco da Gama 181
villagisation 37, 109, 197
violence 77, 154, 228, 241, 243–245, 270
climate change-induced 243
gender-based 174. *See also* gender *and* sexual harassment

Wakikuyu 106
Wanderobo people 109
war
World War I 38
Word War II 75, 111, 113
WASCAL vi, x, xiii, xvi, xx, 199, 204, 209, 211–221, 224
waste
disposal 180, 185, 190, 265, 277
reduction 191–192
water
conservation 67, 166
drinking 41, 86, 184, 234, 237
management 250, 252, 254, 260–261, 263
running 52, 130
shortage/scarcity 85, 166, 184, 237
supply 66, 84, 97, 110, 166, 184–185, 247
system 111, 185
weather
annual 5
conditions 25, 34, 40, 44, 48, 164–165, 167, 173
data 10–11, 13, 59
effect/impact of 57, 62, 73, 79, 151–152, 161, 183, 237
extreme w. event 1, 11, 51, 98, 118–120, 149, 153–156, 159, 164, 171, 177, 180–181, 238, 243–244
god. *See* Eng'ai
prediction 44
stations 8, 53, 56–57
wildlife 14, 36, 38, 41, 60, 68, 109, 234, 243
World Bank 50, 255, 259
World Commission on Environment and Development (WCED) 178

Yaoundé, Cameroon 269

Zambia xx, 60, 164, 170, 256, 260–261, 268–269
Zimbabwe v, xiv, xviii, xx, 18, 48–55, 58–60, 63–65, 70, 164, 173, 260–261, 269

www.ingramcontent.com/pod-product-compliance
Lightning Source LLC
Chambersburg PA
CBHW071231230926
51849CB00001B/68
9789004410831